CONCRETE FORMWORK

AMERICAN TECHNICAL PUBLISHERS, INC.
HOMEWOOD, ILLINOIS 60430

Leonard Koel

1 2 3 4 5 6 7 8 9 -88- 9 8 7 6 5

Printed in the United States of America

Library of Congress Cataloging-in-Publication Data

Koel, Leonard.
Concrete formwork.

Includes index.
1. Concrete construction—Formwork. I. Title.
TA682.44.K64 1988 624.1'834 88-6278
ISBN 0-8269-0704-0 (soft)

Contents

Acknowledgments

The author and publisher are grateful to the following companies and organizations for providing technical information and assistance.

American Concrete Institute
American Plywood Association
American Society for Testing and Materials
Associated Pile and Fitting Corporation, Clifton, NJ
Barclay and Associates
William Brazley and Associates
The Burke Company
Calweld, Inc.
Caterpillar, Inc.
Concrete Construction Publications, Inc.
Deere and Company
Economy Forms Corporation
The Garlinghouse Company
Gomaco Corporation
Kaiser Cement Corporation
Laser Alignment, Inc.
Metal Forms Corporation
Morrow Crane Co., Inc.
Occupational Safety and Health Administration
Portland Cement Association
Power Curbers, Inc.
Richmond Screw Anchor Company
Simpson Strong-Tie Company, Inc.
Symons Corporation
Chris P. Stefanos Associates, Inc.
Walsh Construction Company of Illinois
Western Forms, Inc.
Wire Reinforcement Institute
David White Instruments

Introduction

CONCRETE FORMWORK provides information on the safe construction of formwork for residential, light commercial, and heavy construction projects. Various aspects of form construction are covered, ranging from site preparation to concrete placement to stripping forms.

Chapters 1 and 2 provide information on the building site and methods and materials of forming walls. Chapters 2 through 5 cover residential foundation construction, flatwork, and heavy construction. Chapter 6 includes information on precast concrete construction and tilt-up construction. Chapter 7 describes concrete mix and placement and stripping forms. Chapter 8 is divided into three sections: Math Fundamentals, Printreading, and Estimating Form Materials and Concrete.

The appendices and glossary provide additional information. Appendix A includes general formulas and tables used by a form builder or estimator. Appendix B, Construction Materials, provides information on lumber dimensions, reinforcement, and quantities of concrete required for concrete walls, footings, and floor slabs. Appendices C and D include the American Concrete Institute Recommended Practices and the Occupational Safety and Health Administration Concrete and Shoring Regulations. Appendix E describes leveling instruments and their use, including the builder's level, transit-level, and laser transit-level. The illustrated glossary includes approximately 275 terms commonly used by a form builder.

Review Questions are included at the end of Chapters 1 through 7. In chapter 8, Review Questions are provided at the end of Sections 1 and 3. Review Questions test the understanding of the students of the material presented in the chapter. Answers to Review Questions involving math applications are rounded to 2 places after the decimal point. Section 2 of chapter 8 consists of Printreading Exercises related to the prints on pages 233 through 247. Printreading Exercises test knowledge of prints acquired in previous experience. Types of questions included in the Review Questions and Printreading Exercises are Completion, Multiple Choice, and Identification. Always record the answer in the space provided. Answers for all questions are in the Instructor's Guide for CONCRETE FORMWORK.

Completion

Write the appropriate response in the space provided.

steel **1.** High tensile ________ cables are placed in the form when pretensioning concrete.

Multiple Choice

Select the response that correctly completes the statement, and write the appropriate letter in the space provided.

B **1.** The recommended slope for deep excavations under average soil conditions is ________°.

A. 33
B. 45
C. 75
D. 90

Identification

Write the letter of the answer in the space provided.

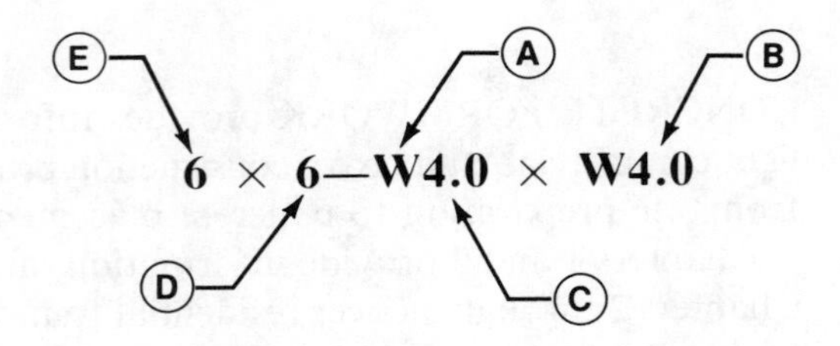

B 1. Cross-sectional area of transversal wire

C 2. Cross-sectional area of longitudinal wire

D 3. Transversal spacing in inches

E 4. Longitudinal spacing in inches

A 5. Smooth wire

Plans

Five prints are included on pages 233 to 247 and are to be used with the Printreading Exercises in Chapter 8, Section 2. Prints included are the Plot Plan, Slab-on-Grade Foundation, Crawl Space Foundation, Full Basement Foundation, and Heavy Construction Foundation. The Full Basement Foundation print is also used for the Review Questions for Chapter 8, Section 3. The size of each print has been modified and should not be scaled.

CHAPTER 1

The Building Site

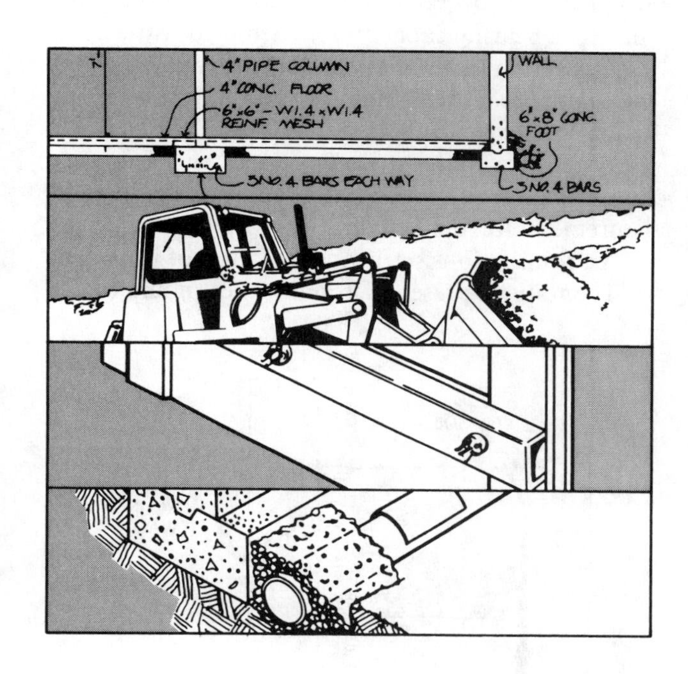

The building site is the area in which construction occurs. A building site may be a small residential lot or a large area used for a heavy construction project. Property lines indicate the boundaries of a building site and are a reference for groundwork and building layout.

The foundation design of a structure and amount of groundwork required are based on the soil conditions of the building site. The soil characteristics determine the bearing capacity of the soil and amount of settlement that can be expected. Sandy soil contains larger soil particles than soil with a higher percentage of clay. Sandy soil has a greater bearing capacity, and less soil and foundation settlement.

Preliminary groundwork is completed before constructing a foundation. A set of prints is used to determine the location of a structure and the amount of groundwork to be completed. Carpenters establish building lines based on the prints. Building lines indicate the location of the foundation of the structure. Operating engineers perform the groundwork using earth-moving equipment such as backhoes, bulldozers, and motor graders. The groundwork may range from a small amount of grading to massive trenching and excavating. Carpenters verify grade levels during the groundwork and shore earth walls of heavy excavations.

The foundation of a structure is constructed after the groundwork is completed. The size and shape of a foundation is based on the prints and local building codes and zoning ordinances.

SOIL CONDITIONS

Soil conditions determine the type of foundation design required for a building. One of the most important factors influencing foundation design is the type of soil found beneath the structure. Some soils have a higher *bearing capacity* than others. Bearing capacity is the ability of the soil to support weight. A number of different earth layers (*strata*) often exist below the ground surface of the building site. The bearing capacity for each layer varies. *Excavation* (removal) of some of the earth layers may be required so the foundation footings rest on soil of adequate bearing capacity. See Figure 1-1. In heavy construction work, piles are often driven deep into the ground to adequately support the load requirements of the building.

All foundations settle over a period of time. The soil conditions present determine the amount of settlement. Excessive or uneven settlement can cause cracks in the foundation, resulting in structural damage to the building. Foundations built in proper soil conditions can reduce settlement problems.

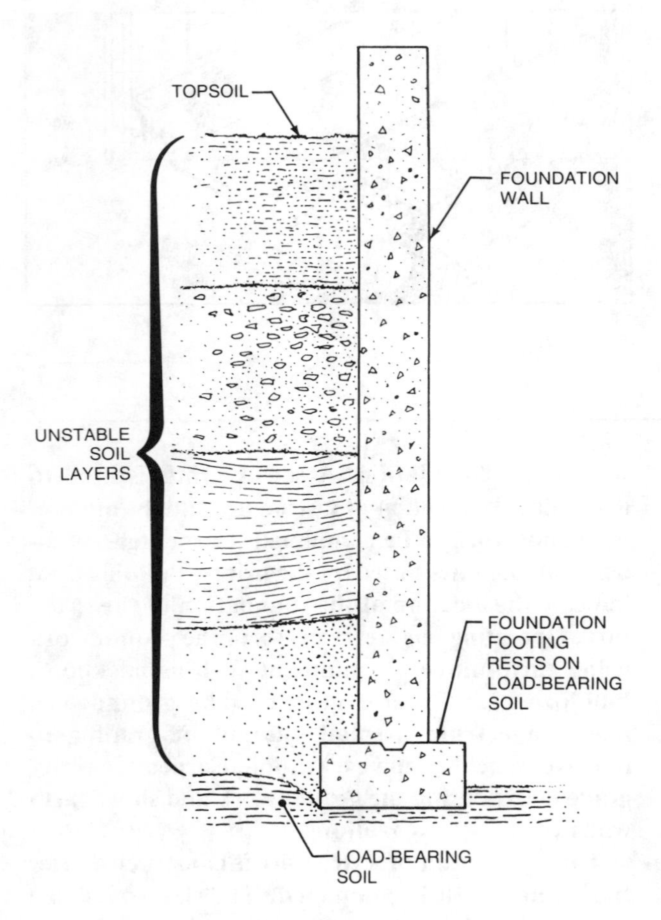

Figure 1-1. Excavation for a foundation footing requires that the foundation footings rest on soil with adequate bearing capacity.

Soil Composition

All soils consist of particles that originate from the breakdown and decomposition of solid rock. The major factor in differentiating soils is the size of the soil particles. Soils are generally classified as *granular* or *cohesive*. Granular (coarse-grained) soils such as gravel and sand have large particles that can be easily seen by eye. Cohesive (fine-grained) soils such as silts and clays are comprised of very small particles and in some cases can only be seen through a microscope. See Figure 1-2.

Samples taken from a building site often show mixed soils. The predominant soils found in the mixture determine the final analysis of the sample and its expected bearing capacity. Classification charts that give the approximate bearing capacities of various soil mixtures common to the location are

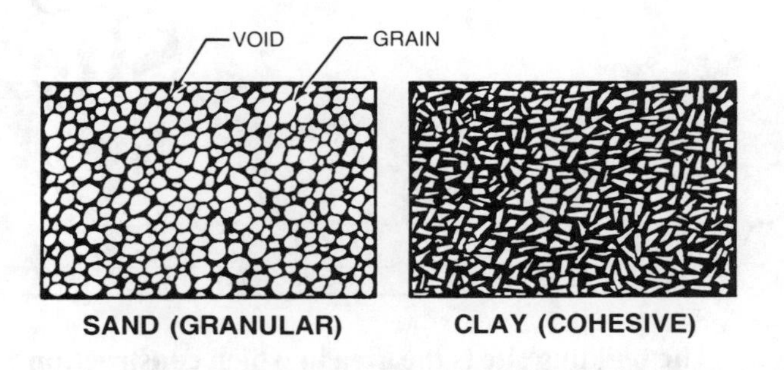

SOIL TYPES	APPROXIMATE SIZE LIMITS OF SOIL PARTICLES
Boulders	Larger than 3″ diameter
Gravel	Smaller than 3″ diameter but larger than #4 sieve
Sand	Smaller than #4 sieve[1] but larger than #200 sieve[2]
Silts	Smaller than 0.02 mm diameter but larger than 0.022 mm diameter
Clays	Smaller than 0.022 mm diameter

[1] Approximately 1/4″ in diameter.
[2] Particles less than #200 sieve not visible to the naked eye.

Figure-1-2. A major factor in soil classification is the size of the soil particles. Sand grains are larger than clay grains and have fewer air and water voids between grains.

SOIL CLASSIFICATION[1]				
Major Division			**Typical Names**	**Presumptive Bearing Capacity[2], Tons Per Sq Ft**
Coarse-Grained Soils (More than half of material is larger than the smallest particle visible to the naked eye)	Gravels (more than half of coarse fraction is larger than ¼″)	Gravels with Fines (appreciable amount of fines	Well-graded gravels, gravel-sand mixtures, little or no fines.	5
			Poorly graded gravels or gravel-sand mixtures, little or no fines.	5
		Clean Sands (little or no fines)	Silty gravels, gravel-sand-silt mixtures.	2.5
			Clayey gravels, gravel-sand-clay mixtures.	2
	Sands (more than half of coarse fraction is smaller than ¼″)	Clean Gravels (little or no fines)	Well-graded sands, gravelly sands, little or no fines.	3.75
			Poorly graded sands or gravelly sands, little or no fines.	3
		Sands with Fines (appreciable amount of fines)	Silty sands, sand-silt mixtures.	2
			Clayey sands, sand-clay mixtures.	2
Fine-Grained Soils (More than half of material is smaller than the smallest particle visible to the naked eye)	Silts and Clays (liquid limit is less than 50)		Inorganic silts, very fine sands, rock flour, silty or clayey fine sands or clayey silts with slight plasticity.	1
			Inorganic clays of low to medium plasticity, gravelly clays, sandy clays, silty clays, lean clays.	1
			Organic silts, and organic silty clays of low plasticity.	
	Silts and Clays (liquid limit is greater than 50)		Inorganic silts, micaceous or diatomaceous fine sandy or silty soils, elastic silts.	1
			Inorganic clays of high plasticity, fat clay.	1
			Organic clays of medium to high plasticity, organic silts.	
Highly Organic Soils			Peat and other highly organic soils.	

[1]Based on ASTM D2487—Classification of Soils for Engineering Purposes.
[2]National Building Code, 1976 Edition, American Insurance Association.

Figure 1-3. Bearing capacity of soil is expressed in tons per square foot for common soil mixtures.

available. These soil classifications apply to normal conditions. Specific job sites may require further testing for an accurate analysis. See Figure 1-3.

Soil Compressibility. *Compressibility* (pressing together) of the soil below the foundation determines the amount a foundation will settle. The amount of compressibility is determined by the reduction of the spaces (voids) between the soil particles that contain air and/or water. Sandy soil contains larger soil particles than silt or clay, which results in less compression and foundation settlement. Silt or clay containing small soil particles compresses more, resulting in more foundation settlement.

Soil Types. The types of soil are *rock, virgin soil,* and *fill.* Rock has the greatest soil-bearing capacity if it is level and free of faults (cracks). Virgin soils include, in the order of their strengths, gravel, sand, silt, and clay. Most construction takes place in virgin soils. Fill consists of soil brought from some other location and deposited at the building site. Fill does not provide a dependable foundation base. When constructing on a site with fill, foundation footings must rest on ground excavated to firm, undisturbed virgin soil, or additional support must be provided.

Soil Testing

Soil testing determines the bearing capacity and compressibility of the soil. Soil samples taken from the building site are analyzed in the testing laboratory. Soil testing information makes it possible to calculate the amount of bearing capacity per square foot of soil area. This data is used in determining the type of foundation that must be constructed over the soil.

A variety of methods is used to test subsurface (below the surface) soil conditions. Some of these methods employ geophysical instruments that give information about below-grade conditions without earth samples having to be dug up. When subsurface soil samples are required, they are brought to the surface by equipment such as power augers and other types of test drilling rigs. Open test pits at 10′ intervals, dug by hand or mechanical digging equipment, are considered one of the more reliable methods for shallow exploration. This method makes it possible to examine the earth layers in their natural condition. See Figure 1-4.

As a general rule, more extensive laboratory soil testing is done for heavy construction projects that require deeper excavations. It is seldom required for residential or other light construction buildings where the foundation excavation is based on established practice in the area. In addition, soil surveys have been made of much of the land area in the United States. This information has been compiled by soil engineers of the Soil Conservation Service, U.S. Department of Agriculture, and is available to the general public. Local building codes may also list bearing capacities for different types of soil in the area.

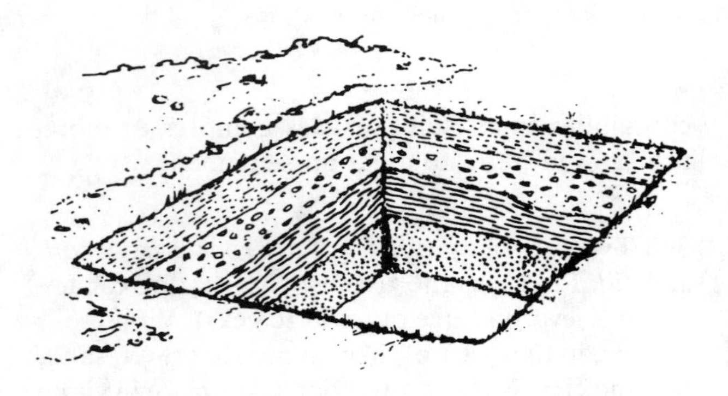

PIT METHOD OF DETERMINING SOIL CONDITIONS

Figure 1-4. Subsurface soil conditions can be determined by using instruments, digging, or drilling.

Soil Moisture

Soil moisture is water in the soil, which affects soil conditions. The type and amount of soil moisture must be considered in the design and construction of a foundation. If water collects in an enclosed space such as the area beneath the floor of a residential crawl space foundation, it can cause odors, mold, and wood decay. Water penetrating the walls of a full basement foundation can make the basement area unsuitable for a storage, work, or living space. Methods commonly used to prevent potential surface and groundwater problems include proper grading around the perimeter of the building, drain tiles, and vapor barriers. See Figure 1-5.

Soil moisture is caused by surface water, groundwater, and capillary action. Surface water results from rain, downspout discharge, and melting snow.

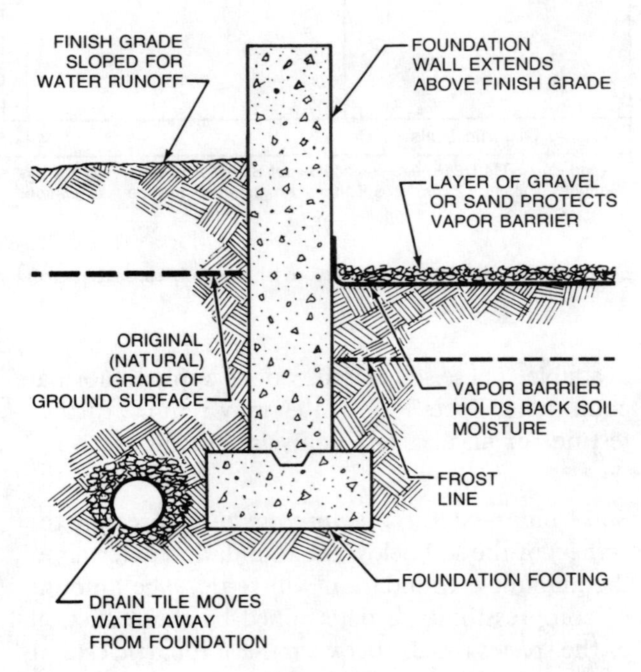

Figure 1-5. Methods to control surface and groundwater include sloping the finish grade outside of the building, and placement of drain tiles and a vapor barrier.

By sloping the finish grade away from the foundation walls, surface water can be directed away from the building. For nonpaved areas a recommended slope is 6″ in 10′ (5%) away from foundation walls. If the area around the foundation is paved, a slope of 1/8″ in 1′ (1%) is usually adequate.

Groundwater is caused by the water table level at the job site. The water table is the highest point below the surface of the ground that is normally saturated with water. Water table levels vary in different geographical areas. In addition, water tables tend to rise during wet seasons because of water penetration from rain and melting snow. During dry seasons, water tables subside to their normal levels.

The amount of moisture in the soil is also affected by *capillary action* of the soil. Capillary action is a physical process occurring in all types of soil that causes water and vapor to rise from the water table and move up toward the surface of the ground. However, water and vapor rise higher in porous fine-grained soils such as silt and clay than in coarse-grained soils such as gravel and sand. See Figure 1-6.

Capillary action can create dampness at the surface of basement floor slabs and slabs that rest directly on the ground. Moisture accumulation must also be avoided in the enclosed areas of crawl space foundations.

Vapor barriers (ground covers) are widely used to control ground surface dampness caused by capillary action. Four mil polyethylene film is commonly used because of its resistance to decay and insect attack. Concrete may be placed directly on the polyethylene film to form a slab. When concrete is not placed directly on the polyethylene film, such as in a crawl space, a covering layer of pea gravel or sand is recommended to protect the film from damage and retain the desired position. The amount of capillary action can also be controlled by removing the more porous soil adjoining the foundation walls and replacing it with gravel.

Drain tile or *drain pipe* around the foundation provides the best means of directing groundwater away from the foundation. Some of the more traditional drain tile methods use round clay or concrete sections of pipe with a 1/4″ space between each section. The groundwater enters the tile through this open space and then flows along the tiles. A strip of asphalt-saturated paper is placed over the top of the space between the tiles to prevent soil from falling inside the tile. Another type of drain tile has holes at the bottom of the tile sections. The groundwater seeps into the holes and flows through the inside of the tiles. See Figure 1-7.

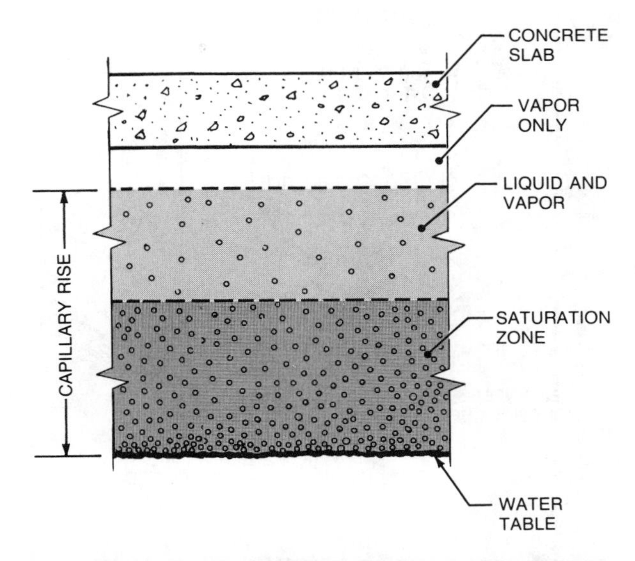

CAPILLARY RISE	SOIL TYPE	SATURATION ZONE
11.5′	Clay	5.7′
11.5′	Silt	5.7′
7.5′	Fine Sand	4.5′
2.6′	Coarse Sand	2.2′
0.0′	Gravel	0.0′

Figure 1-6. Capillary action of water from the water table is greater in silt and clay than in sand and gravel.

Another kind of porous drain pipe recently developed is made of a mixture of portland cement, basaltic trap rock, and sand. Water flows into the pipe through tiny channels that are too small to allow the passage of anything but water. This type of pipe features a self-sealing slip joint that does not require mortar or wrapping. Most of the types of drain pipe range in diameters from 4″ to 24″, and lengths from 2′ to 3′.

Plastic corrugated pipe is another type of drain pipe widely used today. It is manufactured in diameters ranging from 3″ to 18″. The smaller diameter pipes are available in coils of 100′ to 300′. The large diameters (10″ to 18″) are available in 20′ lengths. Plastic drain pipe is also available as perforated and nonperforated tubing. The perforated tubing is used for the same purpose as the clay and concrete drain pipes. It offers the advantage of fast installation and lower labor cost. Nonperforated

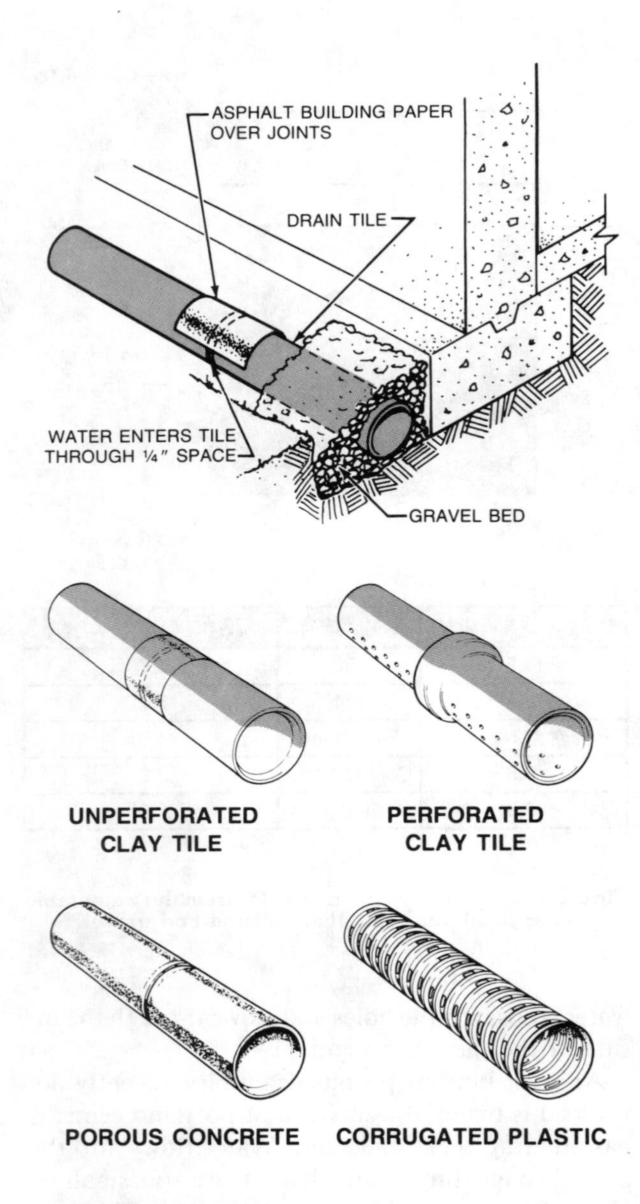

Figure 1-7. Drain tile systems move groundwater away from the foundation.

tubing is used for such purposes as downspout runoff, window well drainage, and floor and driveway drains.

Drain tiles or pipes extend around the perimeter of the building and are placed alongside the base of the footing and surrounded by a layer of gravel. Drain tile or drain pipe should be positioned at a minimum slope of 1″ in 20′ of length. Collected water drains toward a lower elevation and is absorbed by the ground or one or more *catch basins*. A catch basin (*dry well*) is an area dug out and filled with gravel. See Figure 1-8. In some areas where storm sewers exist below the street levels, water drainage systems may be connected to the storm sewers.

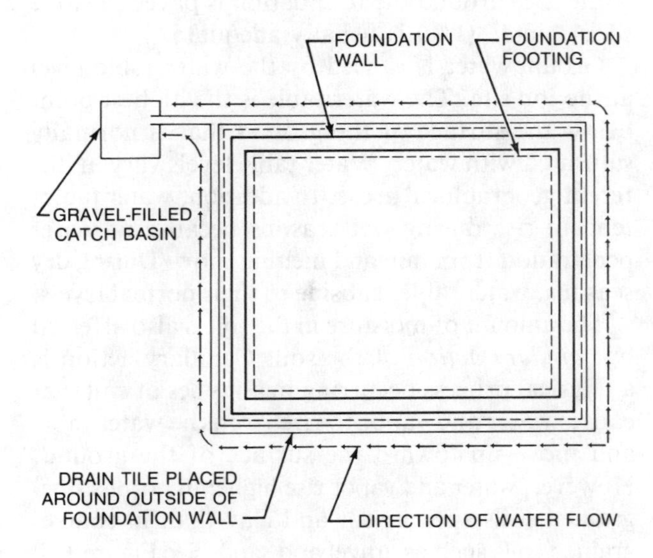

Figure 1-8. Drain tile placed around the perimeter of the foundation wall should have a minimum slope of 1″ in 20′ so that groundwater flows toward the gravel-filled catch basin.

Frost Line

The *frost line* is the depth to which soil freezes in a particular area. Soil freezing is caused by temperature drop and surface water penetrating the soil. The frost line varies with different climate areas ranging from 0″–5″ on the West Coast to 72″–108″ in some of the far northern sections of the country. See Figure 1-9. In addition, the amount of frost penetration is affected by the type of soil underlying the building site. For example, clay and silt tend to absorb and hold moisture, therefore allowing greater frost penetration. Coarse sands and gravels drain well, resulting in shallower frost penetration. Increasing water drainage through the use of gravel and drain tiles helps limit the depth of frost penetration in more porous soils.

The frost line must be considered in excavation work and the design and construction of a foundation. The footings of any foundation system should always be placed below the frost line. Footings placed above the frost line are subject to *soil-heaving* action. Soil heaving occurs as the soil freezes and expands, resulting in the building structure moving

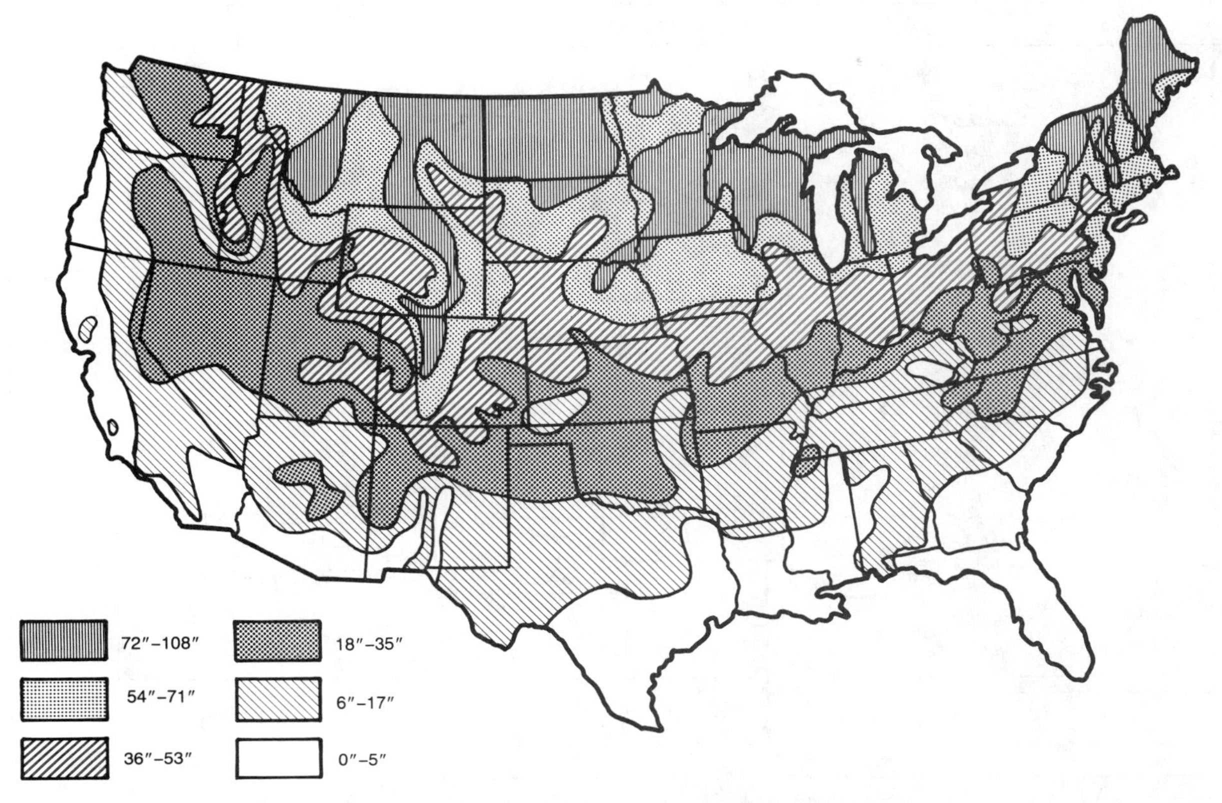

Figure 1-9. The frost line varies according to climate in the United States.

upward. When the soil thaws, the structure drops down again. This up-and-down movement can result in structural damage to the foundation.

SITE WORK

Site work includes layout work and groundwork that are completed before the foundation construction begins. Groundwork is the preliminary grading and excavation required at the job site. Additional grading and backfilling is often required after the foundation work has been completed. The initial layout for the groundwork is determined by information provided in the plot plan of the prints. Additional data may be contained in the foundation plan and the written specifications of the prints. All this information must conform to the local building codes and zoning regulations.

The Plot Plan

The *plot plan* (site plan) in the prints provides information regarding preliminary site work. Before drawing the plot plan, the architect has a survey performed on the lot to establish and mark the legal boundaries of the property, and record the existing grades. Additional information provided by the plot plan is the setback, compass direction, bench mark and elevations, utility hookups, roads, sidewalks, terraces, tree locations, and easements. See Figure 1-10.

Property Lines. Property lines define the boundaries of the building lot. Building lots are usually mapped and recorded by local building or zoning authorities. The lot surveyor studies the zoning maps and records to determine the exact boundaries of the property. The surveyor establishes the two front corners of the lot by measuring from existing reference points. Reference points are established by surveys that locate the marks. In some localities street curbs are used as reference points. In other localities property corners are found by measuring from the center of a road or markers placed at intervals in sidewalks. The two rear lot corners are normally laid out with a transit-level by sighting at a right angle from the two established front corner stakes.

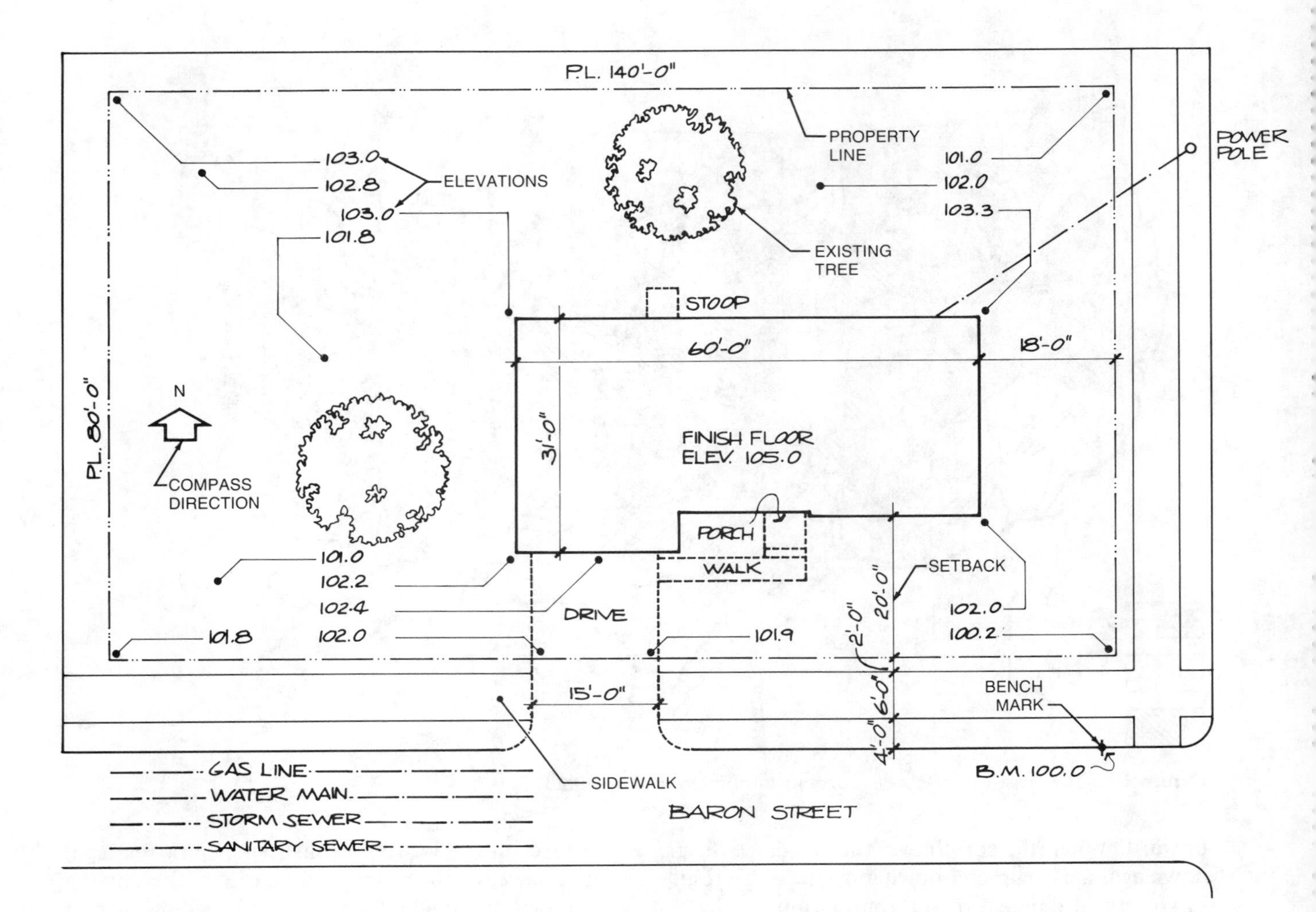

Figure 1-10. The plot plan provides information related to the preliminary site work.

To identify the lot corners, the surveyor places corner stakes at each corner. Corner stakes, also called *hubs,* are often wood stakes driven into the ground so that the top of each stake is flush with the surface of the ground. Hubs may also consist of a pipe with a cork or lead plug. Small nails driven into the top of the corner stakes or hubs mark the exact corners of the property. The property lines and measurements shown on the plot plan represent lines extending from the four corners of the lot. In Figure 1-10 the lot measurements are 140'-0" × 80'-0".

Grades and Elevations. Grades and elevations for the job site are specified in the plot plan. The plot plan gives the grade levels at all four corners of the lot. Grade levels are shown at the building corners as well as at different points within the interior of the lot. Grade levels marked on the plot plan are usually given in engineering measurements of feet and tenths or hundredths of a foot. Carpenters and other construction workers use rules with feet and inch measurements; therefore, it is often necessary to convert engineering measurements to inch and fractional equivalents. A table can be used for this purpose, or the conversions can be done mathematically. (See Chapter 8.) The grade levels show how the ground surface will be sloped and contoured. Some plot plans include *contour lines* to help clarify the desired slopes of the ground. Contour lines show the slope of the ground with lines extending from identified grade levels. See Figure 1-11.

The plot plan normally shows a *bench mark* (job datum) established by the surveyor at a convenient spot close to the property. The bench mark may be a mark chiseled at the street curb, a plugged pipe driven into the ground, a brass marker, or a wood

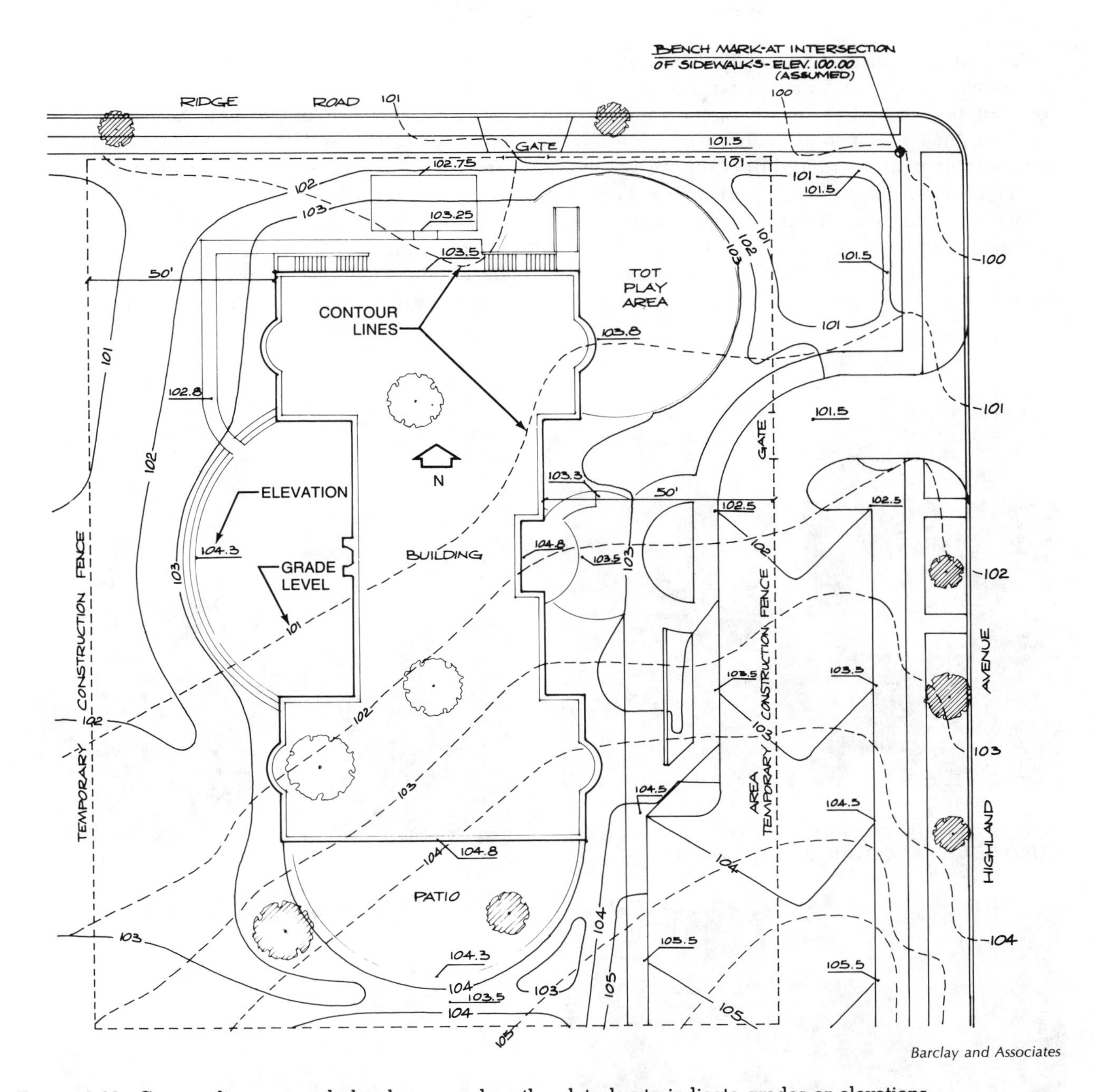

Barclay and Associates

Figure 1-11. Contour lines or grade levels are used on the plot plan to indicate grades or elevations.

stake placed near one corner of the lot. In addition, bench marks may be established on adjoining buildings or power poles.

The bench mark shown on the plot plan may be identified as the number of feet the ground is above sea level at that point or it may be shown as 100.0 ′ as shown in Figure 1-11. The bench mark establishes a reference for all other elevations on the plot plan. For example, the elevation along the south end of the building is 104.8 ′; therefore, the surface at that point is 4.8 ′ (104.8 ′ – 100.0 ′) higher than the bench mark. The grade level at the southwest edge of the lot is 103 ′; therefore, the elevation is 3 ′ (103.0 ′ – 100.0 ′) higher at that point.

Groundwork

Groundwork includes operations such as grading, excavation, and trenching that is completed before the foundation construction begins. The amount of groundwork required depends on the existing contours of the lot and the depth of the foundation footings. Most groundwork is done with earth-moving equipment such as bulldozers, motor graders, power shovels, backhoes, and other heavy equipment. See Figure 1-12. Heavy equipment is run by operating engineers. Carpenters often work with operating engineers by setting up lines to guide the earth-moving work, and assisting in checking grade levels and depth of excavation.

Grading. The amount of grading required is determined by the condition of the ground. Lots that are fairly flat and level or only slightly sloped may require very little grading. Lots that are steeply sloped or have very uneven surfaces require considerable grading. In order for the surface to conform to the grade levels shown on the plot plan, soil may have to be removed from or added to different

Deere and Company

BULLDOZER USED TO START EXCAVATIONS AND STRIP ROCKS AND TOPSOIL AT SURFACE OF SITE

Deere and Company

HYDRAULICALLY POWERED EXCAVATOR USED FOR DEEPER EXCAVATIONS, TRENCHING, AND LOADING

Caterpillar, Inc.

BACKHOE LOADERS USED FOR TRENCHING, BACKFILLING, AND SMALL LOADING OPERATIONS

Deere and Company

MOTOR GRADERS USED FOR FINAL GRADING OPERATIONS

Figure 1-12. Heavy earth-moving equipment is used for grading, excavation, and trenching.

areas of the lot. The finished surface should be sloped away from the building so that surface water caused by rain and melting snow will flow away from the building.

During the grading operations, the levels of grading can be measured and checked by using a rod and leveling instrument such as a builder's level or transit-level. (See Appendix E.)

Residential Excavation and Trenching. Crawl space foundations require trenches or shallow excavations dug down to the proper footing level. The trenches for a crawl space foundation must be deep enough to place the footing below the frost line and at depths prescribed by local building codes. Residential buildings or other types of small buildings may require a large excavation for a below-grade full basement foundation. See Figure 1-13.

If a footing form is required, the trench must be wide enough to allow room for the construction of the form. In both types of foundations, lines must be set up showing the boundaries of the building

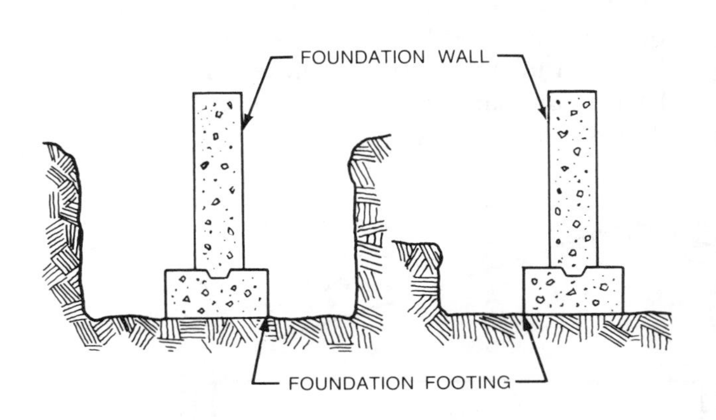

CRAWL SPACE FOUNDATION

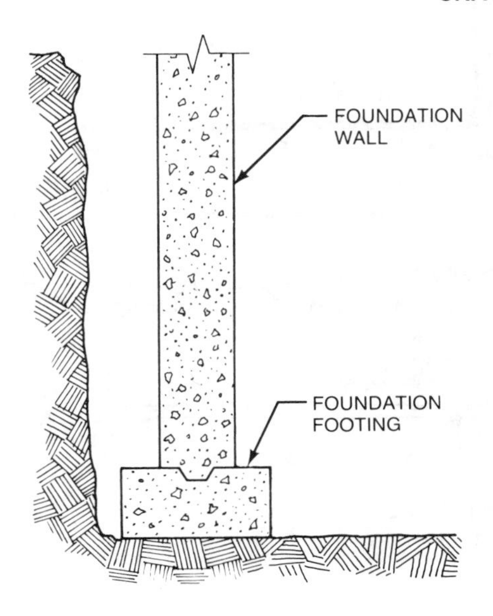

FULL BASEMENT FOUNDATION

Figure 1-13. Crawl space foundations require trenches dug down to the proper footing level. Full basement foundations require complete excavation of the area required by the basement.

in order to establish the perimeters of the excavation. The walls of the excavation should be at least 2′ outside the building lines to allow enough room for the formwork. However, if the soil is loose and unstable, the excavation for the foundation walls should extend farther back.

The depth of an excavation must extend to firm and stable soil and be below the frost line. A complete set of prints usually provides section view drawings that give the necessary information needed to determine the depth of the excavation. For a full basement, the depth of excavation can be calculated by adding the slab-to-joist height, slab thickness, and the footing height, less the distance the wall projects above the finish grade. See Figure 1-14. In warm climate areas where frost lines are not a factor, many codes only require that the *topsoil* (a soft surface layer of earth in which vegetation grows) be removed and the trenches dug a minimum of 6″ into natural undisturbed soil.

Backfilling. *Backfilling* is the replacing of soil around the outside foundation walls after the walls have been completed. Backfilling is performed after the forms have been stripped (removed) from the walls and the waterproofing and drain pipe work have been finished. The soil used for backfill should be free of wood scraps and any other type of waste material. Backfill must be placed carefully against the foundation wall and well compacted to avoid future shrinkage. Gravel is trucked in and used for backfill on many jobs since it allows better water drainage around the building. See Figure 1-15.

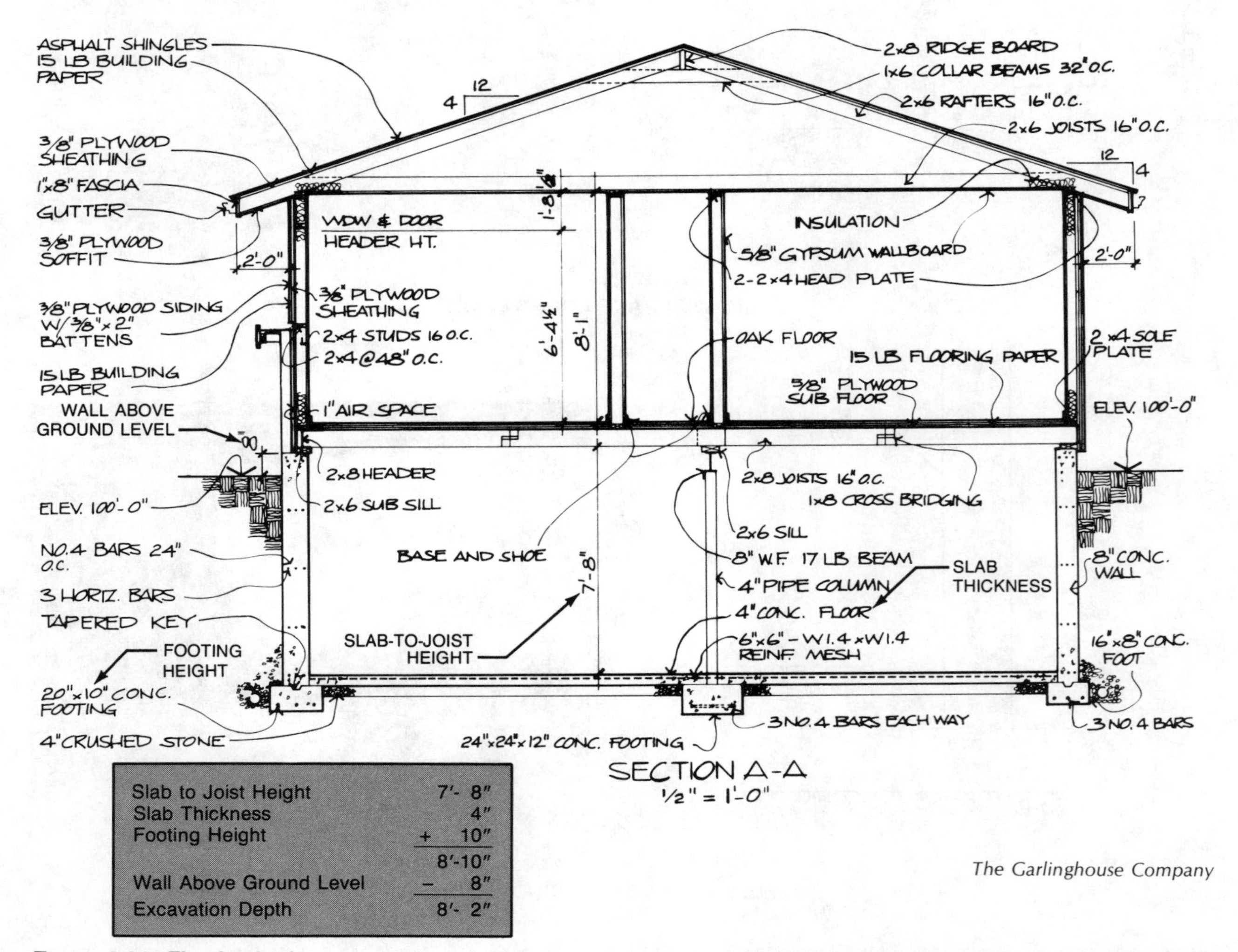

Slab to Joist Height	7′- 8″
Slab Thickness	4″
Footing Height	+ 10″
	8′-10″
Wall Above Ground Level	− 8″
Excavation Depth	8′- 2″

The Garlinghouse Company

Figure 1-14. The depth of excavation for a full basement foundation is calculated by adding the footing height and the foundation wall height, and then subtracting the wall projection above the finish grade.

Figure 1-15. Gravel backfill is placed around the outside of a completed foundation wall.

Heavy Construction Excavation

Heavy construction excavation is required in the construction of large buildings. When deep and extensive trench and excavation work is necessary, safety becomes a major factor. Heavy construction excavation work is subject to strict local, state, and national code regulations. Precautions must be taken to ensure that the excavation banks do not collapse and cause injury or death to persons working in the excavation. The method used to protect excavation banks from collapsing depends on the type of of soil in the area, the depth of the excavation, the type of foundation being built, and the space around the excavation.

Before beginning the excavation, the builder must secure all possible information regarding any underground installations in the area including sewer, water, fuel, and electrical lines. Precautions must be taken not to disturb or damage any utility while digging, and to provide adequate protection after they have been exposed. All owners of underground facilities in the area of the excavation should be advised of the work prior to the start of the excavation.

Many safety codes also require that the excavation be inspected by a qualified person after a rainstorm or any other hazardous natural occurrence. Earth bank cave-ins or landslides may be averted by increasing the amount of shoring and other means of protection.

Convenient and safe access to excavated areas must be provided for workers. This may consist of stairways, ladders, or securely fastened ramps.

Some soil types pose greater problems than others during excavation. Sandy soil is always considered dangerous even when allowed to stand for a period of time after a vertical cut. The instability can be caused by moisture changes in the surrounding air or changes in the water table. Vibration from blasting, traffic movement, and material loads near the cut can also cause earth to collapse in sandy soil.

Clay soils present less risk than sand; however, clay can also be dangerous if it is soft. A simple test of clay conditions can be accomplished by pushing a 2 × 4 into the soil. If the 2 × 4 can be easily pushed into the ground, it indicates that the clay is soft and may collapse. Silty soils (a combination of sand and clay) are also unreliable and require the same precautions as sand.

Sloping. If there is sufficient space around the construction site, sloping the earth banks may be all that is necessary to prevent collapse. The Occupational Safety and Health Administration (OSHA) code regulations for the construction industry recommends a 45° slope for excavations with average soil conditions. Solid rock, shale, or cemented sand and gravels may require less slope. Compacted sharp sand or well-rounded loose sand may require more than a 45° slope. See Figure 1-16.

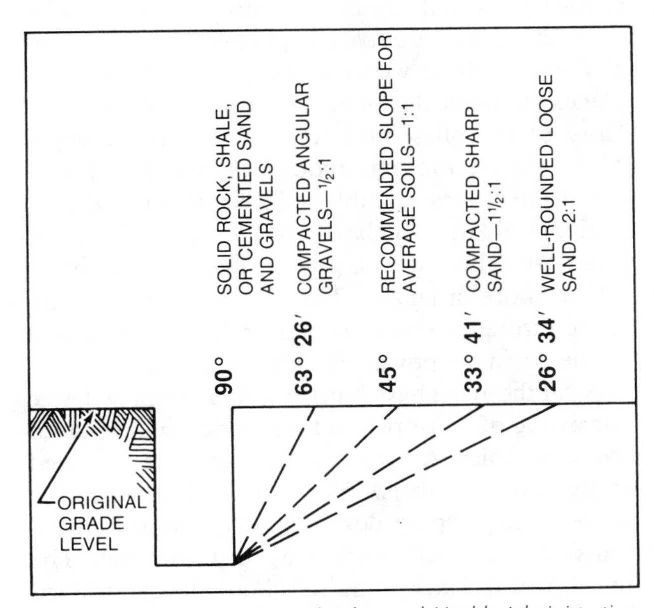

Occupational Safety and Health Administration

Figure 1-16. Soil conditions determine the angle that earth banks are sloped around excavations 5′-0″ or more in depth.

Shoring Vertical Walls. Shoring (supporting) the vertical walls of an excavation is required when sloping is considered unsafe or inadequate. Soil types such as clays, silts, loams, or non-homogenous soils usually require shoring. Shoring may also be required where there is insufficient room for sloped banks. This is particularly true in downtown urban areas where the new construction is right next to existing buildings. In addition to preventing injury from collapse of excavation banks, the stability of the foundation walls of adjoining buildings must be protected. Shoring for high vertical walls is designed by a civil engineer and the installation is supervised by qualified personnel. The shoring system should not be removed until the construction in the excavated area is completed and all necessary steps are taken to safeguard workers. Two methods commonly used to shore high vertical excavation banks are the use of interlocking sheet piling and soldier piles. See Figure 1-17.

Interlocking sheet piling consists of steel pilings that can be reused many times and offers the additional advantage of being watertight. Each individual sheet piling is lowered by crane into a template that holds it in position. The piling is then driven into place with a pile driver. Braces may also be installed to help support the metal sheets.

Soldier piles, also called *soldier beams*, are H-shaped piles that are driven into the ground with a pile driver and are spaced approximately 8′ apart. Three-inch thick wood planks called *lagging* are placed between the flanges or directly against the front of the piles. Soil conditions and the depth of the excavation may require *tie-backs* that consist of steel strand cables placed in holes drilled horizontally into the banks of the excavation. The holes are drilled with a power auger and are often 50′ or more in length. The tie-back cables are inserted through an opening in the pile and are secured in the earth by power grouting the hole.

After the grout has set up, a strand gripping device consisting of a gripper and gripper casing is placed over the cables. A hydraulic tensioning jack is used to tighten the cables. When the jack releases the cables, the gripping device holds them and maintains the required tension against the pile. The number of tie-backs required should be determined by an engineer whose decision will be based on soil conditions and the depth of the excavation. Some soldier pile systems may also include a heavy horizontal steel waler held in place with tie-backs.

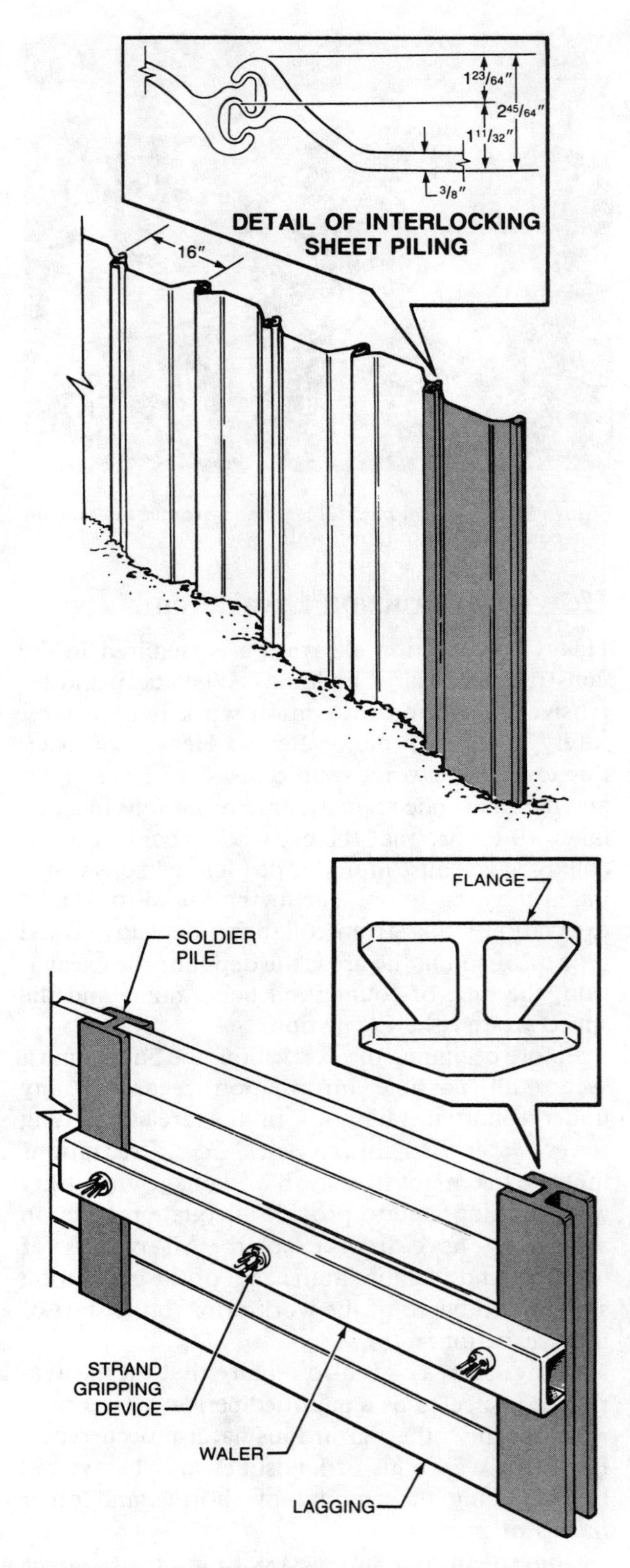

Figure 1-17. Interlocking sheet piling and soldier piles are used to shore the walls of deep excavations for heavy construction projects.

Trenches. Trenches 5 ′ or more in depth, in hard, compact soils are commonly shored by placing vertical timbers on opposite sides of the trench. These timbers are held in place by wood cross braces or screw jacks. The wood cross braces or screw jacks and upright timbers should be spaced a maximum of 5 ′ OC. One cross brace is required for every 4 ′ of the trench depth, with no less than two braces. Trenches dug in loose and unstable soil should be supported with wood sheet piling reinforced with 4 × 4 stringers and cross braces. Maximum spacing between stringers is 5 ′ OC. See Figure 1-18.

Groundwater. The presence of groundwater in a deep excavation can be a serious problem and may hold back the progress of construction work. Water may also temporarily collect because of rain or melting snow. However, more serious problems are caused by underground streams or high water tables in the area. Mechanical pumps are commonly used to remove groundwater accumulating in the excavation. Groundwater in the excavation can also be controlled by lowering the water table in the excavation area. This requires sinking a series of well points and removing the water with a suction pump.

SCHEDULE FOR WOOD HORIZONTAL BRACES FOR HARD, COMPACT SOIL TRENCHES

Trench Width	Minimum Timber Size
1′-0″	4″ × 4″
3′-0″	4″ × 6″
6′-0″	6″ × 6″
8′-0″	Increase Proportionately

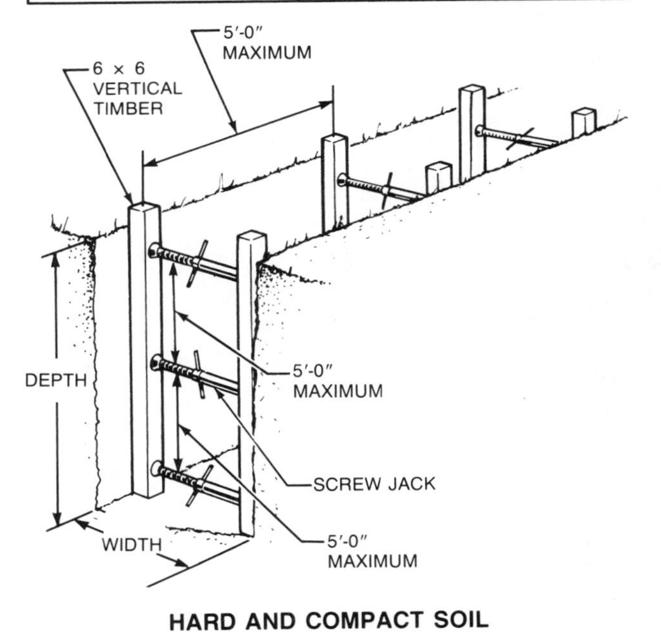

SIZE OF WOOD SHEET PILING FOR LOOSE-SOIL TRENCHES

Trench Depth	Minimum Thickness
4′-0″ TO 8′-0′	2″
OVER 8′-0′	3″

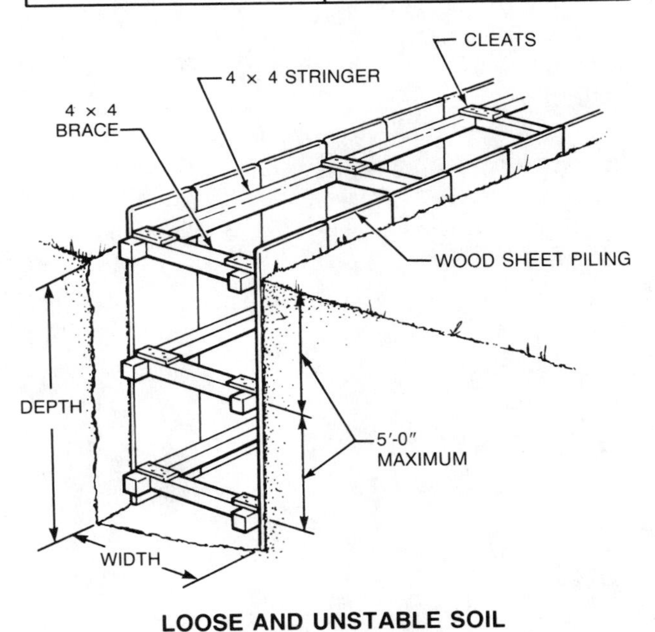

Figure 1-18. Shoring for trenches in hard, compact soil requires vertical timbers and cross bracing. Shoring for trenches in loose, unstable soil requires wood sheet piling in addition to stringers and cross bracing.

Chapter 1—Review Questions

Completion

_____ 1. The size of the soil _____ is the major factor differentiating types of soil.

_____ 2. Soils are generally classified as granular or _____.

_____ 3. Sand and gravel are referred to as _____-grained soils.

_____ 4. Silt and clay are referred to as _____-grained soils.

_____ 5. Sandy soil is subject to less _____ than silt or clay.

_____ 6. The _____ capacity of soil is determined by the weight per square foot of soil area.

_____ 7. Power _____ are often used to bring earth samples to the surface of the ground.

_____ 8. The recommended slope for finish grade around foundation walls is _____″ in _____′ for paved areas.
_____

_____ 9. The recommended slope for finish grade around foundation walls is _____″ in _____′ for nonpaved areas.
_____

_____ 10. The water _____ is the highest point below the surface of the ground that is normally saturated with water.

_____ 11. _____ action is the physical process occurring in soil that causes water to rise from the water table.

_____ 12. Vapor _____ are used for ground covers to control ground surface dampness.

_____ 13. _____ tile is placed around foundation footings to move groundwater away from the foundation.

_____ 14. The _____ line is the depth to which soil freezes in a particular area.

_____ 15. _____ occurs in soil when it freezes and expands.

Multiple Choice

_____ 1. Grade levels on a plot plan are usually expressed in _____.
A. feet and inches
B. fractions of a foot
C. feet and tenths or hundredths of a foot
D. a percentage of a foot

__________ **2.** Contour lines show the __________ of the lot.
A. slope
B. shape
C. location
D. none of the above

__________ **3.** A(n) __________ is established by a surveyor to locate various elevations.
A. hub
B. easement
C. setback
D. bench mark

__________ **4.** When grading a lot, the finished surface should __________.
A. be perfectly level
B. slope from front to back
C. slope away from the building
D. slope toward the center

__________ **5.** The walls of a foundation excavation should be __________ the building lines.
A. 1′ inside of
B. even with
C. at least 2′ outside of
D. 5′ outside of

__________ **6.** Print information regarding the depth of excavation is given in the __________.
A. plot plan
B. foundation plan views
C. foundation section views
D. floor plan

__________ **7.** A rule that applies to excavation of footing trenches requires that the __________.
A. bottom is below the frost line
B. topsoil is removed
C. trench is wide enough to allow for form construction
D. all of the above

__________ **8.** __________ is replacing the soil around the outside of a completed foundation wall.
A. Trenching
B. Excavating
C. Grading
D. Backfilling

__________ **9.** The recommended slope for deep excavations under average soil conditions is __________°.
A. 33
B. 45
C. 70
D. 90

__________ **10.** Soldier piles are used to __________.
A. support the floor of a building
B. shore vertical walls of deep excavations
C. stiffen formwork
D. shore trenches

CHAPTER 2

Wall Form Methods and Materials

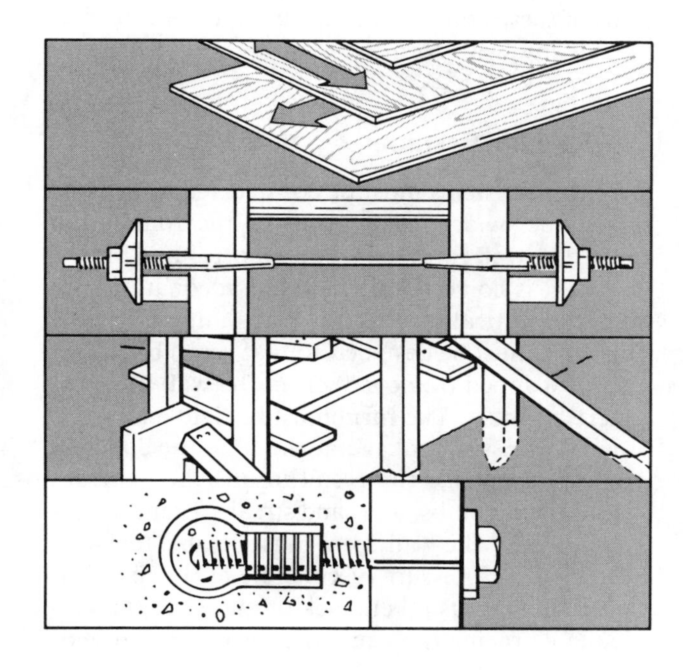

Concrete is used for various types of walls, including foundation walls of residential and heavy concrete structures. A foundation wall rests on a foundation footing and ranges in height from low walls for crawl space foundations to higher and thicker walls for high-rises. The type of foundation footing used depends on the bearing capacity of the soil and type of structure to be constructed.

A concrete wall is formed by placing concrete in a plastic state into wall forms and allowing it to set. Built-in-place wall forms and panel forms are used for residential and light commercial construction. Panel forms and ganged panel forms are used for heavy construction projects. A built-in-place form is constructed in its final position. Panel forms are prebuilt panel sections constructed with studs and a top and bottom plate. Ganged panel forms are large forms constructed by bolting prefabricated panel forms together.

Plywood is commonly used to sheath the form walls. Dimensional lumber is used as stiffeners and supports, and includes base plates, studs, walers, braces, and stakes. Form ties are used to space and tie form walls and prevent them from spreading. Form ties for low wall forms consist of wood cleats nailed to the tops of the form walls. High wall forms are tied together with patented metal ties.

Safe and established construction procedures must be followed when constructing wall forms. Consult the American Concrete Institute (ACI) or Occupational Safety and Health Administration (OSHA) for information regarding safe construction procedures.

WALL FORM CONSTRUCTION

Wall forms are constructed in various shapes, heights, and thicknesses. Two methods are commonly used for high walls. One method consists of sheathing stiffened by studs and double walers. A second method consists of sheathing stiffened by single walers and strongbacks. See Figure 2-1. Both systems require braces to hold the walls in position and devices to tie the opposite walls together. Variations of these two methods are used in the construction of low wall forms.

Wall Form Systems

Wall form systems must be constructed to the correct shape, width, and height of the foundation walls. The wall form must be supported and braced so that it is correctly aligned and adequately supports all vertical and lateral loads imposed on it. The form must be designed so that it can be conveniently stripped (removed) from the wall after the concrete has set. The form must be tight enough to prevent excessive leakage at the time the concrete is placed. Leakage can result in unsightly surface ridges, honeycombs, and sand streaks after the concrete has set. The wall form must be able to safely withstand the pressure of concrete at the time it is placed. Short cuts taken, lack of materials, and inadequate bracing can cause the form to collapse or move during concrete placement.

For residential and other light construction projects, *built-in-place* or *panel* systems are commonly used. Built-in-place forms are constructed in place on the job site. Panel systems consist of prefabricated panel sections framed with studs and a top and bottom plate. Heavy construction projects usually require panel forms or *ganged panel forms.* A ganged panel form consists of a number of prefabricated panel forms tied together to create a much larger single panel.

Built-in-Place Forms. Built-in-place forms are constructed in place over a footing or concrete slab that acts as a platform for the wall form. The most common procedure is to fasten a base plate to the footing or concrete slab for the outside wall with powder-actuated fasteners or concrete nails. Wood studs are nailed to the base plate and tied together with a ribbon board. The sheathing is then nailed to the studs and the walers are placed. See Figure 2-2. After securing the walers, wall ties are inserted through predrilled holes in the sheathing. Wall ties extending from the outside walls are inserted through holes in the inside wall. The inside form walls are constructed in a manner similar to the outside form walls. The inside walls are then aligned and braced.

A simplified built-in-place plank system for forming low walls has become popular in many areas. The forms are constructed of 2″ thick planks (actual size 1½″) and tied together with 2 × 4 stakes or uprights. A flat form tie secured with tapered wedges spaces and holds the opposite walls together. See Figure 2-3. Standard-sized ties can be obtained for wall thicknesses ranging from 6″ to 16″. This type of wall form is commonly used for low foundation walls and footings not exceeding 4′ in height. However, forms for higher walls may also be constructed using this method.

Panel Forms. Panel form systems consist of prebuilt panel sections framed with studs and a top and bottom plate. Panel form systems increase the speed and efficiency of construction. The panel sections can be built in the shop or on the job, and with proper care can be reused many times. If many similar panel sections are required, a template table facilitates the construction of the panels. A template table is constructed of plywood nailed to a frame and supported by legs. The studs are placed between cleats laid out to the correct spacing of the studs, and the sheathing is nailed to the studs. See Figure 2-4.

After the panel sections are prefabricated they are placed into position. Smaller panels can be placed by hand; larger panels require lifting equipment. Panel sections are fastened to the footings by driving concrete nails or powder-actuated fasteners through the base plate of the panel section. The panels are fastened to each other with bolts or 16d duplex nails. A filler panel is placed in any leftover space that is less than a full panel width. Walers are then nailed to the panel studs to keep the panels aligned. Vertical strongbacks, if required, are then fastened to the walers. After securing the walers or strongbacks, wall ties are inserted through predrilled holes in the panels. The inside walls are constructed in a manner similar to the outside walls. As the inside panels are tilted into place, the wall ties are inserted through predrilled holes. Braces are attached to the walers

The Burke Company

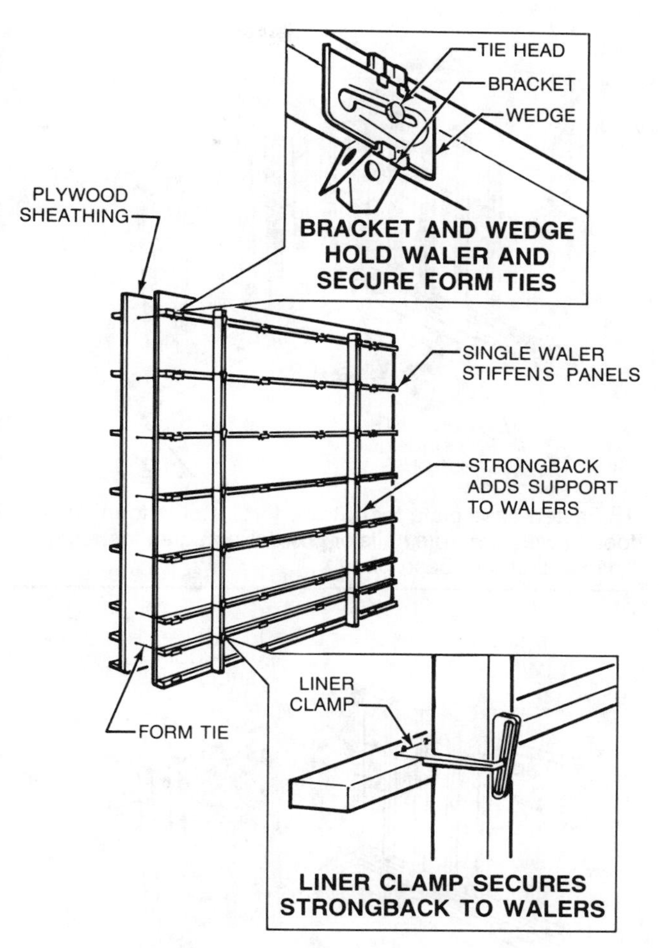

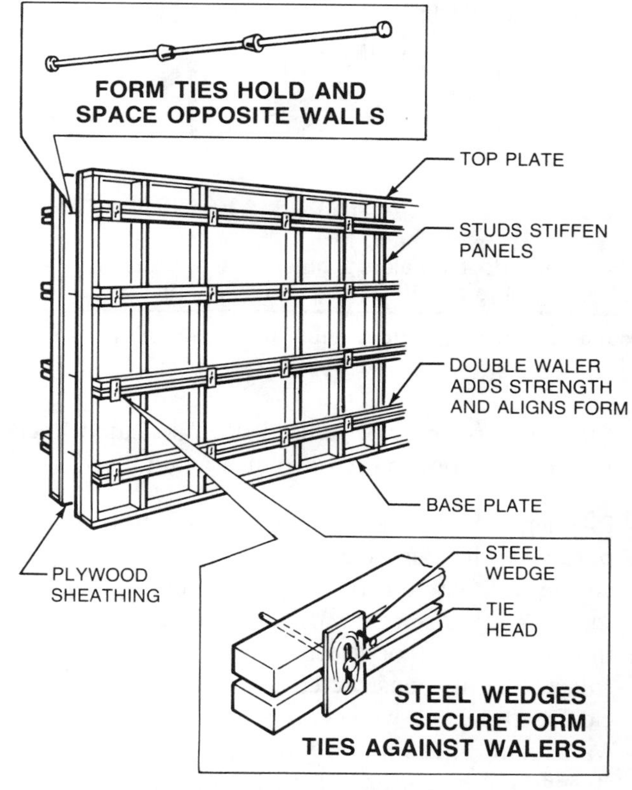

The Burke Company

Figure 2-1. Single or double waler systems may be used to form high foundation walls.

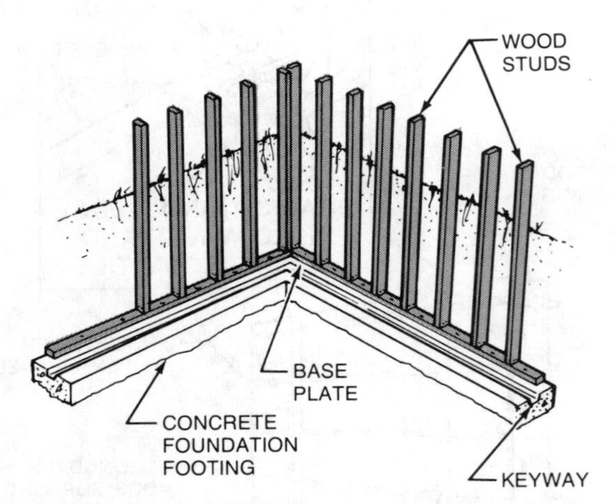

1. Fasten base plate for outside form wall to foundation footing with concrete nails or powder-actuated fasteners. Toenail studs to base plate.

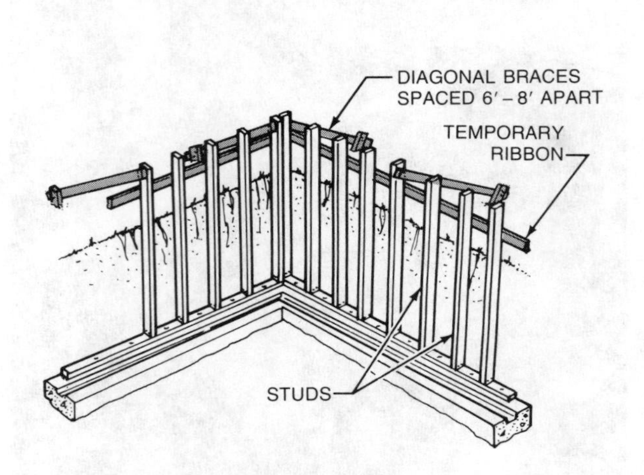

2. Tie studs together with temporary ribbon board. Secure studs and ribbons with diagonal braces every 6′ to 8′.

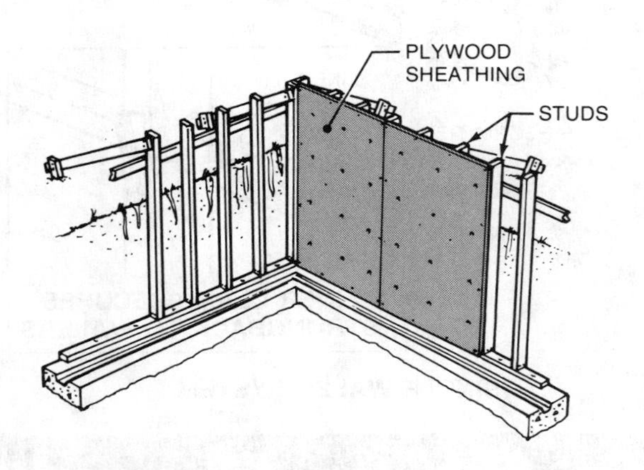

3. Nail plywood sheathing to inner edge of studs.

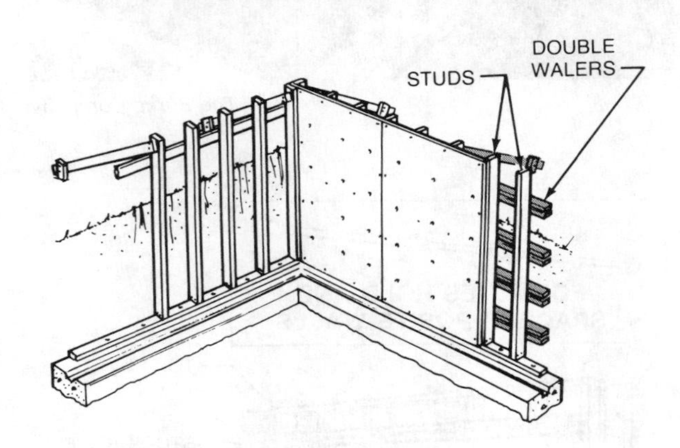

4. Attach double walers to outer edge of studs. Align and brace wall form.

Figure 2-2. Built-in-place wall forms are constructed over a foundation footing or floor slab. Plywood sheathing is reinforced with studs, walers, and braces.

or strongbacks and the wall is aligned and braced. See Figure 2-5.

Single Wall Forms. Single wall forms are built when conditions make it difficult to remove one side of a form after the concrete has set. This might occur where a foundation wall is to be constructed very close to or against the existing foundation of an adjoining building, or where a wall is constructed next to solid rock or extremely hard soil. Single wall forms consist of plywood sheathing, studs, walers, strongbacks, and braces. Single wall forms must be heavily braced to hold them in position. See Figure 2-6. Single wall forms may be built-in-place, or prefabricated panels can be used. Prefabricated steel-framed plywood panels are also available.

Pilasters. A pilaster is a rectangular column incorporated with a concrete wall and is used to strengthen the wall where it receives a concentrated structural load. Pilasters are also constructed to support the ends of beams or girders spanning concrete walls. Pilaster forms constructed of plywood sheathing, cleats, studs, and walers are erected along with the wall form. Kickers and patented ties are used to strengthen the corners of the form. See Figure 2-7. The footing should project out to form a base for the pilasters wherever they are placed in foundation walls.

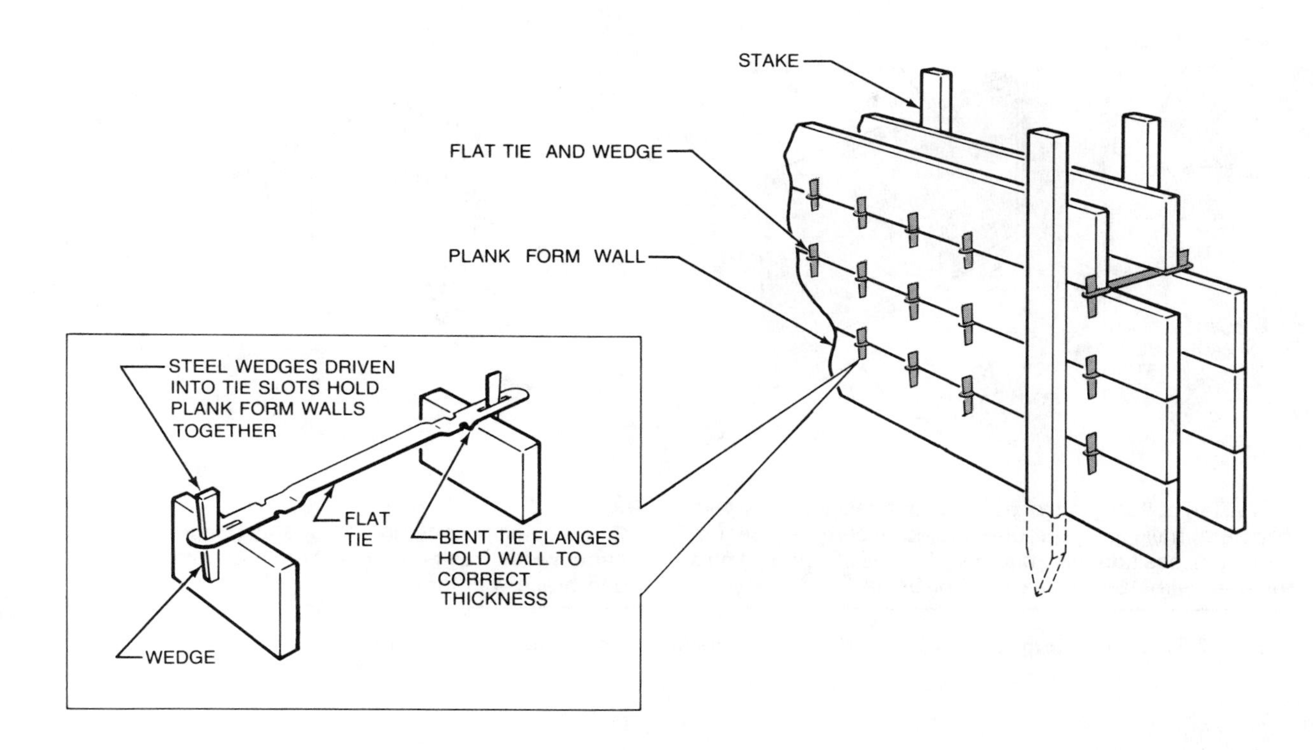

Figure 2-3. The built-in-place plank method is used to build low wall forms.

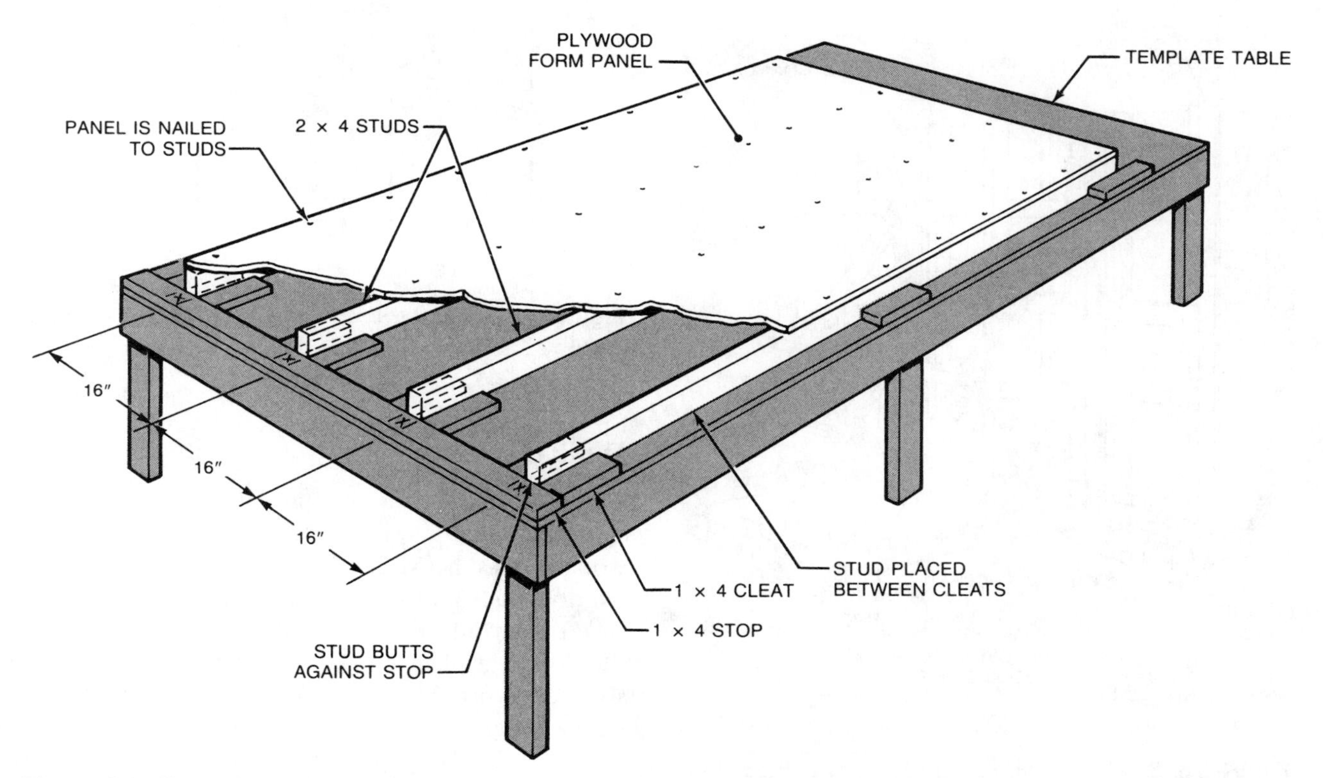

Figure 2-4. A template table facilitates construction of prefabricated panel form sections.

DIAGONAL BRACES
PANELS NAILED TOGETHER
STAKE
BOTTOM PLATE FASTENED TO FOOTING
FOUNDATION FOOTING
KEYWAY

1. Fasten bottom plate of panel to foundation footing with concrete nails or powder-actuated fasteners. Secure panels in position with diagonal braces. Fasten panels together with 16d duplex nails or bolts.

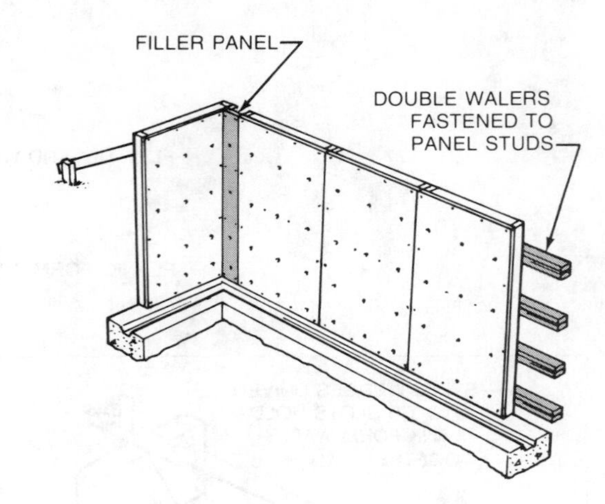

2. Place filler panel in any openings that are not a full panel width. Fasten double walers to panel studs. Align and brace wall form.

Figure 2-5. A panel form system is a cost-effective method of constructing wall forms.

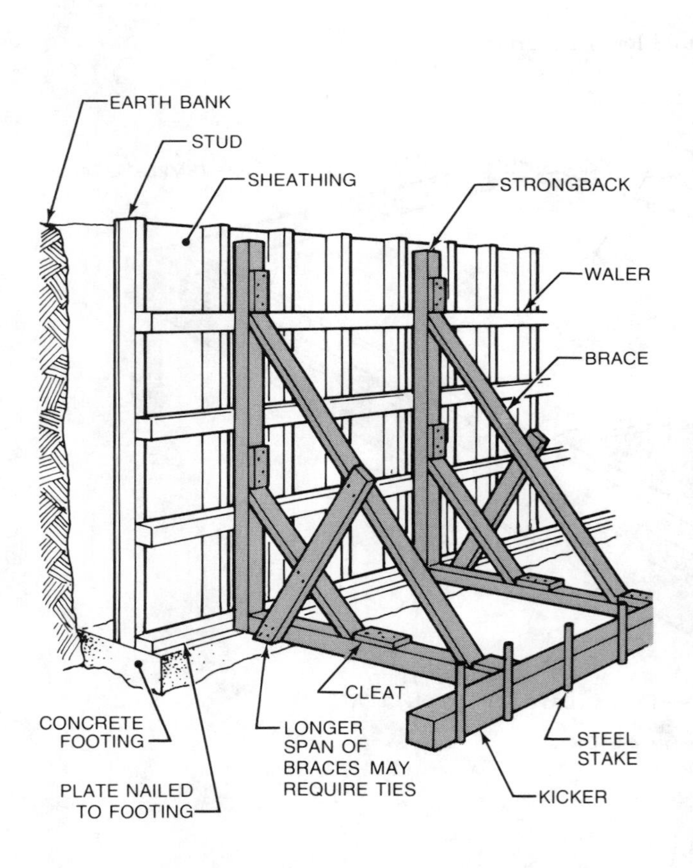

Figure 2-6. Single wall forms must be heavily braced to secure them in position.

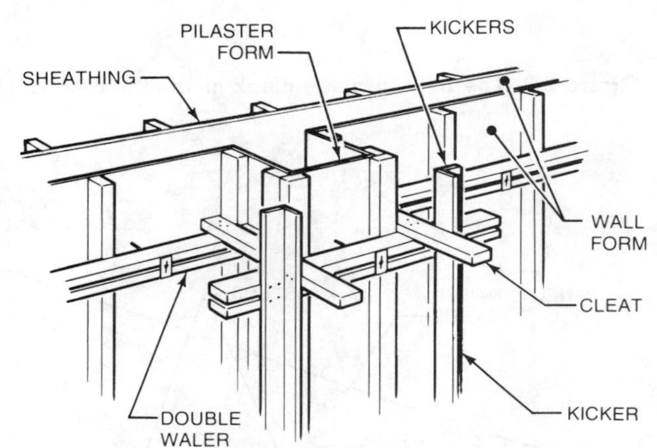

Figure 2-7. A pilaster is a rectangular column joined to a wall.

Wall Openings

Wall openings for doors and windows are created by placing *bucks,* which are rigid and well-braced wood frames, between the two form walls. Door and window bucks must be rigid enough to resist pressure that would result in distortion at the time of the concrete placement. The bucks must be constructed to provide an opening in the concrete form large enough to accommodate the door and window frame, and also allow some clearance for plumbing and leveling the frame. A typical buck design consists of an outside frame reinforced by 2 × 4s placed on edge and horizontal cross braces. The frame may also be built of 2″ thick planks or ¾″ plywood. More bracing is required with thinner material. Arches are formed over door and window openings by fastening an arched section at the top of the door or window buck. The arched section is laid out on two pieces of plywood that are then secured to the top of the buck. The arched form is then enclosed with ¼″ thick plywood. See Figure 2-8. If a recess is required to receive the frame, a *recess strip* is nailed to the outside of the buck's frame. If a wood nailing strip is to be provided in the concrete, a wedged piece is attached to the frame. An inspection pocket may be cut out at the bottom of a window buck to observe the flow and consolidation of the concrete beneath the buck. When the concrete reaches the bottom of the buck, the inspection pocket is replaced and cleated down.

Door and window bucks are usually placed after the outside form walls have been erected. The bucks can be laid out and built against the wall. See Figure 2-9. Many form builders, however, prefer to prefabricate the bucks and then attach them to the form wall.

Wood or metal door and window units are used for the finished openings in concrete walls. The wood units, and some types of metal units, are commonly placed in the openings after the concrete sets. Other metal door and window units are set in the form prior to placing the concrete. In this case, the metal door jamb and window frame also serve as bucks and must be carefully plumbed, aligned, and well-braced. See Figure 2-10.

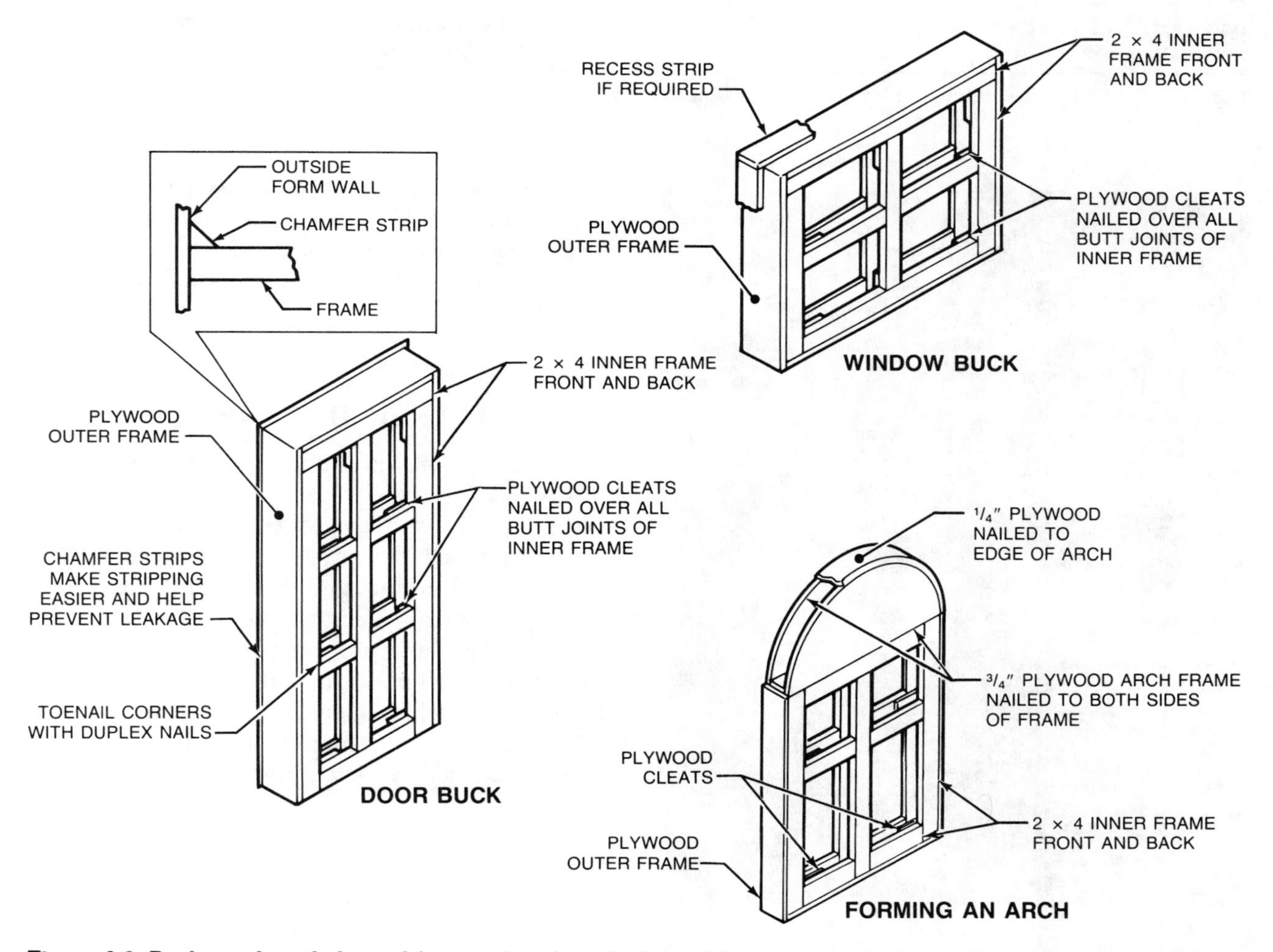

Figure 2-8. Bucks are heavily braced frames placed inside the wall form to provide door and window openings in concrete walls.

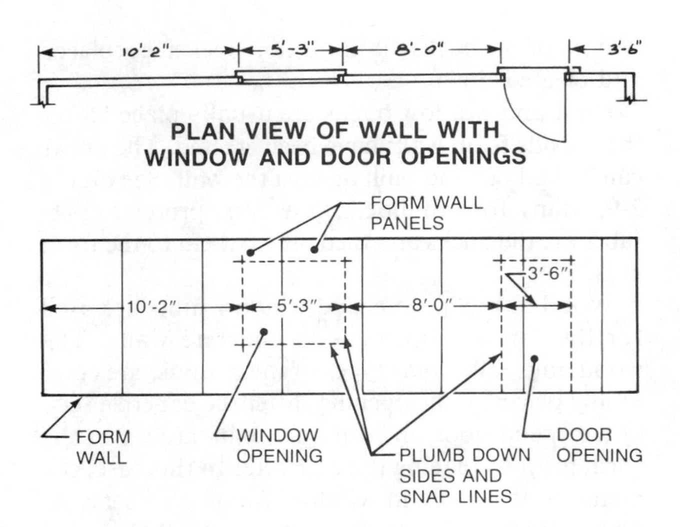

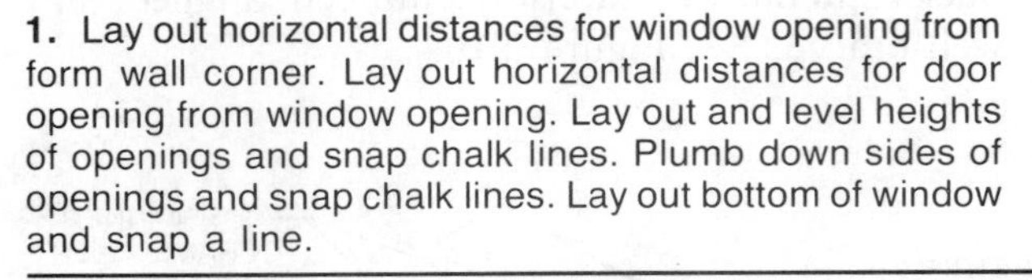

1. Lay out horizontal distances for window opening from form wall corner. Lay out horizontal distances for door opening from window opening. Lay out and level heights of openings and snap chalk lines. Plumb down sides of openings and snap chalk lines. Lay out bottom of window and snap a line.

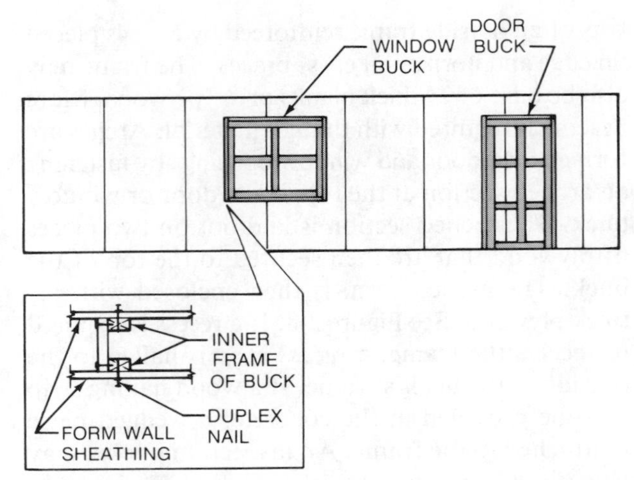

2. Set door and window bucks to snapped lines. Drive duplex nails through the form wall sheathing into the inner frames of the bucks.

Figure 2-9. Door and window buck placement is determined by using the plan view.

Western Forms, Inc.

Figure 2-10. Permanent metal frames placed in the wall forms must be carefully plumbed, aligned, and braced.

Small openings for large diameter pipes, vents, heating and ventilating ducts, and utilities may also be required. Small vent, duct, and pipe openings can be formed by constructing wood box frames to the size of the openings. Cellular plastic blockouts and fiber and sheet metal sleeves are also used.

MATERIALS AND COMPONENTS

Sheathing (usually plywood) and the lumber (2 × 4, 4 × 4, 2 × 6, etc.) used for structural form components such as studs, walers, strongbacks, and bracing are used to construct wall forms. The thickness of the sheathing and the size and spacing of the structural components are determined by the anticipated loads and pressures exerted on the form. Forms for small and light wall forms are designed based on established procedures. However, wall forms for larger and more complicated structures may be designed by a structural engineer.

Plywood

Plywood sheathing is most commonly used to surface wall forms. Plywood is available in large panels, which reduces labor costs involved in constructing and stripping forms. In addition, plywood produces fewer joint impressions on the finished concrete walls, thus reducing the cost of finishing and rubbing the surface.

Plywood is a manufactured panel product made up of *veneers* (thin layers of wood) that are glued together under intense heat and pressure. The panels are always made up of an odd number of layers, allowing the front and back veneer grain to run the same direction. Each layer is placed with the grain running at a right angle to the adjacent layer. This process, known as *cross-lamination*, increases the strength of the panel and minimizes shrinking and warpage. The outside surfaces of a plywood panel are the *face veneer* and *back veneer*. The inner veneers are the *crossbands* and the center layer is the *core*. See Figure 2-11.

Most plywood panels used in formwork are manufactured from softwood lumber species such as Douglas fir. Other woods used are pine, spruce, larch, and hemlock. The most common form panel thicknesses are ½″, ⅝″, and ¾″. Standard size 4′ × 8′ sheets are used most often; however, sheets 5′ wide and lengths ranging from 5′ to 10′ are also stocked by suppliers. Some builders prefer 5′ × 10′ panels because they can be cut to 2′-6″ × 10′ sections and handled more easily by one person.

The manufacturing of plywood is governed by strict product standards established by the U.S. Department of Commerce in cooperation with industry associations such as the American Plywood Association (APA). The two major types of plywood are *exterior* and *interior*. Exterior plywood is bonded with waterproof glue and interior plywood is bonded with water-resistant glue. Each type is further broken down into grades that are based on the condition and appearance of the outer veneers. Grade-use guides that provide information about plywood grades are published by various industry associations.

Plyform®. Plyform® is a plywood product specifically designed for concrete formwork. It is the most highly recommended panel for most forming operations. The Grade-Use Guide for Concrete Forms identifies three grades of Plyform® as B-B Plyform®, High Density Overlaid (HDO) Plyform®, and Structural 1 Plyform®. Additional information regarding Plyform®, such as typical trademarks and recommendations for usage, may also be found in the guide. See Figure 2-12.

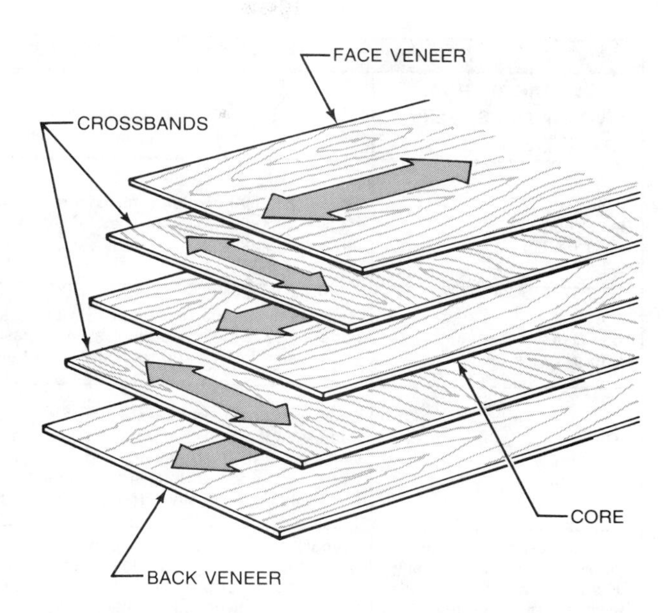

Figure 2-11. Plywood consists of an odd number of veneers. The grain direction of adjacent layers runs perpendicular to each other.

Textured Plywood. Textured plywood is used to produce special surface effects such as wood grain and boards. See Figure 2-13. A textured wall surface is produced by using ¼″ textured plywood as a liner for a thicker structural panel or a thicker textured form panel without a liner. By using a thicker textured form panel, labor is minimized. However, the textured panels can only be reused a limited number of times because the textured surfaces damage easily when the panels are stripped from the concrete walls.

Deflection and Panel Strength. Deflection (bending) of a wall form occurs when panels of inadequate thickness are used, or when the stiffeners (studs or walers) are spaced too far apart. A wall form must be constructed to hold the concrete to a straight and true surface. See Figure 2-14. The required panel thickness for a particular form design must take into consideration the maximum concrete pressure anticipated for that form, as well as the spacing and type of stiffeners used.

An important factor when positioning plywood in the wall form is the direction of the grain in relation to the panel stiffeners. Plywood has much greater strength when the face grain is placed perpendicular rather than parallel to the stiffeners.

GRADE-USE GUIDE FOR CONCRETE FORMS*					
Use These Terms When You Specify Plywood	Description	Typical Trademarks	Veneer Grade		
			Faces	Inner Plies	Backs
APA B-B Plyform Class I & II**	Specifically manufactured for concrete forms. Many reuses. Smooth, solid surfaces. Mill-oiled unless otherwise specified.	APA PLYFORM B-B CLASS I EXTERIOR 000 PS 1-83	B	C	B
APA High Density Overlaid Plyform Class I & II**	Hard, semi-opaque resin-fiber overlay, heat-fused to panel faces. Smooth surface resists abrasion. Up to 200 reuses. Light oiling recommended between pours.	HDO•PLYFORM 1• EXT•APA•PS1-83	B	C-Plugged	B
APA Structural 1 Plyform**	Especially designed for engineered applications. All Group 1 species. Stronger and stiffer than Plyform Class I and II. Recommended for high pressures where face grain is parallel to supports. Also available with High Density Overlay faces.	APA STRUCTURAL I PLYFORM B-B CLASS I EXTERIOR 000 PS 1-83	B	C or C-Plugged	B
Special Overlays, Proprietary panels and Medium Density Overlaid plywood specifically designed for concrete forming**	Produces a smooth uniform concrete surface. Generally mill treated with form release agent. Check with manufacturer for specifications, proper use, and surface treatment recommendations for greatest number of reuses.				
APA B-C EXT	Sanded panel often used for concrete forming where only one smooth, solid side is required.	APA B-C GROUP 1 EXTERIOR 000 PS 1-83	B	C	C

*Commonly available in 19/32″, 5/8″, 23/32″, and 3/4″ panel thicknesses (4′ × 8′).
**Check dealer for availability in your area.

American Plywood Association

Figure 2-12. The grade-use guide is used to determine the type of plywood to use in the construction of forms. The trademarks are usually stamped on the back veneer.

Symons Corporation

Figure 2-13. Special surface designs, such as a brick pattern, are produced with textured plywood.

Curved Forms. Plywood is also used to sheath curved forms. The bending capacity of dry plywood is based on the *minimum bending radius*. The minimum bending radius is the smallest radius that the plywood can be subjected to without structural damage. See Figure 2-15. A 1/4″ thick panel bent across the grain has a 2′ bending radius and can be shaped to a curve having a 2′ radius. Shorter radius curves can be produced by bending plywood across the grain rather than with the grain.

Gradual curves can be made with dry panels ranging in thickness from 1/4″ to 3/4″. Shorter radius curves can also be constructed by wetting or steaming the panel, or cutting spaced kerfs in the panel. Another curving method is to use two thinner panels instead of a single thick panel.

Preparation and Maintenance. Proper preparation and maintenance allows plywood panels to be used several times. Oiling the faces of plywood panels before use reduces moisture penetration that may

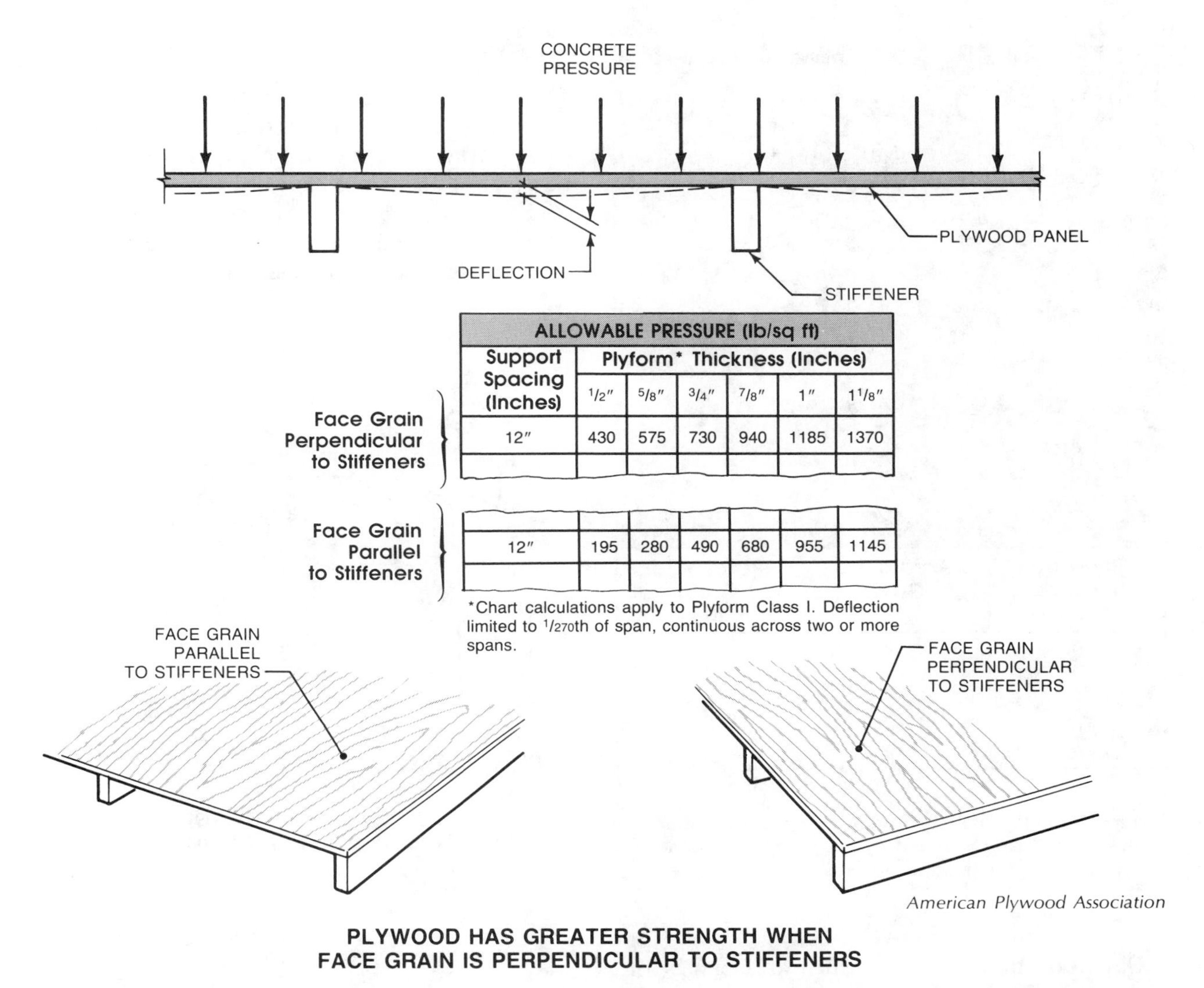

ALLOWABLE PRESSURE (lb/sq ft)							
	Support Spacing (Inches)	Plyform* Thickness (Inches)					
		1/2"	5/8"	3/4"	7/8"	1"	1 1/8"
Face Grain Perpendicular to Stiffeners	12"	430	575	730	940	1185	1370
Face Grain Parallel to Stiffeners	12"	195	280	490	680	955	1145

*Chart calculations apply to Plyform Class I. Deflection limited to 1/270th of span, continuous across two or more spans.

Figure 2-14. Deflection of a wall form can occur at the time of concrete placement if the plywood panels are not thick enough or the stiffeners are placed too far apart.

damage the panel. A liberal amount of oil should be applied a few days before the plywood panel is to be used, and then wiped clean so only a thin layer remains. The oil also acts as a release agent, making it easier to strip the form panels from the concrete walls. Other release agents including waxes, oil emulsions, cream emulsions, and water emulsions can also be sprayed on the panel.

After a panel has been stripped from the concrete wall, it should be inspected for wear and cleaned with a fiber brush. Necessary repairs such as patching holes with patching plaster or plastic wood should be made. Tie holes may be patched on the inside of the forms with metal plates. The panel should then be lightly oiled. Panels should be laid flat and face-to-face when stored, and kept out of the sun and rain. If this is not possible, the panels should be protected by a loose cover that allows air circulation without heat buildup.

Framework, Stiffeners, and Bracing

Framework, stiffeners, and bracing lumber for concrete forms should be straight and structurally sound. Partially seasoned (dry) stock is recommended because fully dried lumber tends to swell when it becomes wet and creates distortions when aligning the forms. Completely unseasoned (green) lumber dries out and warps during hot weather, which also causes distortions.

MINIMUM BENDING RADII		
Plywood Thickness (in.)	Across the Grain (ft.)	Parallel to Grain (ft.)
1/4	2	5
5/16	2	6
3/8	3	8
1/2	6	12
5/8	8	16
3/4	12	20

American Plywood Association

Figure 2-15. Plywood is bent to form curves. Shorter radius curves are produced by bending plywood across the grain rather than with the grain.

Softwood lumber species such as Douglas fir, southern pine, spruce, and hemlock are generally utilized for structural purposes. When ordering stock, the estimate of lumber should be based on minimal waste because much of this material is reused after the forms are stripped from the finished concrete walls.

The most frequently used stock is 2 × 4s; however, 2 × 3s, 2 × 6s, 4 × 4s, and 2″ planks of various widths are also common. *Actual* lumber sizes differ from *nominal* lumber sizes; for example, a 2 × 4 is actually $1\frac{1}{2}'' \times 3\frac{1}{2}''$. This difference is a result of the amount of waste removed from the rough lumber during surfacing in the planing mill, and the anticipated shrinkage while it is drying. (See Standard Lumber Size table in Appendix B.)

The key factor in determining the size and spacing of stiffeners for a form wall is the pressure exerted when the concrete is being placed into the form. Other factors that determine the size and spacing of stiffeners include the rate of placement, ambient temperature, and consistency of the concrete.

Various methods are used to stiffen and brace low wall forms. One method requires studs or stakes, a single waler placed toward the top edge of the form, wood ties, and braces. Another low wall forming method requires 2″ thick planks, stakes, wood ties, and braces. See Figure 2-16.

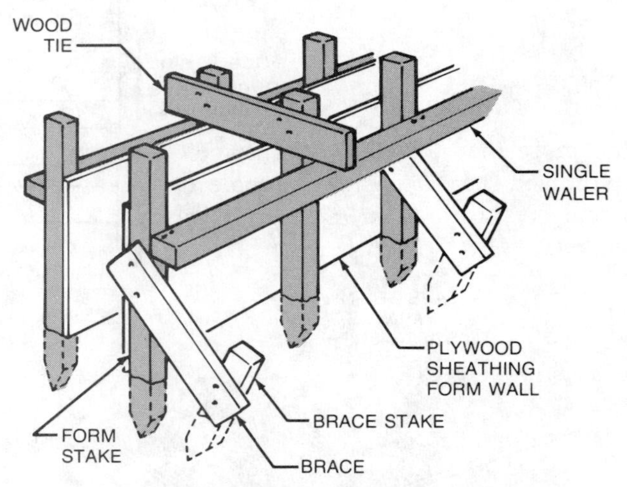

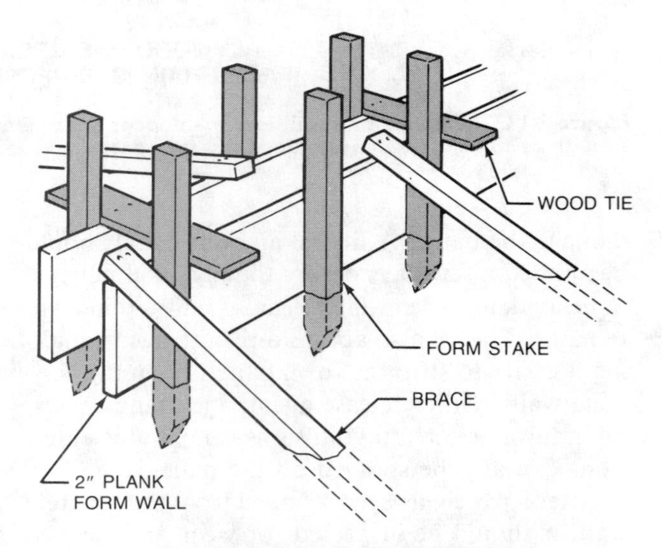

Figure 2-16. Low wall forms are stiffened by using stakes, studs, or wood ties.

Studs, Walers, and Strongbacks. Most high wall forming systems are constructed and stiffened with a combination of studs and walers, or walers and strongbacks. An older but still widely used method for stiffening higher form walls consists of vertical studs and double walers (usually 2 × 4s). The studs give rigidity to the form panel and are spaced 12″, 16″, or 24″ apart, depending on the anticipated concrete pressure. The double walers (wales) are positioned horizontally and are fastened to the studs with nails, clips, or brackets. The double walers reinforce the studs, align and tie together the form panels, and provide a wedging surface for patented wall ties. Typical spacing for double walers is to place walers within 12″ of the top and bottom of the wall, and no more than 24″ OC for the intervening rows. However, the spacing may vary, depending on concrete pressure and the size and spacing of other materials. For walls where precise alignment is required, vertical strongbacks may be fastened behind the double walers.

A newer forming method eliminates the studs and uses only single walers strengthened by vertical strongbacks. Eliminating the studs requires the walers to be spaced closer together. Single walers are placed within 8″ of the top and bottom of the wall, and no more than 16″ OC for the intervening rows. See Figure 2-17. The spacing of the single walers may be closer to accept greater concrete pressure. Under normal circumstances, vertical strongbacks placed over single walers are spaced every 6′, or closer if conditions require.

Corner Ties. The corners of wall forms are subjected to extreme pressure during the concrete placement. Corners must be tied and braced so that a form failure does not occur. An effective method to tie corners together is to overlap the walers at the corners and nail kickers against the walers. Patented metal devices are also available for tying the corners. See Figure 2-18.

Bracing Wall Forms. Lateral pressure against a wall form is caused by the movement and force exerted when placing the concrete, wind load, and pressures resulting from the weight of workers and materials on scaffolds attached to the form. Adequate lateral bracing must be provided to keep the walls straight and prevent collapse.

The two forces acting on a form brace are *tension* (pulling away) and *compression* (pushing against). Some braces are designed to handle both tension and compression and are only required on one side of the form. Other braces are only effective for compression and may be required on the two opposite sides of the form.

Wood form braces are 2 × 4s set at approximately a 45° angle. Braces set at less than a 45° angle are subject to greater force. Braces placed at more than a 45° angle are adequate for walls 6′ or less in height. A brace placed at more than a 45° angle with higher walls presents a danger of bending and becoming ineffective. Long braces (12′ or more) should be strengthened at the midpoint by nailing stiffeners across the braces. The stiffeners should

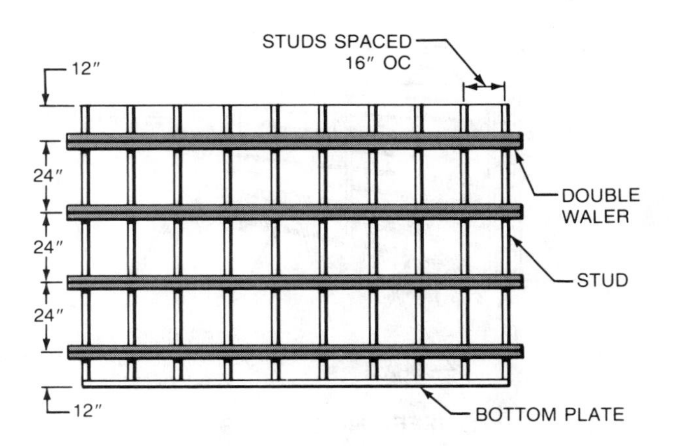

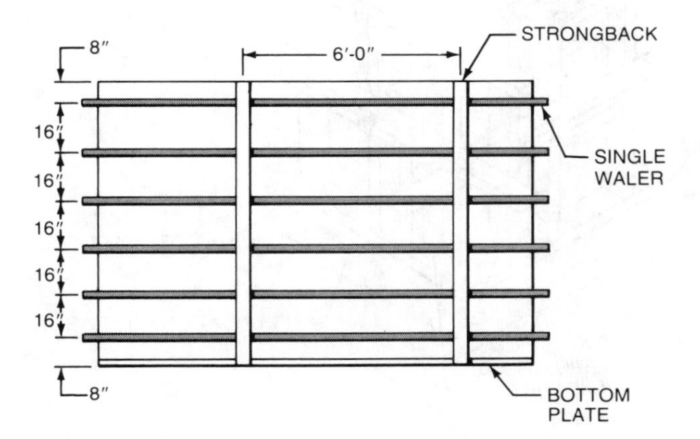

Figure 2-17. Minimum waler spacing must be maintained to accept the pressure of the concrete during placement.

also be braced to the ground to resist bending. See Figure 2-19. For very high walls with long braces, two rows of stiffeners, one high and one low, may be required. Typical spacing for braces is every 8′ for form walls up to 8′ in height, and every 6′ for walls up to 10′ in height.

Braces on higher walls should be fastened to a waler or strongback. See Figure 2-20. Braces should be attached at the lower end to the top stakes or pads. The method used depends on the type of surface present.

Patented metal devices are also available to facilitate bracing operations. One commonly used device features a turnbuckle that is fastened at either end of a 2 × 4 wood brace by driving nails or screws through a metal angle bracket. When placed at the upper end of the wood brace, a metal brace plate

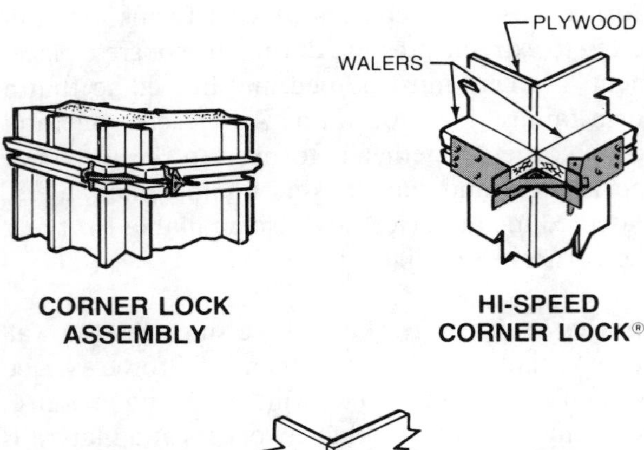

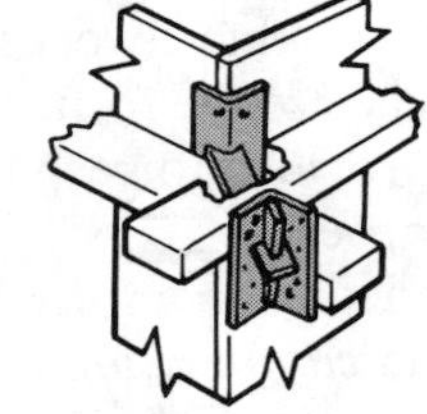

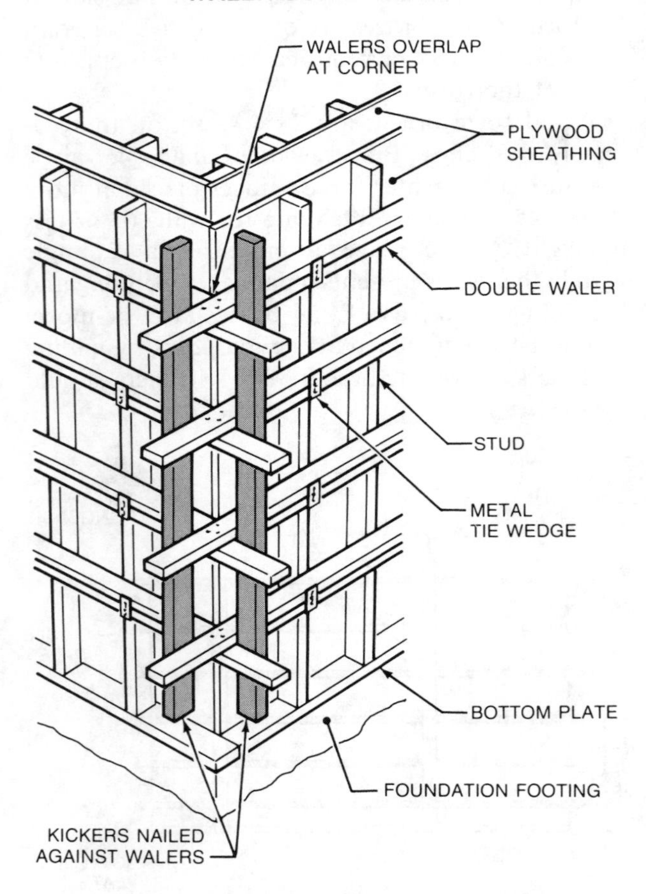

Figure 2-18. Wall form corners must be tied together to withstand the extreme pressure of the concrete.

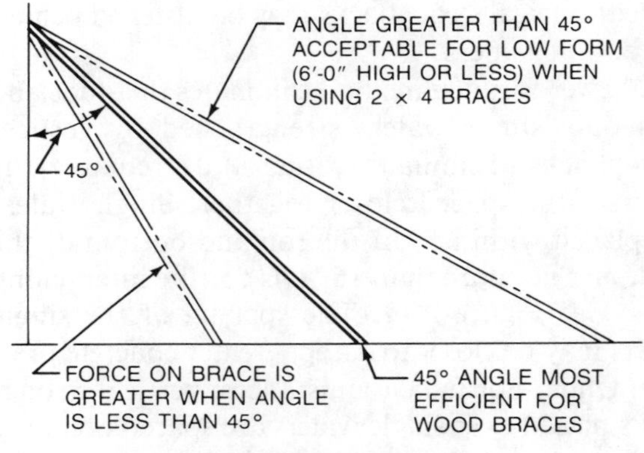

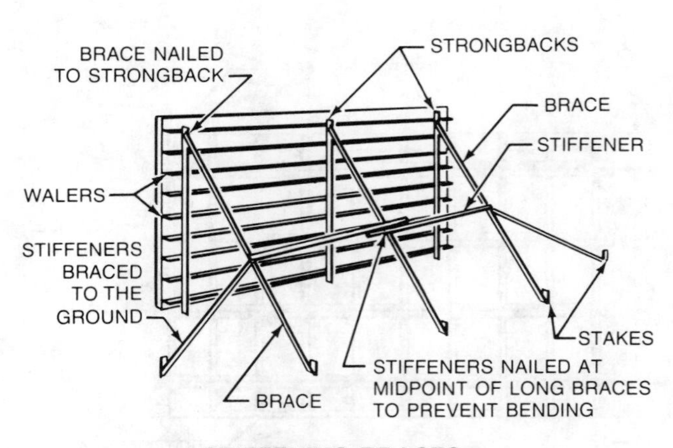

Figure 2-19. Forces affecting wall forms must be counteracted by using proper bracing angles and methods.

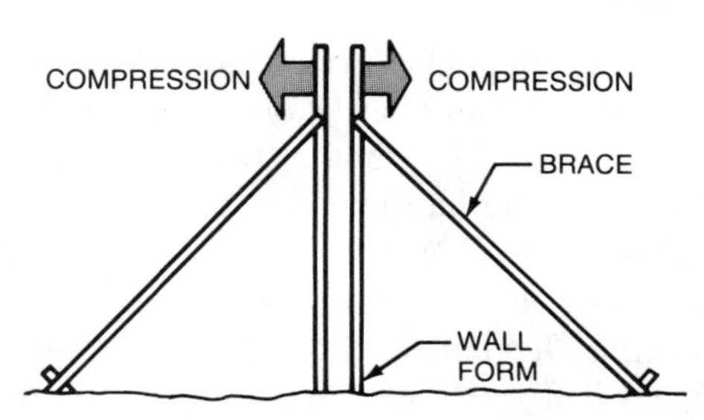

Compression forces require braces on two opposite sides of form. Braces designed for tension and compression strength are attached to one side of form only.

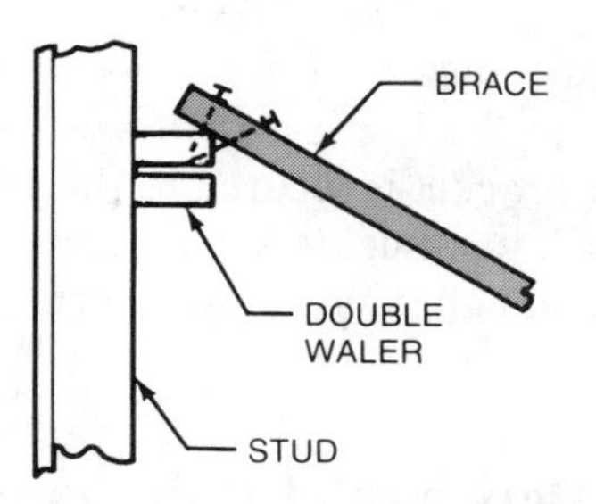

Brace attached to waler when strongback is not used. Brace can be attached to stud on low wall forms.

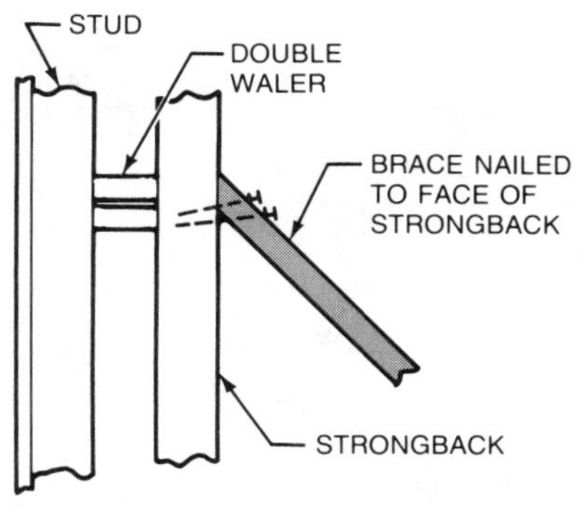

Brace nailed to face of strongback. Good compression strength.

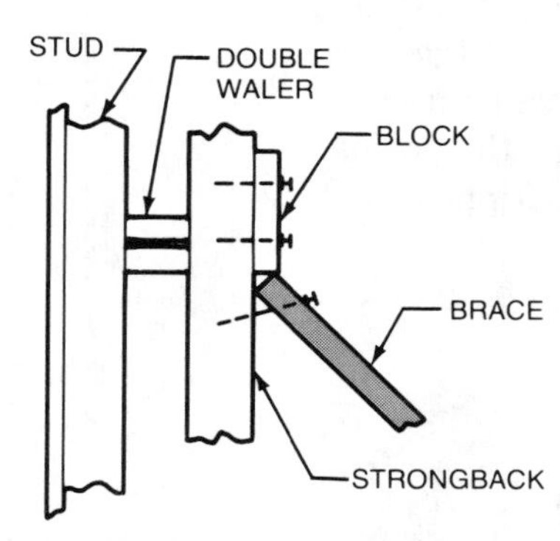

Brace nailed to strongback and block. Good compresion strength.

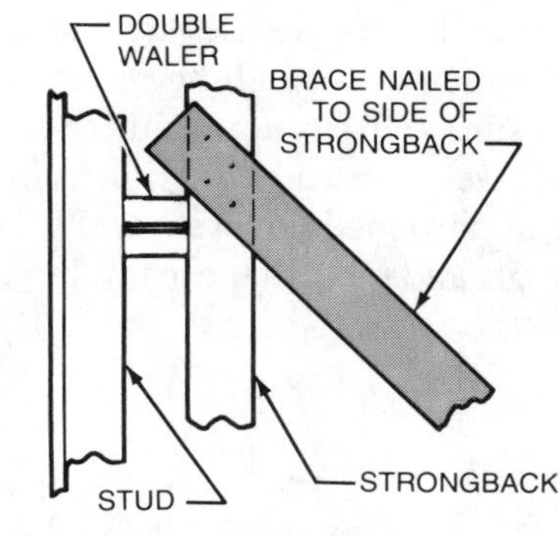

Brace nailed to side of strongback. Good tension and compression strength.

ATTACHING BRACES TO WALL FORMS

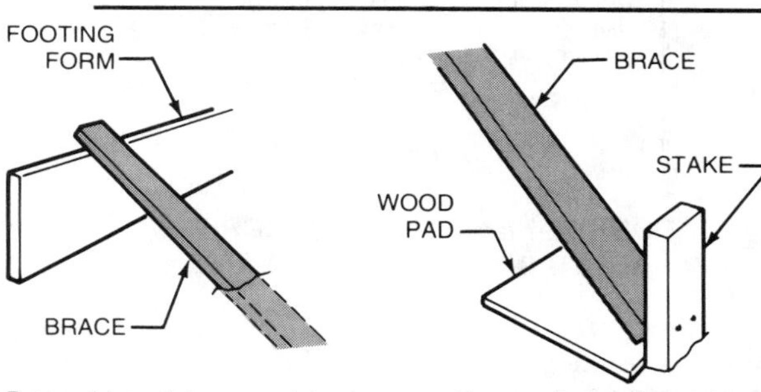

Brace driven into ground for low footing form. Good compression strength.

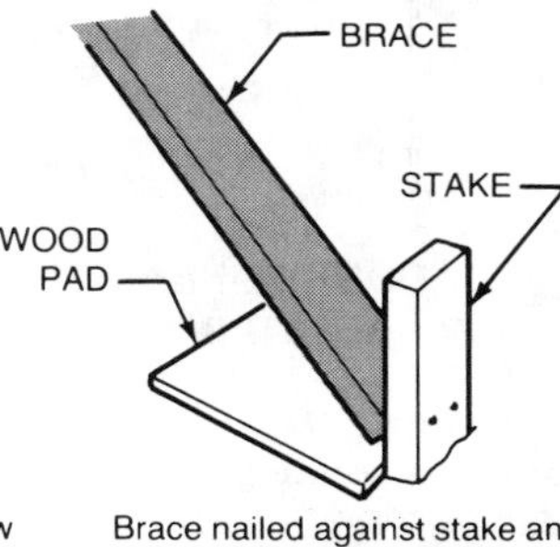

Brace nailed against stake and rests on wood pad. Good compression strength.

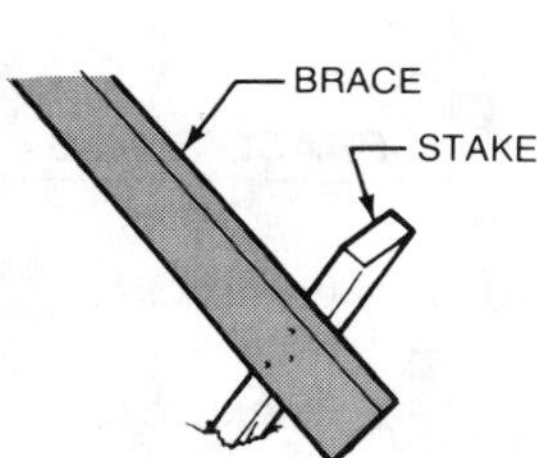

Brace nailed against edge of stake. Use for loose soil. Good compression and tension strength.

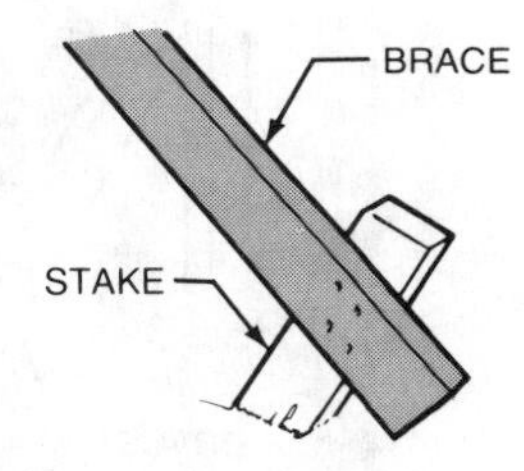

Brace nailed against face of stake. Use for firm soil. Good compression and tension strength.

ANCHORING BRACES TO THE GROUND

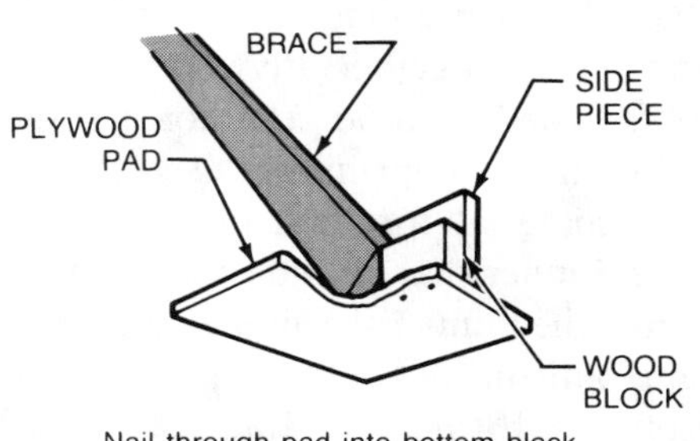

Nail through pad into bottom block.

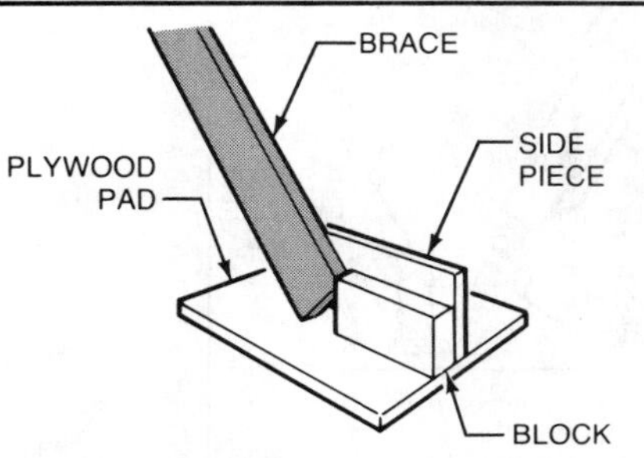

Nail side piece to block. Fasten pad to concrete with concrete nails.

ANCHORING BRACES TO CONCRETE SLAB

Figure 2-20. Braces are attached to strongbacks or walers for maximum strength.

fastens the turnbuckle to a waler or strongback. When placed at the lower end of a wood brace, a metal anchor bracket is secured to the ground with a steel stake. A turnbuckle allows easy adjustment for plumbing and aligning the wall form. See Figure 2-21.

Fastening Devices and Procedures

Nails and spikes are the most common fasteners used in light form construction. Bolts, lag screws, and other devices are used in the construction of heavier forms. Proper nailing procedures must be used to ensure the strength and duribility of the forms. However, because the forms must be stripped from the hardened concrete walls, the use of too many nails and nails that are too large should be avoided.

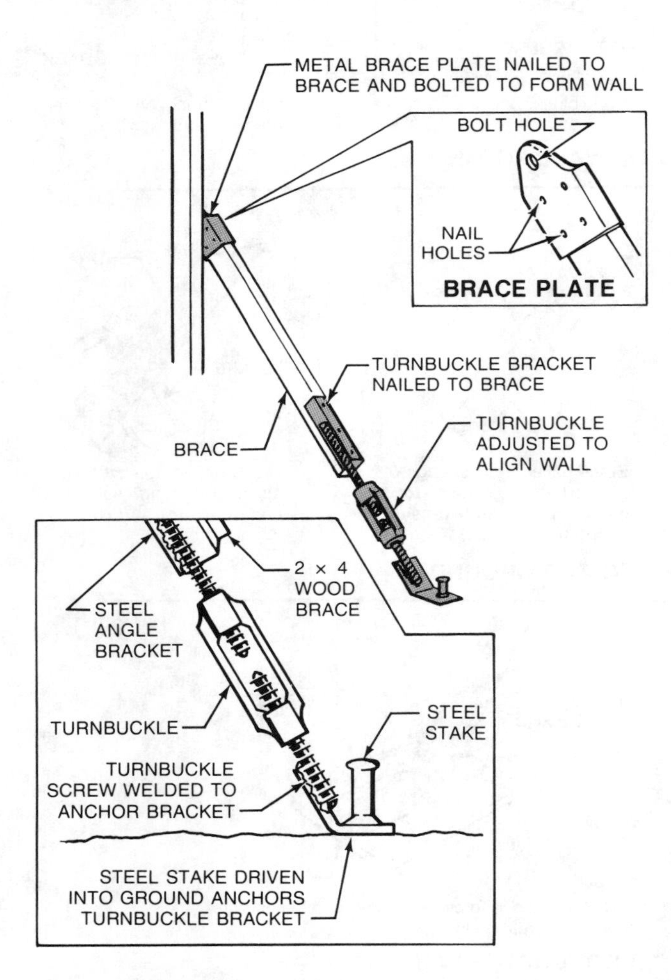

Figure 2-21. Patented metal devices are used with wood form braces.

Nails and spikes are available in a variety of types and sizes. Nail lengths are designated by a number and the letter *d*, which is the designation for *penny*. Some of the nail lengths commonly used in formwork are 6d (2″), 8d (2½″), 10d (3″), and 16d (3½″). Spikes are 16d and larger nails. *Common, box, double-headed,* and *concrete nails* are commonly used in form construction. See Figure 2-22.

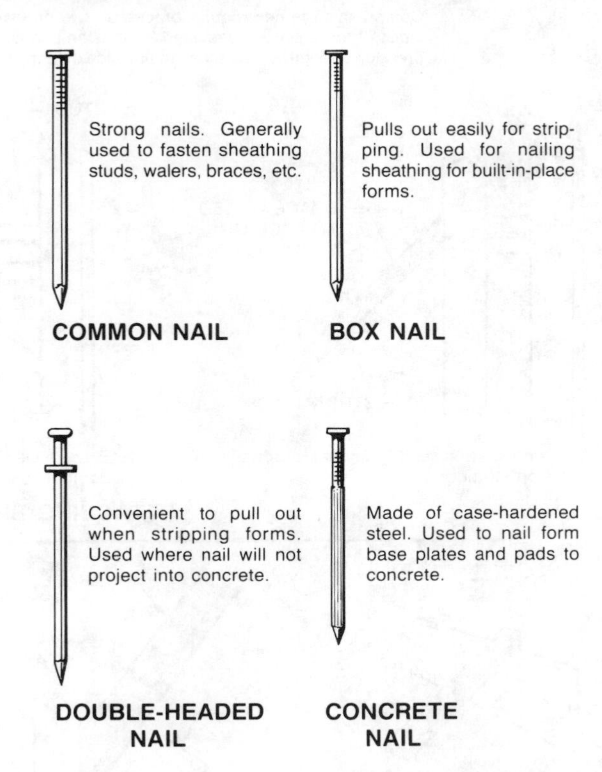

Figure 2-22. Many types of nails are commonly used in form construction. The type of nail used depends on the application.

Common nails are used to fasten sheathing to the studs or walers and to fasten form frames, stiffeners, and bracing together. Plywood ⅝″ or thicker should be fastened with 6d common nails.

Box nails are primarily used to nail sheathing or liner material for built-in-place forms. Box nails have thinner shanks and flatter heads than common nails. The thinner shank is an advantage when stripping built-in-place forms because it is easier to pull from the forms. The flat head of a box nail also leaves less of an impression on the finished concrete wall.

Double-headed (duplex) nails are used extensively to nail walers, strongbacks, kickers, blocks, and other form components. The second head makes it convenient to pull the nail with a claw hammer or wrecking bar during stripping operations. As a result, there is less bruising and damage to reusable form lumber. Double-headed nails are recommended anywhere they do not protrude into the concrete at the time of the concrete placement.

Concrete (masonry) nails are used to fasten form base plates to concrete slabs or footings. Concrete nails are made of special case-hardened steel to resist bending. They must be driven straight to avoid chipping or breaking out the concrete.

Nailing Methods. Proper nailing methods are essential to good form construction. The strength of a nailed joint depends on the lateral and/or withdrawal resistance of the nail and the nailing procedure used. A few general rules should be followed for maximum withdrawal resistance. See Figure 2-23.

1. The withdrawal resistance of a nail is much greater when driven into the edge grain rather than the end grain of wood.
2. When fastening pieces of wood of different thicknesses together, nail through the thinner piece into the thicker piece.
3. When nailing plywood to solid wood, the pointed end of the nail should go into the solid wood.
4. Toenailing is often a better alternative to nailing into end grain. For best results, toenails should be driven at a 30° angle with one-third of the nail penetrating the piece being fastened.
5. Maximum withdrawal resistance can be accomplished by clinching a nail across the grain.
6. Drive nails straight into the pieces being joined. This facilitates removal when stripping forms.

Form Ties

Form ties are used to tie opposite form walls together so they do not shift or spread while the concrete is being placed. Low form walls are tied together with wood cleats nailed to the tops of the walls or to stakes extending above the walls. Higher form walls are held together with metal ties. A variety of *pat-*

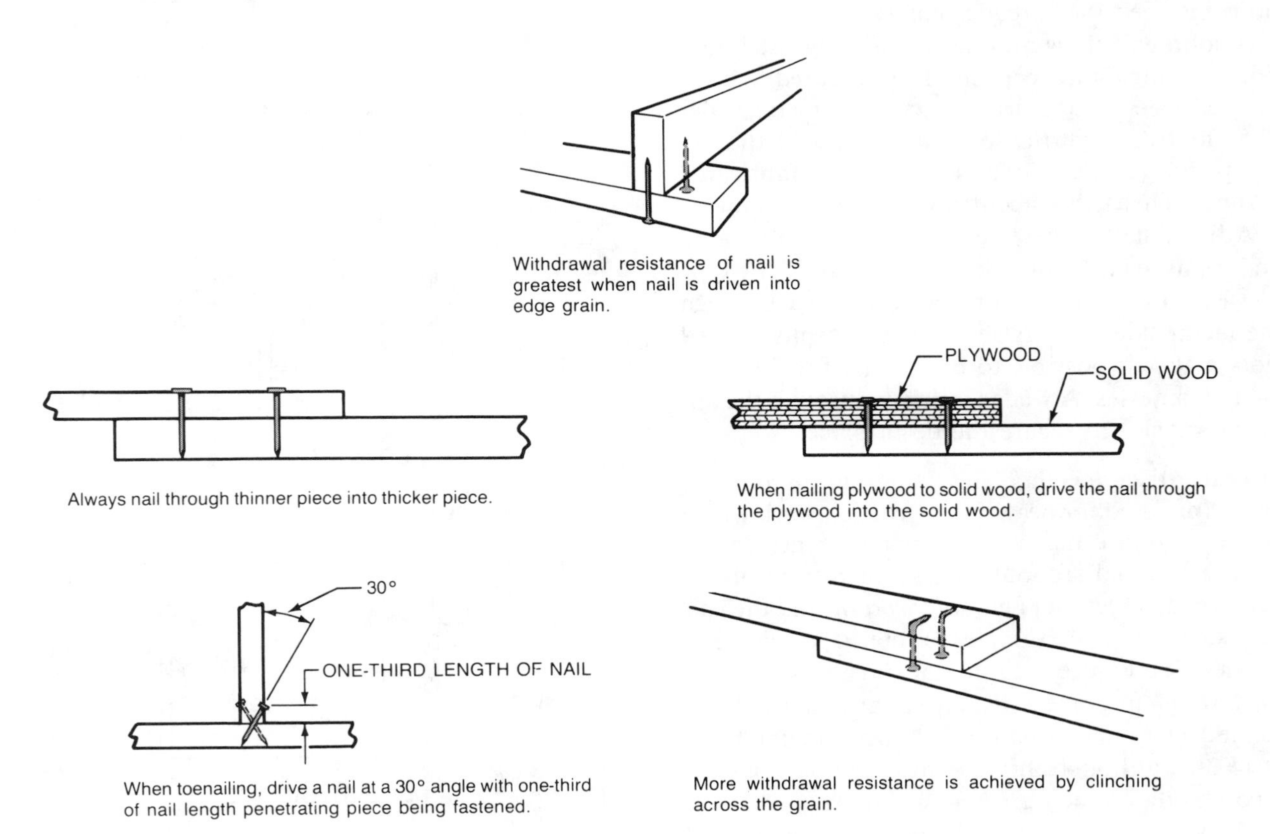

Figure 2-23. Proper nailing methods are essential when building forms.

ented ties are available for securing and spacing form walls. Patented ties consist of a rod that passes through the wall with holding devices at each end. The two basic patented tie designs are the *continuous single-member* types and *internal disconnecting* types. Their working loads range from 1000 to 50,000 pounds.

Continuous Single-Member Ties. A *snap tie* is the most common type of continuous single-member tie. Snap ties are used with both double and single waler systems and are available for wall thicknesses ranging from 6″ to 26″. Various wedge and wedge-bracket devices are used to secure the snap ties. A snap tie is a metal rod extending from the outside surfaces of the opposing walers. The ties are tightened by driving slotted metal wedges behind buttons at the ends of the ties. Small plastic cones or metal washers placed in the section of the tie passing between the form walls act as spreaders holding the walls the correct distance apart. After the forms are stripped from the completed wall, the sections of the snap tie protruding from the wall are snapped off at the *breakbacks*. Breakbacks are grooved sections between the spreader cones.

A loop end tie is another commonly used continuous single-member tie. It is secured with a tapered steel wedge driven against a metal waler plate and through the loop at the end of the tie. Loop end ties are frequently used in prefabricated forms such as the steel-framed plywood panels.

Adjustable flat ties are used with prefabricated and plank wall forms. See Figure 2-24. This type consists of a flat piece of metal set on edge between the metal side rails. A series of uniformly spaced slots makes it possible to use the tie for different wall thicknesses. A wedge is driven through the appropriate slot to secure the tie in place.

Internal Disconnecting Ties. Internal disconnecting ties are used for heavier construction work where greater loads are anticipated. These form tie systems feature external sections that screw into an internal threaded section. The *waler rod tie* and the *coil tie* systems are two examples of internal disconnecting ties. See Figure 2-25. Metal spreaders hold the form walls the correct distance apart. Internal disconnecting ties are used for wall sizes ranging from 8″ to 36″, although they can also be assembled to accommodate much greater thicknesses.

The waler rod systems (she bolts) are composed of an inner rod threaded at each end that screws into two waler rods. The inner rod comes in various lengths for different wall thicknesses. The waler rods are fastened to the walers with large hex nut washers. The waler rod tapers, which facilitates removal. After the concrete sets, the waler rods are unscrewed and removed, and the inner rod remains in the concrete.

The coil tie system features external bolts that screw into an internal device consisting of metal

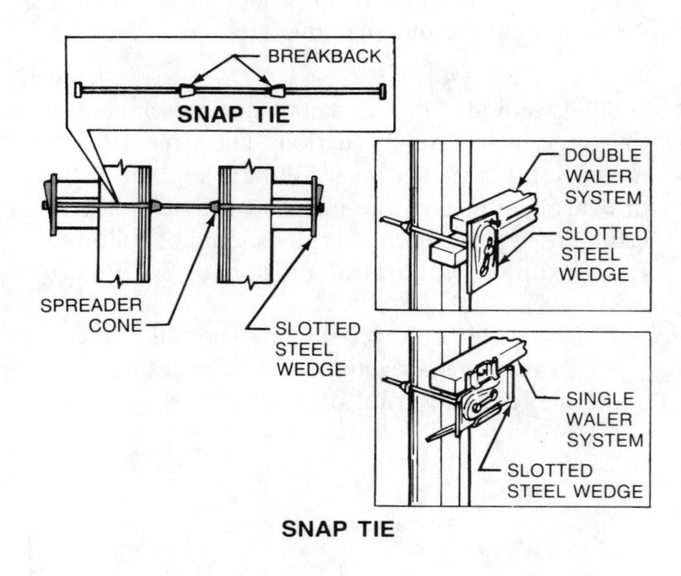

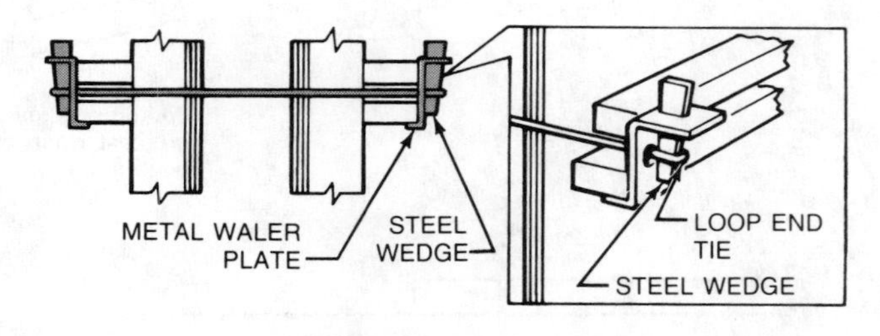

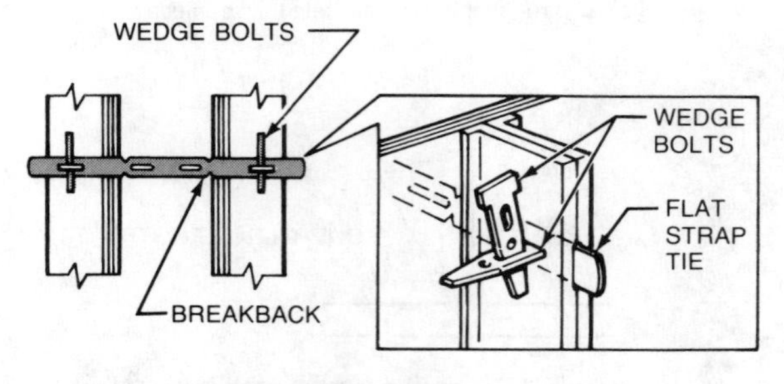

Figure 2-24. Continuous single-member ties are used with single and double waler systems.

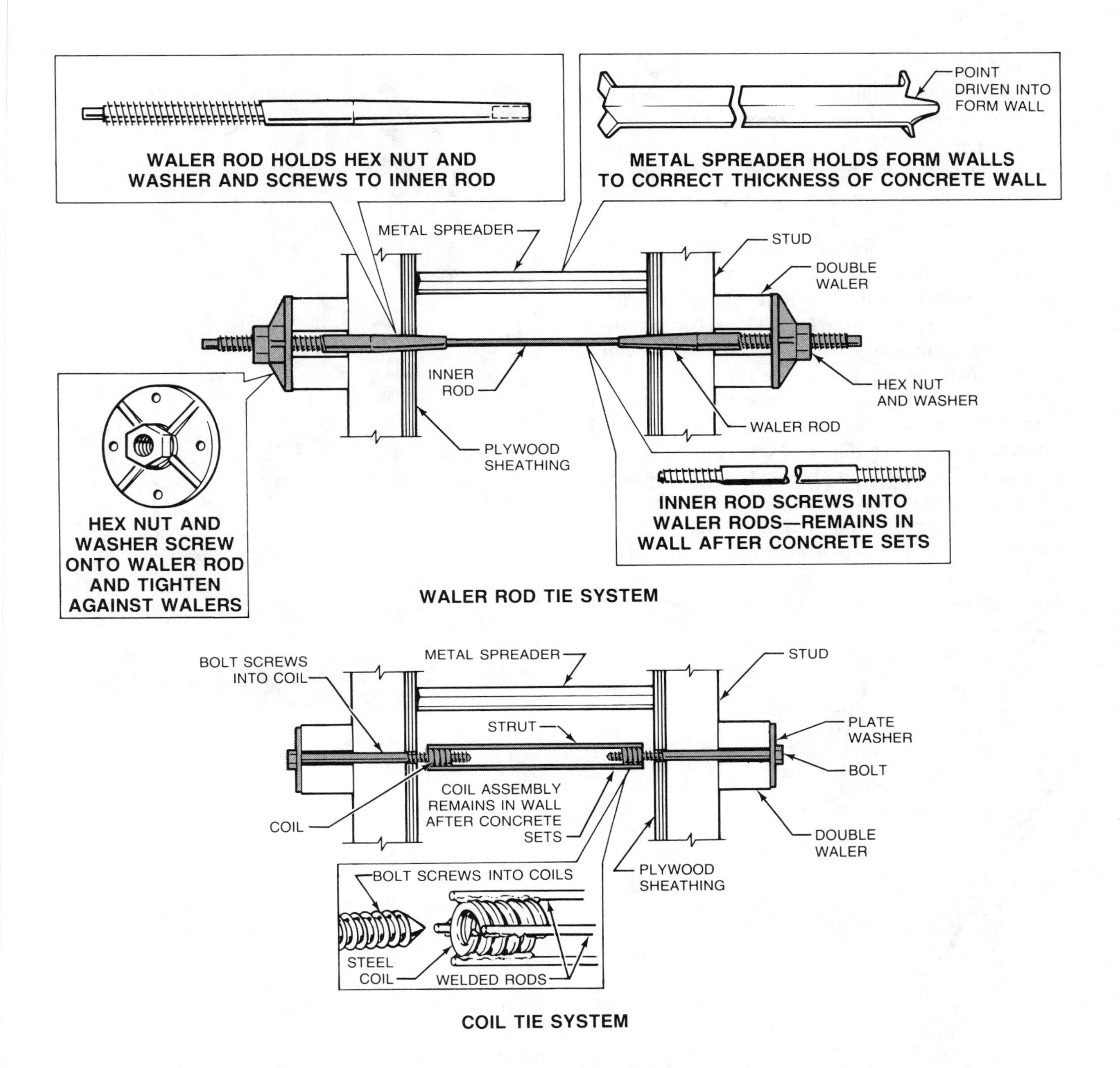

Figure 2-25. Internal disconnecting ties are used with heavier wall forms.

struts with helical coils at each end. The coil assembly remains inside the concrete after the bolts are removed.

Placing Wall Ties. The type of tie to be used and the spacing between the ties are determined by several factors:

1. Anticipated concrete pressure during placement;
2. Size and spacing of studs;
3. Size and spacing of walers, if necessary.

The load capacity of the tie must be considered in the spacing of the ties. The total load on a tie is determined by the contributing area of form around the tie. The contributing area equals one-half the vertical distance between ties (which is the distance between walers) times one-half the horizontal distance between the ties. Standard horizontal tie spacing is 24″ OC; however, greater pressure and other structural factors may require a shorter distance between the ties.

Form panels are predrilled when tie holes are required by drilling a number of panels at one time with an electric drill. The holes must be drilled square to the plywood surface. Holes for snap ties should be slightly larger than the end buttons, but smaller than the spreaders. See Figure 2-26.

Snap ties are usually placed after the outside wall forms have been constructed. Each tie is slipped through a predrilled hole, and one end of the tie is secured to the walers with brackets and/or wedges. The inside wall panels are set in place (doubled up) by tilting each panel and guiding the snap ties into the holes as the panel is straightened into its final position. The ties are then tightened against the inside walers with brackets and/or wedges. See Figure 2-27.

Figure 2-27. The ties are guided into the inside form panel as it is tilted into place.

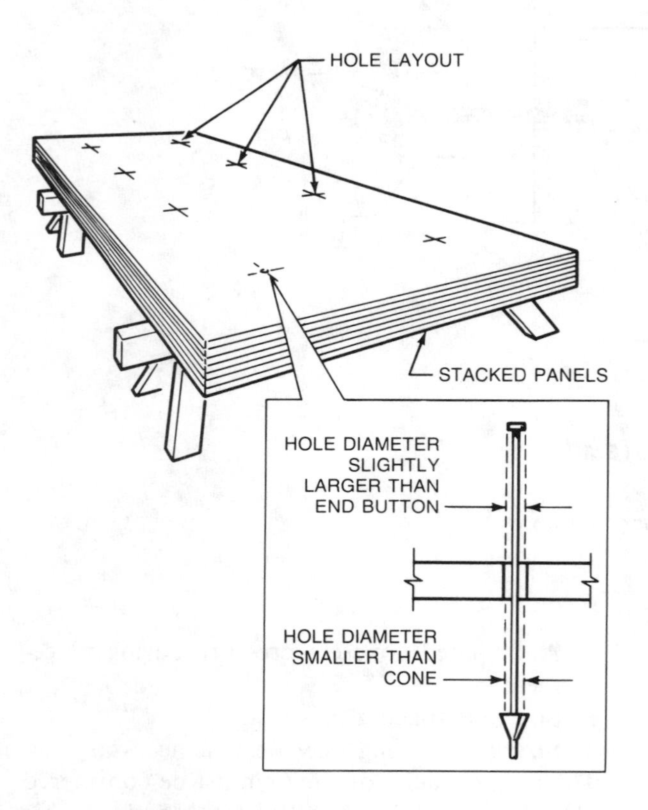

Figure 2-26. A number of panels may be predrilled at the same time to ensure proper placement of the ties.

In another method for placing wall ties, the outside wall form is constructed in the conventional manner with full panels, and the inside wall form is made up of smaller panels laid horizontally. The first row of tie holes in the outside wall form are laid out and drilled 12″ from the bottom with the following rows 24″ OC. When constructing the inside wall form, a 12″ wide bottom panel is placed horizontally and 24″ wide panels are placed horizontally on top of it. As the inside panels are set in place, the snap ties are inserted into the tie holes that have been drilled in the outside wall form and hammered into the top edges of the inside form panels. See Figure 2-28.

When waler rod or coils tie systems are used, the waler rods or bolts are slipped through predrilled holes after both the inside and outside wall forms have been erected The rods or bolts must engage the threads of the mating section enough to provide maximum strength.

A few rules regarding form tie placement are as follows:

1. Drill tie holes so that the holes in the panels are directly across from each other when the inside and outside forms are in place. Slanted ties lose considerable holding power.
2. When using stud and waler stiffeners, place the ties close to studs to avoid panel deflection caused by the tightened tie.
3. Maintain a uniform tightness for all wall ties. If one tie is tighter than others it carries more of the concrete pressure. This could cause the tie to break, which in turn could cause other ties to break because of the pressure increase. Although uneven tightness may not cause any ties to break, the form panels may deflect, producing bulges in the finished concrete walls.

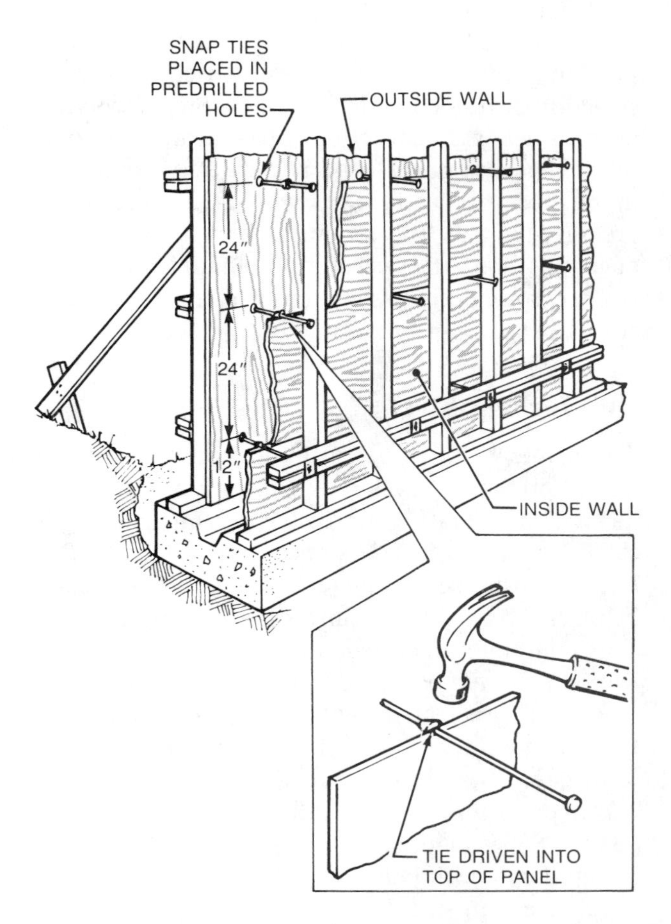

Figure 2-28. The inside form wall may be constructed by using 24″ panels laid horizontally. The ties are driven into the top edge of the panel as the form wall is being erected.

Patching Tie Holes. When removing forms that have been reinforced with continuous single-member ties or internal disconnecting ties, a shallow hole in the surface of the wall will be present. The hole must be patched and sealed in order to give the concrete wall a finished appearance and prevent moisture from reaching the tie ends. Moisture penetrating the tie ends eventually causes rust stains to appear on the surface of the concrete.

Holes are patched with a nonshrink moisture-resistant grout mixture or dry-pack mortar. Some manufacturers offer precast cement compound plugs shaped to fill the holes created by snap tie spreader cones. The plugs are secured in place with a fast tack waterproof neoprene adhesive. Another method is to inject a pressurized epoxy resin into the tie hole, then place a plastic cap insert in the hole. See Figure 2-29. Both the cement and plastic plugs can be placed either flush with or recessed from the surface of the concrete wall.

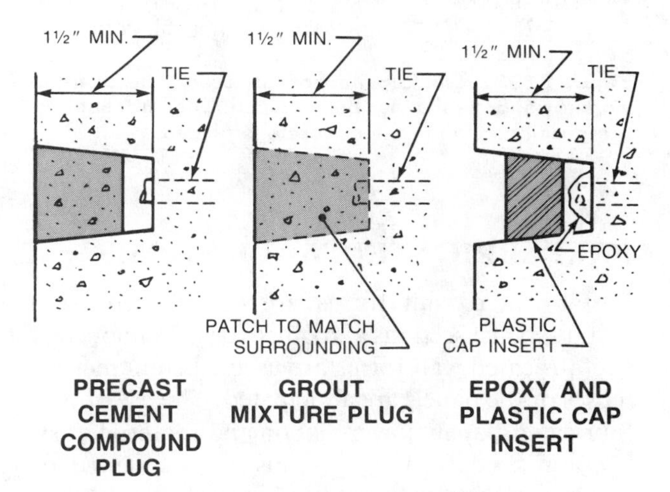

Figure 2-29. Holes created by the snap tie spreader cones must be filled or plugged after the snap ties have been broken off.

Preset Form Anchors. Preset form anchors are used to fasten formwork to previously placed concrete. A straight coil loop insert that receives a coil bolt, or a ferrule loop insert that receives a threaded machine bolt are common form anchors. See Figure 2-30. The insert, along with a temporary bolt and washer, is set in place at the time of concrete placement. After the concrete sets, the insert remains embedded in the concrete. It will later be used to bolt down additional formwork to the existing concrete.

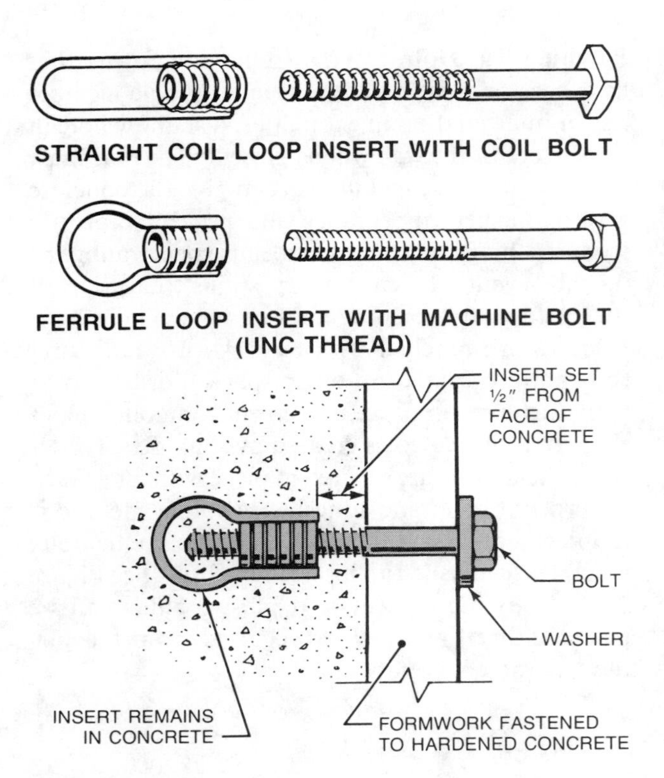

Figure 2-30. Preset form anchors are used to fasten formwork to concrete that has been placed. The form anchors are positioned when the concrete is placed.

PREFABRICATED WALL FORMS

Prefabricated wall forms are constructed from prebuilt panel sections and other form components. Prefabricated wall forms are used when numerous reuses of the panels are anticipated. The use of prefabricated panels lowers labor and material cost. Prefabricated forming systems can be rented or purchased from various manufacturers. Special-purpose custom-made forms, also produced at prefabricating plants, are available for special forming operations.

Panel Systems

The prefabricated wall forms most commonly used consist of modular panel sections usually 2 ′ to 4 ′ in width and 2 ′ to 8 ′ in height. Smaller filler pieces of various sizes are also available. Panel manufacturers also provide the accessories (ties, walers, wedges, braces, etc.) for their particular systems. Although metal framed plywood panels are the most common type of prefabricated panel sections used today, there is a growing use of all-metal panel sections.

Metal-framed Plywood Panels. A metal-framed plywood panel consists of ½″ or ⅝″ Plyform® set in an aluminum or steel frame. Horizontal metal stiffeners spaced approximately 1 ′ apart provide additional support. See Figure 2-31. The frames are designed so the Plyform® panel can be easily replaced when worn or damaged. The panel sections often called *hand-set forms*, are light and easy to handle. Slots in the metal side rails are for wedge bolts that join the panels together. Walers are secured with metal waler ties. Wire, flat, or round ties space and hold the walls together. Braces are secured to the walls with wedge-shaped metal plates.

All-Metal Panels. Prefabricated all-metal panels are made of aluminum or steel. Aluminum hand-set forms consisting of an aluminum face stiffened with an aluminum frame are frequently used in the construction of residential foundations. Steel forms are used in heavy construction. Steel forms are combined in ganged panel forms and are also widely used in the precast industry. See Figure 2-32. Accessories are provided by the manufacturers to assemble, align, and brace the steel and aluminum forms. With proper care and maintenance, steel and aluminum forms can be used indefinitely.

Reinforcing Steel

A concrete wall is subject to both compressive and lateral pressures. Concrete without reinforcement has a great deal of compressive resistance to vertical loads, but far less resistance to lateral loads. The lateral resistance of concrete walls is strengthened by placing *rebars* (steel reinforcing bars) in the walls. This combination is generally referred to as *reinforced concrete construction.*

Rebars are steel bars with ridges and a rough surface. The uneven surface helps bond the concrete to the steel. Standard size rebars range from ⅜″ to 2¼″ in diameter and are identified by numbers from #3 to #18. See Figure 2-33. The diameter of the rebar is found by multiplying the number designation by ⅛″. For example, a #6 rebar is 6/8″, or ¾″, in diameter. The size of the rebars required for a wall, as well as their placement and spacing, is shown in section view drawings of prints.

STEEL FRAME

DADO SLOT FOR WALL TIES

STEEL SIDE RAIL

SLOTS FOR WEDGE BOLTS

STEEL HORIZONTAL STIFFENER

PLYFORM® PANEL

METAL WA

SINGLE WALER

BRACE PLATE

SCAFFOLD BRACKET

BRACE

WEDGE BOLTS SECURE FORMS

METAL FRAMED PLYWOOD PANEL COMPONENTS

PLACING METAL FRAMED PLYWOOD PANELS

Symons Corporation

Figure 2-31. Metal framed plywood panels may be reused many times. The Plyform® panel may be replaced when worn or damaged.

Western Forms, Inc.

ALUMINUM HAND-SET FORMS

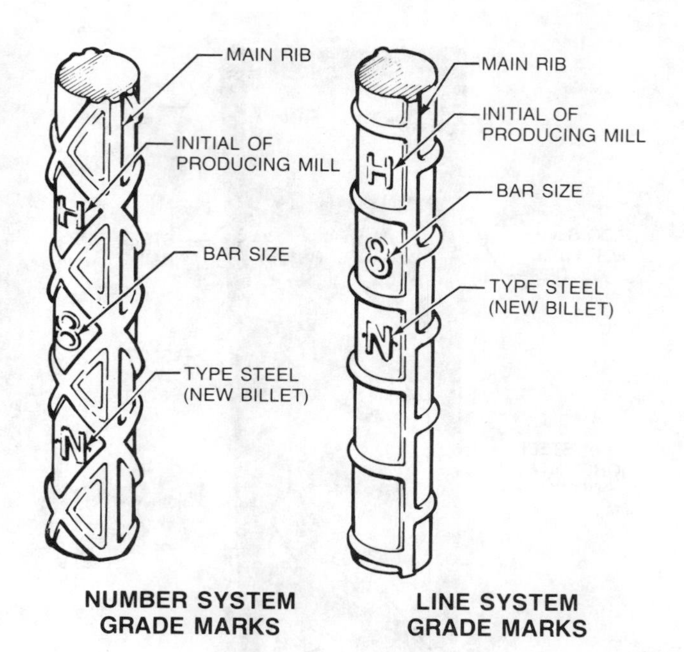

Symons Corporation

STEEL FORMS

Figure 2-32. Metal prefabricated forms are made of aluminum or steel. The steel panels are commonly used on heavy construction projects.

STANDARD REBAR SIZES						
Bar Size Designation	Weight Per Foot		Diameter		Cross-Sectional Area Squared	
	LB	KG	IN.	CM	IN.	CM
#3	0.376	0.171	0.375	0.953	0.11	0.71
#4	0.668	0.303	0.500	1.270	0.20	1.29
#5	1.043	0.473	0.625	1.588	0.31	2.00
#6	1.502	0.681	0.750	1.905	0.44	2.84
#7	2.044	0.927	0.875	2.223	0.60	3.87
#8	2.670	1.211	1.000	2.540	0.79	5.10
#9	3.400	1.542	1.128	2.865	1.00	6.45
#10	4.303	1.952	1.270	3.226	1.27	8.19
#11	5.313	2.410	1.410	3.581	1.56	10.07
#14	7.650	3.470	1.693	4.300	2.25	14.52
#18	13.600	6.169	2.257	5.733	4.00	25.81

American Society for Testing and Materials

Figure 2-33. Steel reinforcing bars (rebars) are used to strengthen lateral resistance of walls. Rebars are available in sizes #3 to #18.

Concrete buildings and other large concrete projects are heavily reinforced throughout the structure. Residential and other light construction foundations may not require rebars if the buildings are not located in a seismic (earthquake) risk zone. In areas where earthquakes occur, local building codes require steel reinforcement. The U.S. Department of Housing and Urban Development *Minimum Property Standards for One and Two Family Dwellings* states the following regarding foundation walls:

1. Where earthquake design is required, and in seismic zone 2 or 3, reinforce concrete walls under the following conditions:
   concrete walls when height exceeds 6 times thickness;
   masonry walls when height exceeds 4 times thickness.
2. Size and spacing of reinforcement shall be in accordance with accepted engineering practice.

Placing Rebars. Proper rebar placement is critical to ensure maximum resistance to lateral loads. A low foundation wall may need only a few rebars placed horizontally. High walls require horizontal and vertical rebars. The horizontal and vertical rebars are tied together to form a steel curtain. Larger and thicker walls may require two or more curtains of steel within the wall. Where only a few rows of rebars are required in low forms, the steel is placed by the carpenters constructing the forms. However, large quantities of rebars are commonly installed by steelworkers specializing in this type of work. The placement of rebars requires careful coordination with other trades. The rebars are usually positioned after the outside wall forms have been set and the openings framed. Other formwork, such as positioning conduit, sleeves, anchors, straps, and inserts, must be completed prior to placing rebars. After the rebars have been placed, the inner wall forms are constructed. See Figure 2-34. In the case of walls that are part of a core, such as an elevator shaft or stairwell, the inner walls are constructed first.

Figure 2-34. Rebars are placed after the outside form walls have been constructed.

Rebars should be free of loose rust, scale, paint, oil, grease, mortar or other foreign matter that may weaken the ability of the concrete to grip the steel. The steel curtains must remain in their proper position within the form while the concrete is being placed. Rebars near the surface of the walls must be protected against corrosion and fire by an adequate layer of concrete. Rusting can occur if the steel is too close to the surface, thus producing cracks in the concrete. See Figure 2-35. Several methods are used to maintain proper spacing between the wall forms and the rebars. These include spacer blocks and plastic snap-on devices, as well as wood strips that are removed as the concrete is being placed.

REQUIRED CONCRETE PROTECTION FOR REBARS		
Application		**Minimum Cover**
Against ground without forms		3″
Exposed to weather or ground but placed in forms	Greater than 5/8″ diameter rebars	2″
	Less than 5/8″ diameter rebars	1 1/2″
Slabs and walls (no exposure)		3/4″
Beams, girders, columns (no exposure)		1 1/2″

Figure 2-35. Rebars must be covered with an adequate layer of concrete to protect them from fire and corrosion.

Chapter 2—Review Questions

Completion

_________ 1. A(n) _________ panel system consists of a number of prefabricated panel forms tied together.

_________ 2. The construction of many similar panel sections can be facilitated by a(n) _________ table.

_________ 3. A(n) _________ is a rectangular column joined to a concrete wall.

_________ 4. _________ are braced frames placed between form walls to create door and window openings.

_________ 5. _________ is a wood by-product commonly used for form sheathing consisting of an odd number of layers.

_________ 6. The most common form panel thicknesses are _________″, _________″, and _________″.

____

____

_________ 7. The standard size plywood panel most commonly used in form construction is _________′ × _________′.

____

_________ 8. Interior plywood is bonded with water-_________ glue.

_________ 9. _________ is the plywood product especially manufactured for form construction.

_________ 10. The _________ capacity of dry plywood is based on the minimum bending radii.

_________ 11. The size of lumber most frequently used for form stiffeners and bracing is the _________″ × _________″.

____

_________ 12. The actual size of a 2 × 6 piece of lumber is _________″ × _________″.

____

_________ 13. The most efficient angle for form braces is _________°.

_________ 14. _________′ is the recommended spacing of braces for walls 10′ high.

_________ 15. A 10d nail is _________″ long.

_________ 16. _________ plywood is used to produce special surface effects on concrete walls.

_________ 17. _________ or bending of a wall form occurs when stiffeners are placed too far apart.

_________ 18. A liberal amount of _________ should be applied to plywood panels to reduce moisture penetration.

_______ 19. Plywood that is ⅝″ thick or more should be fastened with _______d nails.

_______ 20. _______ nails are made of case-hardened steel to resist bending.

Multiple Choice

_______ 1. The _______ nail is commonly used in form construction.
A. common
B. box
C. double-headed
D. all of the above

_______ 2. _______ is a softwood species used for form lumber.
A. Douglas fir
B. Spruce
C. Pine
D. all of the above

_______ 3. Patented ties _______.
A. tie the base plate to the footing
B. secure the braces
C. tie the studs to the form walls
D. space and hold form walls together

_______ 4. The type of continuous single-member tie used most often is a _______ tie.
A. snap
B. loop end
C. flat metal
D. spreader

_______ 5. Waler rods and coil ties are _______.
A. types of snap ties
B. broken off after the concrete hardens
C. internal disconnecting ties
D. secured with wedges

_______ 6. Snap ties are most often placed after constructing _______ wall(s).
A. the inside form
B. the outside form
C. both form
D. none of the above

_______ 7. Waler rods and coil ties are secured to the walers with _______.
A. wedges
B. screws
C. double-headed nails
D. large hex nut washers

_______ 8. The prefabricated panel sections most commonly used are constructed of _______.
A. plywood panels and wood studs
B. plywood panels and metal frames
C. metal panels and wood frames
D. all metal

__________ **9.** __________ are grooved sections between the spreader cones of a snap tie.
A. Helical coils
B. Patented ties
C. Breakbacks
D. none of the above

__________ **10.** The main purpose of rebars in concrete is to __________.
A. increase the compressive strength
B. resist vertical loads
C. increase the lateral strength
D. all of the above

__________ **11.** The diameter of a #4 rebar is __________".
A. 1/4
B. 3/8
C. 1/2
D. 5/8

Identification

__________ **1.** Ribbon
__________ **2.** Stud
__________ **3.** Footing
__________ **4.** Stake
__________ **5.** Keyway
__________ **6.** Brace
__________ **7.** Base plate
__________ **8.** Sheathing
__________ **9.** Waler

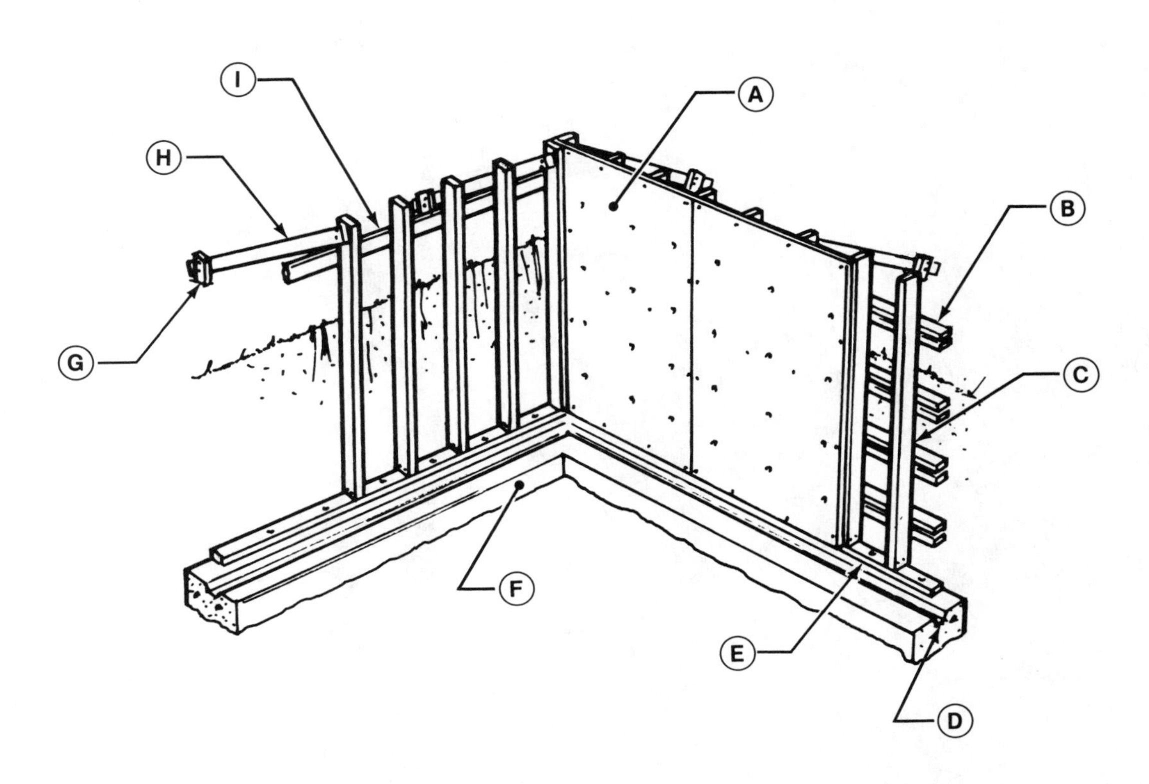

_______________ **10.** Steel stake

_______________ **11.** Base plate

_______________ **12.** Sheathing

_______________ **13.** Stud

_______________ **14.** Brace tie

_______________ **15.** Footing

_______________ **16.** Kicker

_______________ **17.** Waler

_______________ **18.** Strongback

_______________ **19.** Cleat

_______________ **20.** Brace

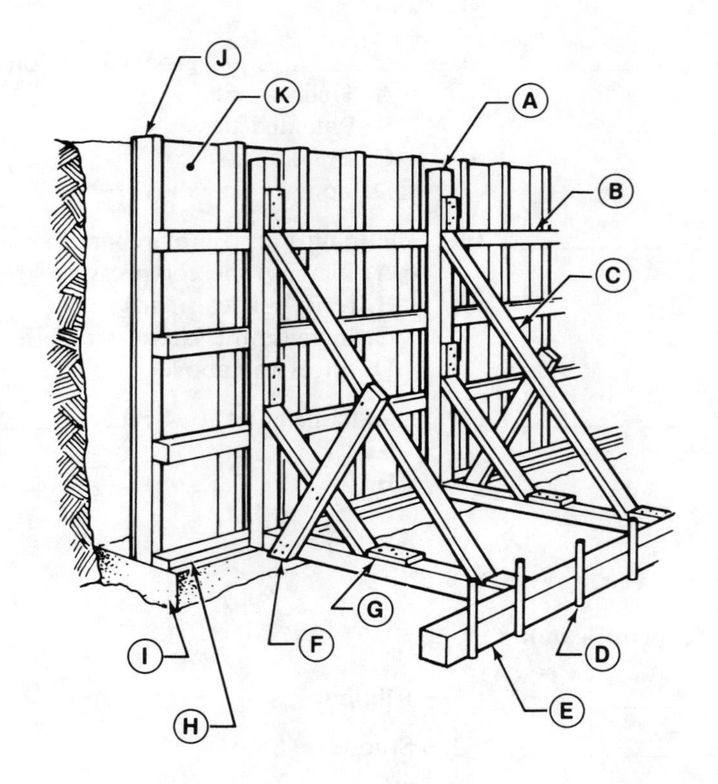

CHAPTER 3
Residential Foundations

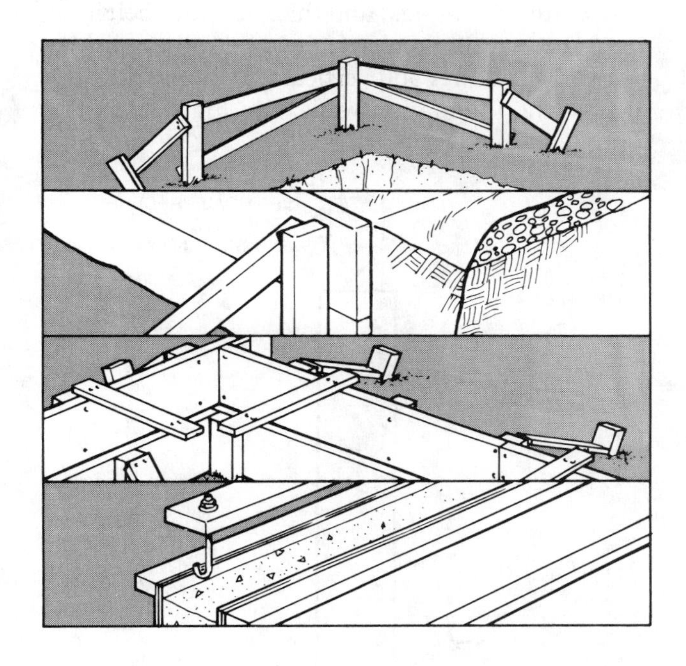

Foundations support residential structures such as one-family and multifamily dwellings. Many types of foundations used for residential structures are also used for light commercial structures. Most foundations are constructed of cast-in-place concrete. The dimensions and shape of a foundation must conform to local building codes and accepted engineering practices.

Full basement, crawl space, and slab-on-grade foundations are commonly used for residential and light commercial structures. Full basement and crawl space foundation designs require foundation wall and footing forms to be constructed. Concrete for slab-on-grade foundations is placed monolithically, or the slab and the foundation are placed separately.

Various foundation shapes are used for residential and light commercial structures, including T-, L-, battered, and rectangular foundations. The design of a foundation is determined by the weight of the foundation, imposed load of the building's superstructure, and bearing capacity of the soil.

Under certain conditions, grade beam or stepped foundations support the superstructure. Grade beam foundations are used with ramped or stepped foundations erected on a sloped surface, or when the bearing capacity of the soil does not provide adequate support for conventional foundation footings. Stepped foundations are commonly constructed on steeply sloped building sites or with crawl space or full basement foundations.

RESIDENTIAL FOUNDATIONS

The major elements in a foundation are *foundation footings* and *foundation walls.* A foundation footing is the part of a foundation that rests directly on the soil and acts as a base for the foundation wall. A *pier footing* acts as a base for wood posts, steel columns, or masonry or concrete piers. Foundation walls are secured to the top of the foundation footings and support the superstructure. The thickness and width of footings, and thickness and height of foundation walls are determined by soil conditions below the footings and vertical and lateral pressure against the foundation walls. See Figure 3-1.

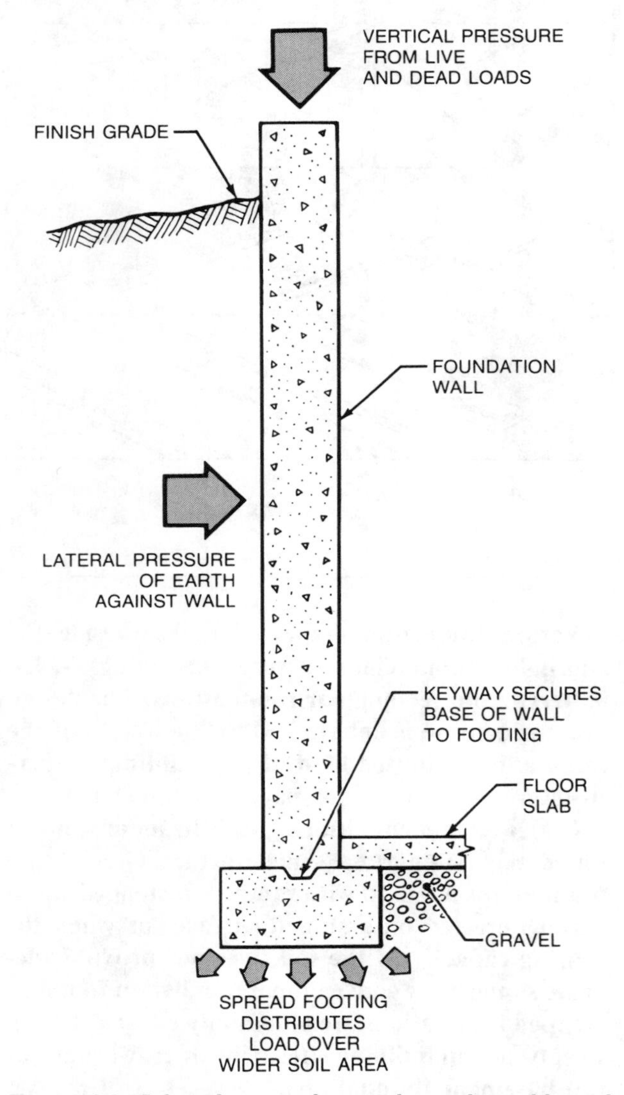

Figure 3-1. A foundation is designed to withstand lateral and vertical pressure. The spread footing distributes the load over a wider soil area.

Vertical pressure on foundation walls is created by *dead* and *live loads* that bear down on the walls. Dead loads are the weight of the entire superstructure. Dead loads are calculated by adding the total weight of materials in the building. Live loads, such as the weight of people, furniture, and snow, are non-constant loads.

Lateral pressure against a foundation wall is created by the force of the earth against the wall. A high foundation wall must be thicker than a low foundation wall to resist greater lateral pressure exerted on it.

Bearing capacity and compressibility of the soil are also considered in foundation design. Foundations are expected to settle a small amount over a period of time. However, too much uneven settlement causes structural damage to the building. Therefore, the foundation must be designed to minimize the amount of settlement.

FOUNDATION CONSTRUCTION

The location of a building on the lot must be determined before constructing the foundation. Dimensions for locating the building are determined from the site or plot plan. After the building location is determined and the building lines have been set, the groundwork is started. The groundwork may involve deep excavation work for a full basement foundation or minor grading and shallow trenching for a crawl space foundation.

For full basement foundations, the footing forms are constructed and the concrete is placed in them. When the concrete in the footings has set sufficiently, the wall forms are constructed and the concrete is placed in them. Low forms used for crawl space foundations may be constructed monolithically. In monolithic construction, an inverted T-shaped form is constructed and the concrete for the footings and walls is placed at the same time.

Foundation Shapes

The *T-foundation* (inverted-T), *L-foundation, rectangular foundation,* and *battered foundation* are commonly used foundation shapes. See Figure 3-2. The T-foundation is the foundation design used most in residential and light commercial construction. In the T-foundation, a wall is placed on a *spread footing* that rests directly on the soil. The spread footing is the base for the wall above and distributes the load of the building over a wider area.

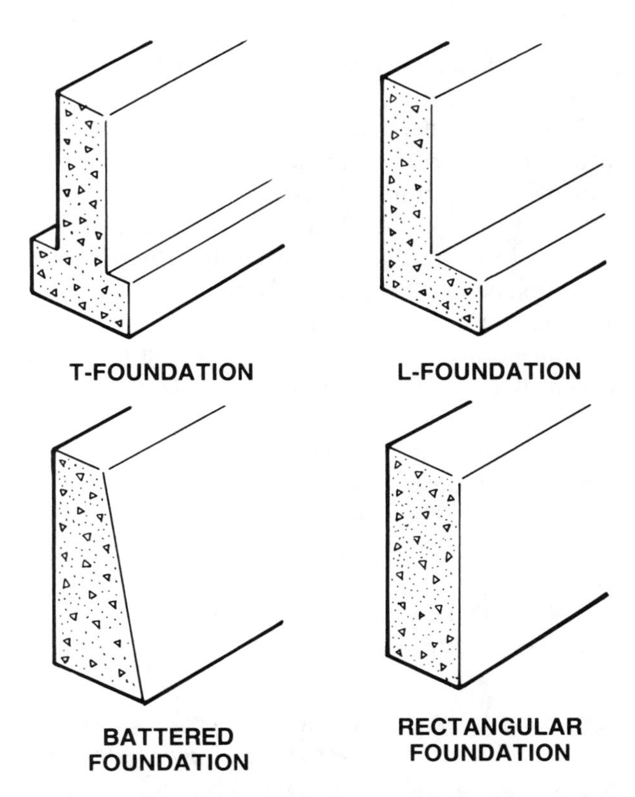

Figure 3-2. Several shapes are commonly used for residential foundation. The T-foundation is used most often.

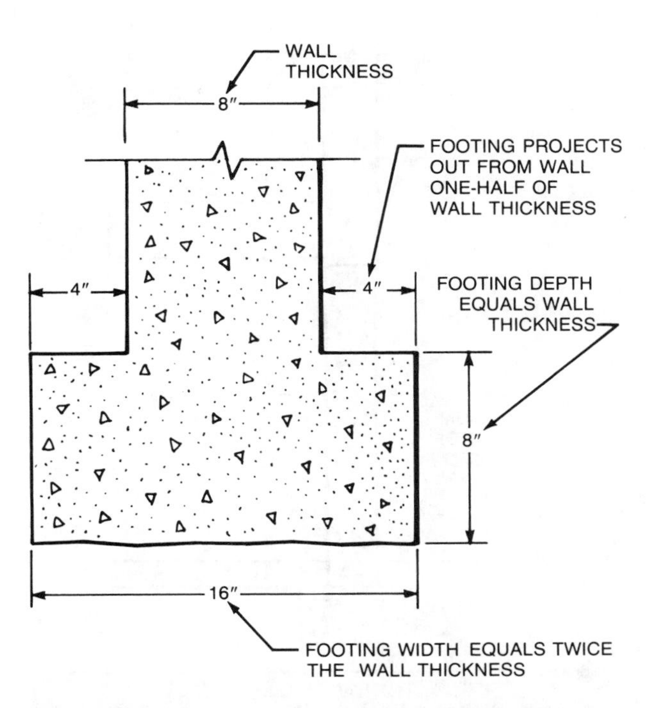

Figure 3-3. A formula is used for calculating the dimensions of a footing resting on soil of average bearing capacity.

Footings for T-foundations should be placed below the frost line, extend below any fill material, and rest on firm, undisturbed soil.

The footings for T-foundations commonly support a cast-in-place concrete wall or a wall constructed of load-bearing solid or hollow concrete masonry units. Thickness and width of footings must conform to local code requirements established for soil conditions in the area.

A formula used to determine dimensions for residential and light commercial foundation footings that rest on soil of average bearing capacity states the following: the depth of footings equals the wall thickness, and the width of footings equals twice the wall thickness. See Figure 3-3. Another general rule is that footings should never be less than 6″ thick and not less than 1½ times the footing projection.

The walls of a T-foundation rest on the center of the spread footing. The base of a foundation wall is secured to the footing by a keyway along the top surface of the footing. In seismic risk areas, rebars extending vertically from the footing are required for both masonry and concrete walls.

A foundation wall must be thick enough to support the vertical load of the building and the lateral pressure against the wall. The thickness of a foundaion wall increases as its height increases. See Figure 3-4. Consult the local building code for specific wall thickness requirements.

Design considerations for the T-foundation also apply to an L-foundation. The main difference between the L-foundation and the T-foundation is that in the L-foundation design the wall rests on one edge of the footing rather than at the center. An L-foundation is used where an existing building foundation does not allow room for a T-foundation. The L-foundation may also be used for concrete retaining walls that hold back earth banks.

Rectangular and battered foundations do not have a spread footing as the base. Small rectangular foundations may be adequate to support light structural loads. They may also be incorporated with slab-on-grade foundation and floor systems. Large rectangular foundations are used as grade beams.

Battered foundations have a vertical exterior surface with a battered (sloped) interior surface. The wide base provides sufficient bearing for the entire wall. Battered foundations may also be used when a new foundation is close to an existing foundation.

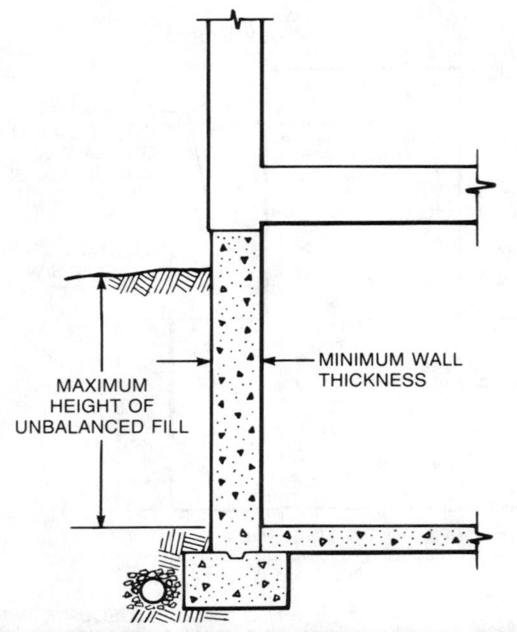

FOUNDATION WALL DIMENSIONS			
Foundation Wall Construction	Maximum Height of Unbalanced Fill	Minimum Wall Thickness	
		Frame Construction	Masonry or Veneer
Hollow Masonry	3′	8″	8″
	5′	8″	8″
	7′	12″	10″
Solid Masonry	3′	6″	8″
	5′	8″	8″
	7′	10″	8″
Plain Concrete	3′	6″	8″
	5′	6″	8″
	7′	8″	8″

Figure 3-4. Concrete and masonry wall dimensions increase as the wall height increases.

Foundation Layout

Foundation layout is based on measurements provided in the plot plan of the prints. A plot plan shows the exact location of the building on the job site. A plot plan commonly indicates the *front setback* of the building. The front setback is the distance from the building to the front property line. The distance from the sides of the building is measured from the side property lines. See Figure 3-5.

Building lines are set up on the job site, based on information in the plot plan, to establish exact boundaries of the foundation walls. Building lines show the area where the ground must be excavated before foundation construction begins. The building lines also indicate where the foundation forms are to be constructed.

Building corners are located using a leveling instrument before establishing building lines. A transit-level is commonly used to establish building corners. (See Appendix E, Leveling Instruments, for information regarding transit-levels.) Building corners must be located accurately because many measurements are taken from these points. When the building corners have been located, wood stakes are driven at each corner of the building and a nail marking the exact position of the corner is driven into the top of each stake. See Figure 3-6.

Building corners may also be established using the *3-4-5 method*. The 3-4-5 method is used to lay out right angles for the building corners. Larger multiples of 3-4-5 (for example, 2 × 3-4-5 = 6-8-10) may also be used when squaring lines over a greater distance. See Figure 3-7.

Batterboards. After building corners have been established, building lines (nylon string or light wire) are stretched to show the exact boundaries of the building. Since it is not practical to stretch building lines by nailing them directly to the tops of the corner stakes, *batterboards* are used to hold the building lines in place. Level 1″ or 2″ thick ledger boards are used as batterboards. Batterboards are usually erected 4′ to 6′ behind each building corner to provide room for excavation or form construction. Batterboards are nailed to 2 × 3 or 2 × 4 wood or metal stakes that have been driven around the corner stakes. See Figure 3-8.

On level lots, the batterboards should be set level to each other at all four corners. The heights of the batterboards are established on the stakes by sighting through a builder's level or transit-level. Stakes are driven solidly into the ground and braced in all directions. The shifting of a batterboard after the building lines have been set up may result in inaccurate foundation layout.

Setting up Building Lines. Building lines are stretched from the four building corners and secured to the tops of the ledger boards. A plumb bob or straightedge and level is used to assure that the building lines intersect directly over the corner stakes of the building. See Figure 3-9. The measurements between the lines should be verified to ensure that they are accurate and conform to the dimensions

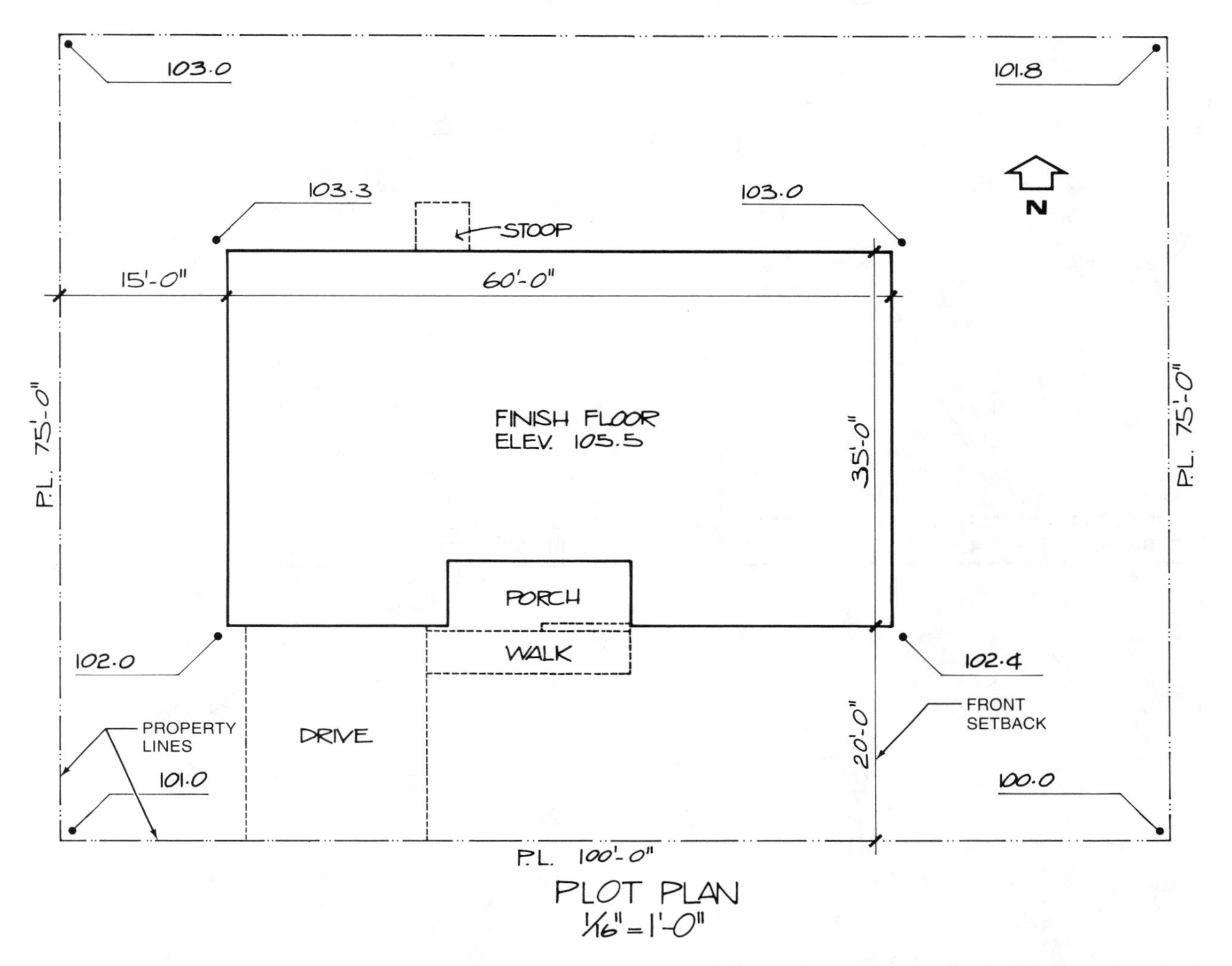

Figure 3-5. The plot plan includes the front setback of the building and the distance from the building to the side property line.

in the foundation plan. Even though the building line dimensions may be accurate, the lines may still be out-of-square.

To verify squareness of building lines, measure diagonally across the building lines. If the diagonal measurements are equal, the building lines are square to each other. If the diagonal measurements are different, the building lines are out-of-square and adjustments must be made. The actual amount that the building lines are out-of-square is one-half the difference between the diagonal measurements. The correction can often be made by shifting the lines at two corners of the building layout a small amount. For example, if the difference in the diagonal measurements is $\frac{3}{4}''$, the line must be shifted $\frac{3}{8}''$ ($\frac{3}{4}'' \div 2 = \frac{3}{8}''$). By moving the building lines a small amount, one of the diagonal measurements is shortened and the other is lengthened. See Figure 3-10.

A $\frac{1}{8}''$ *kerf* (saw cut) is made in the upper edge of the batterboards to secure the building lines. This guarantees that the lines will stay in their proper position and makes it convenient to reset the lines during the foundation construction.

FULL BASEMENT FOUNDATIONS

A full basement foundation provides an area below the superstructure of the building for living space or storage. The basement area is commonly below the ground surface. See Figure 3-11. Concrete pier footings act as bases for posts or columns that support girders. Girders provide central support for the floor directly above. Basement floors have concrete

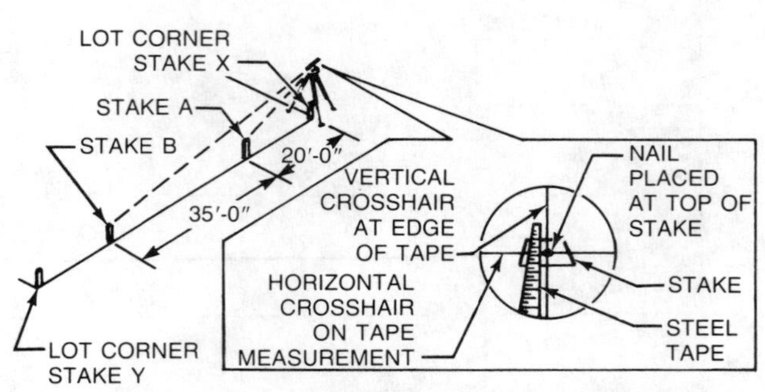

When aligning telescope with tape measurement, the vertical crosshair is at one edge of the tape and horizontal crosshair is at the measurement.

1. Level and plumb transit-level over lot corner stake X. Sight down to lot corner stake Y.

2. Measure front setback (20'-0") using steel tape. Lower telescope until vertical and horizontal crosshairs align with 20'-0" mark. Drive stake A and place a nail. Measure width of building (35'-0"). Raise telescope until crosshairs align with 35'-0" mark. Drive stake B and place a nail.

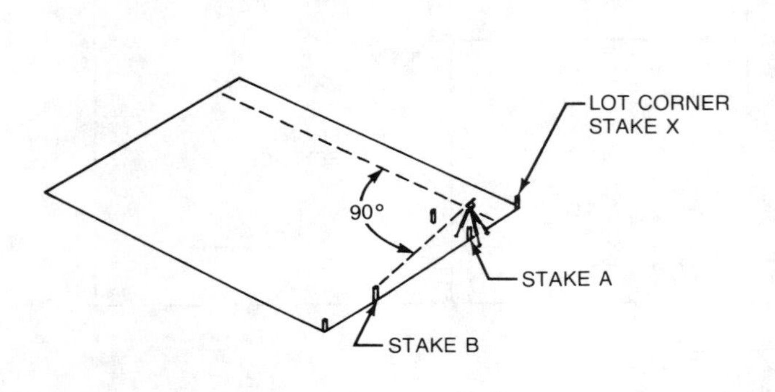

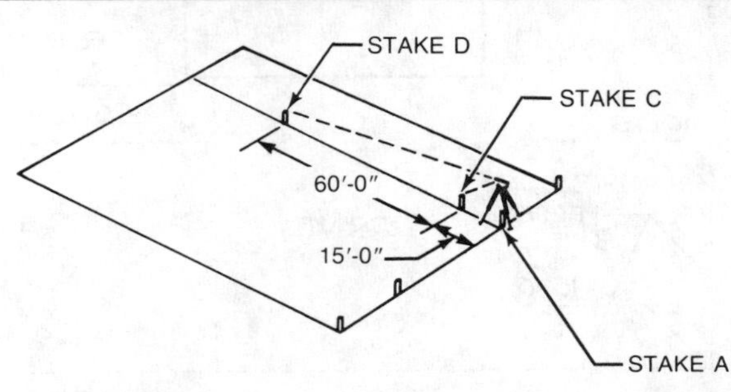

3. Level and plumb transit-level over stake A. Sight down to stake B. Swing telescope 90° to the right.

4. Measure distance from side property line to building (15'-0"). Lower telescope until crosshairs align with 15'-0" mark. Drive stake C and place a nail establishing first building corner. Measure length of building (60'-0") from stake C. Raise telescope until crosshairs align with 60'-0" mark. Drive stake D and place a nail establishing second building corner.

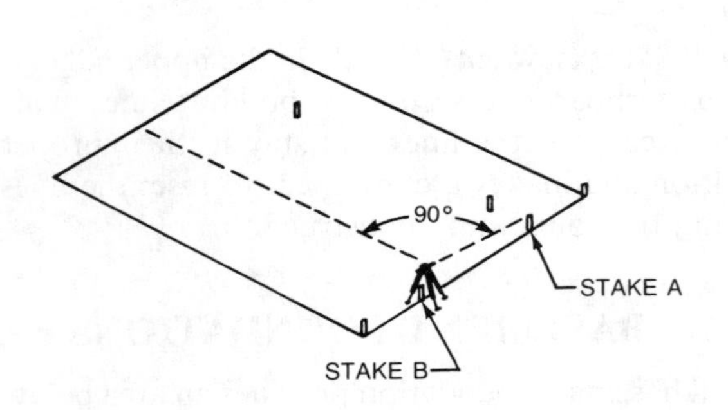

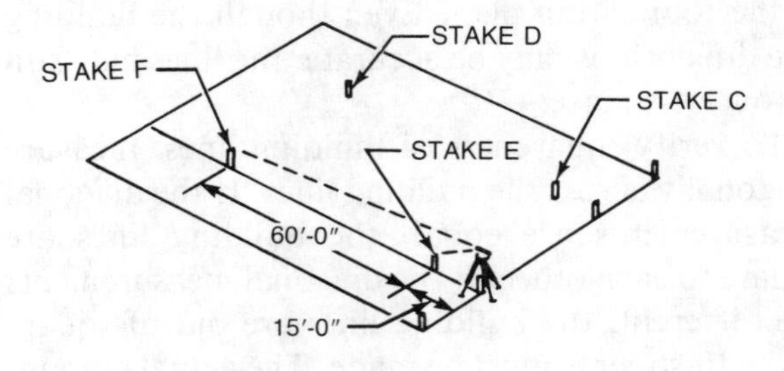

5. Level and plumb transit-level over stake B. Sight back to stake A. Swing telescope 90° to the left.

6. Measure distance from side property line to building (15'-0"). Lower telescope until crosshairs align with 15'-0" mark. Drive stake E and place a nail establishing third building corner. Measure length of building (60'-0") from stake E. Raise telescope until crosshairs align with 60'-0" mark. Drive stake F and place a nail establishing fourth building corner.

Figure 3-6. Building corners may be laid out using a transit-level. The dimensions used are shown on the plot plan in Figure 3-5.

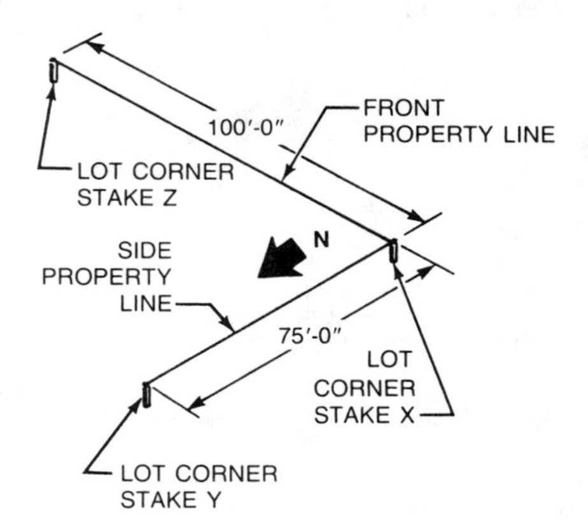

1. Stretch lines from lot corners X, Y, and Z.

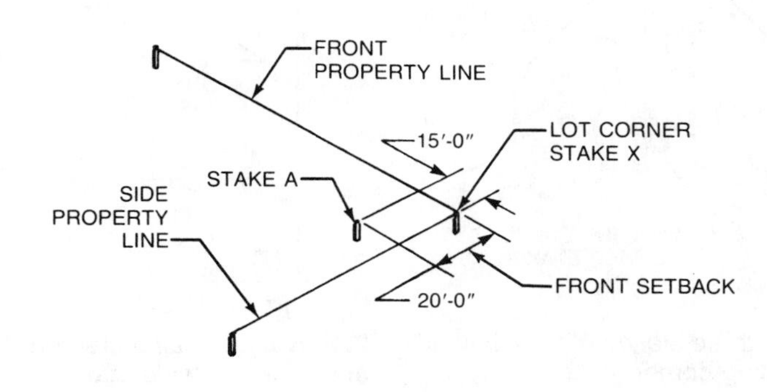

2. Measure the front setback (20'-0") from the front property line and the distance from the side property line to the building (15'-0") at the same time. Drive stake A and place a nail establishing first building corner.

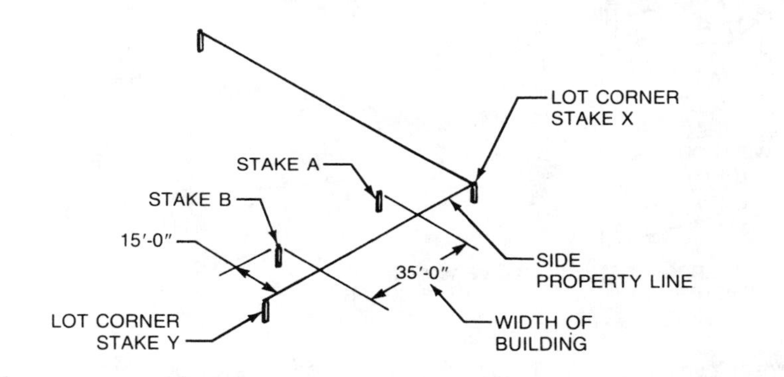

3. Measure the distance from the side property line to the building (15'-0"). Measure the width of the building (35'-0") from stake A. Drive stake B and place a nail establishing second building corner.

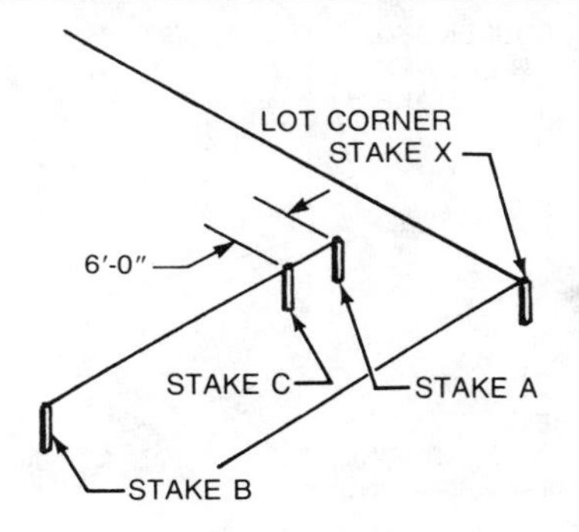

4. Stretch a line between stakes A and B. Drive stake C 6'-0" from stake A and align with stakes A and B. Drive a nail exactly 6'-0" from the nail on stake A.

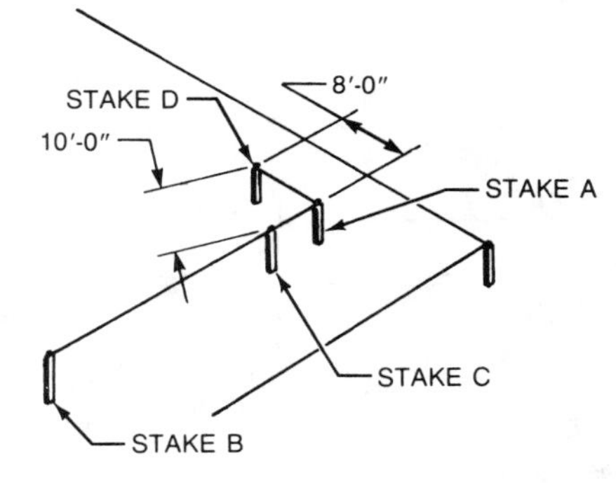

5. Measure 8'-0" from stake A and 10'-0" from stake C. Drive stake D and place a nail exactly where the measurements intersect. Angle DAC is a 90° angle.

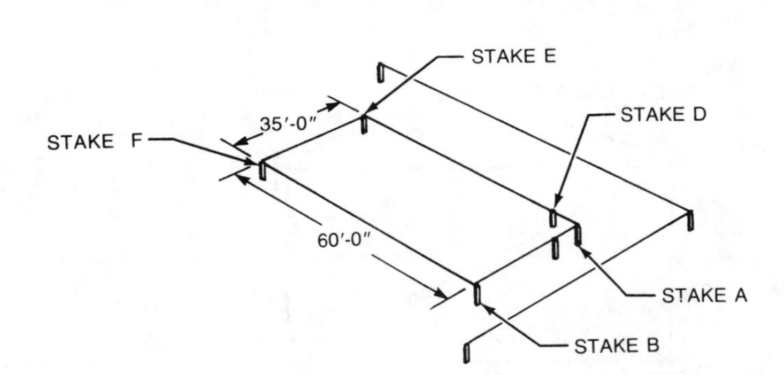

6. Stretch line from stake A and over stake D. Measure length of building (60'-0") from stake A. Drive stake E and place a nail establishing third building corner. Measure length of building from stake B (60'-0"). Measure width of building from stake E and place a nail establishing fourth building corner.

Figure 3-7. The 3-4-5 method can be used to lay out building corners. The dimensions used are shown on the the plot plan in Figure 3-5.

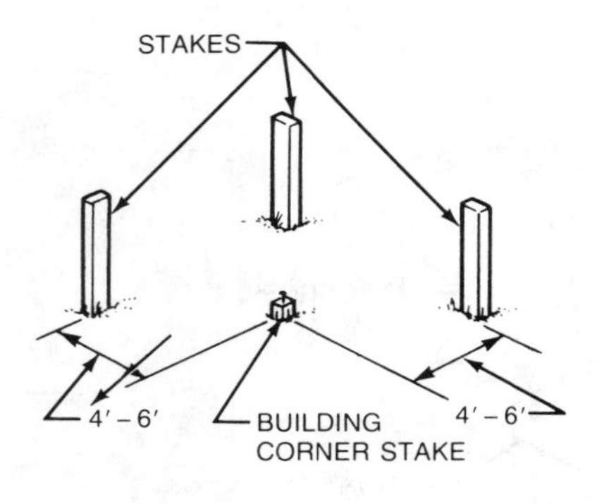

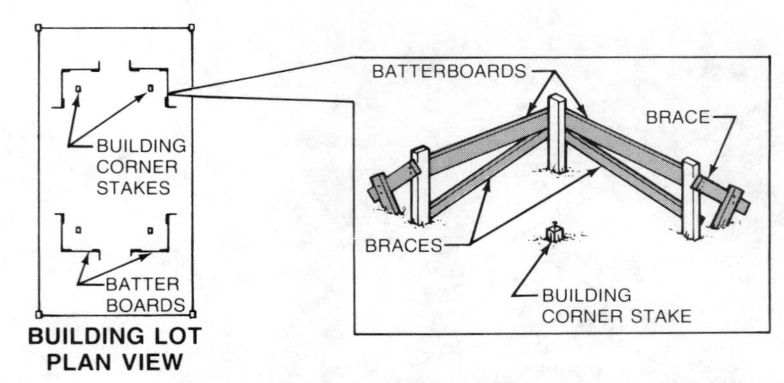

1. Drive three stakes 4' to 6' behind the building corner stakes.

2. Level and nail batterboards to the stakes. Nail braces between the stakes and to the outside stakes.

Figure 3-8. Batterboards hold the building lines that establish the exact position of the foundations.

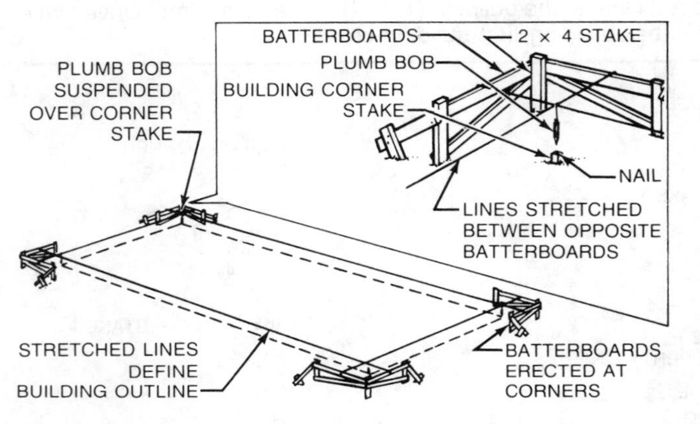

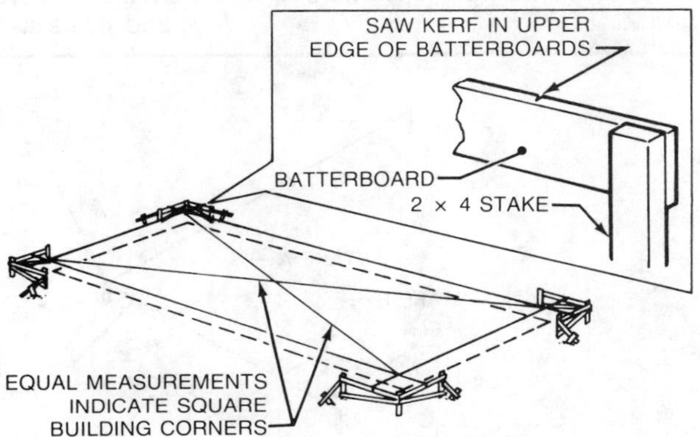

1. Stretch a line between opposite batterboards. Move line at each end until plumb bob aligns with building corner stake in each corner. Fasten line to top of batterboards. Verify building outline measurements against prints.

2. Measure diagonal corners of building lines. Equal diagonal measurements indicate that building lines are square. Cut saw kerfs in upper edge of batterboards to hold lines in place.

Figure 3-9. Building lines must be set up carefully to ensure accurate foundation layout.

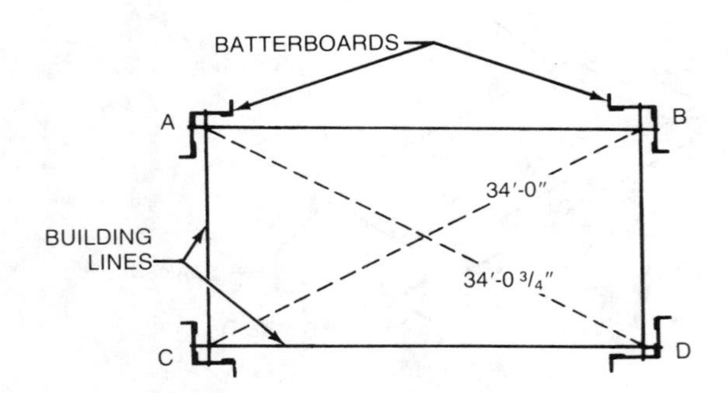

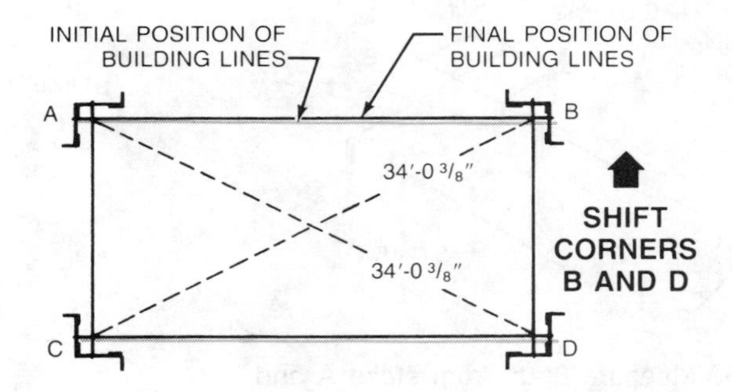

1. Measure distance between diagonal corners. In this example, C-B measures 34'-0" and A-D measures 34'-0 3/4". The difference, 3/4", is twice the amount of adjustment to be made (34'-0 3/4" − 34'-0" = 3/4").

2. Shift lines at B and D 3/8" (3/4" ÷ 2 = 3/8"). This adjustment shortens diagonal A-D and lengthens diagonal C-B to obtain equal measurements (34'-0 3/8").

Figure 3-10. Diagonal measurements are taken to determine if the building lines are square. The corner stakes are adjusted if an error is encountered.

Figure 3-11. The basement area of a full basement foundation extends below the ground level.

slabs that are a minimum of 4″ thick. Information regarding walls, piers, stairways, and window and door openings is provided in the prints. See Figure 3-12.

Building codes require that foundation walls extend above the outside grade level to prevent water from entering the basement. The Council of American Building Officials (CABO) *One and Two Family Dwelling Code* states that the foundation walls should extend at least 8″ above the finish grade. The extended foundation wall also holds the *sill plate* (wood member placed on top of the wall) high enough to minimize damage from moisture and insect attack. The foundation wall should measure a minimum of 7′-0″ from the floor slab to the bottom of the ceiling joists to allow adequate headroom in the basement. Eight-foot high walls above the footing are commonly used to provide a clearance of 7′-8″ above a 4″ floor slab. The basement walls

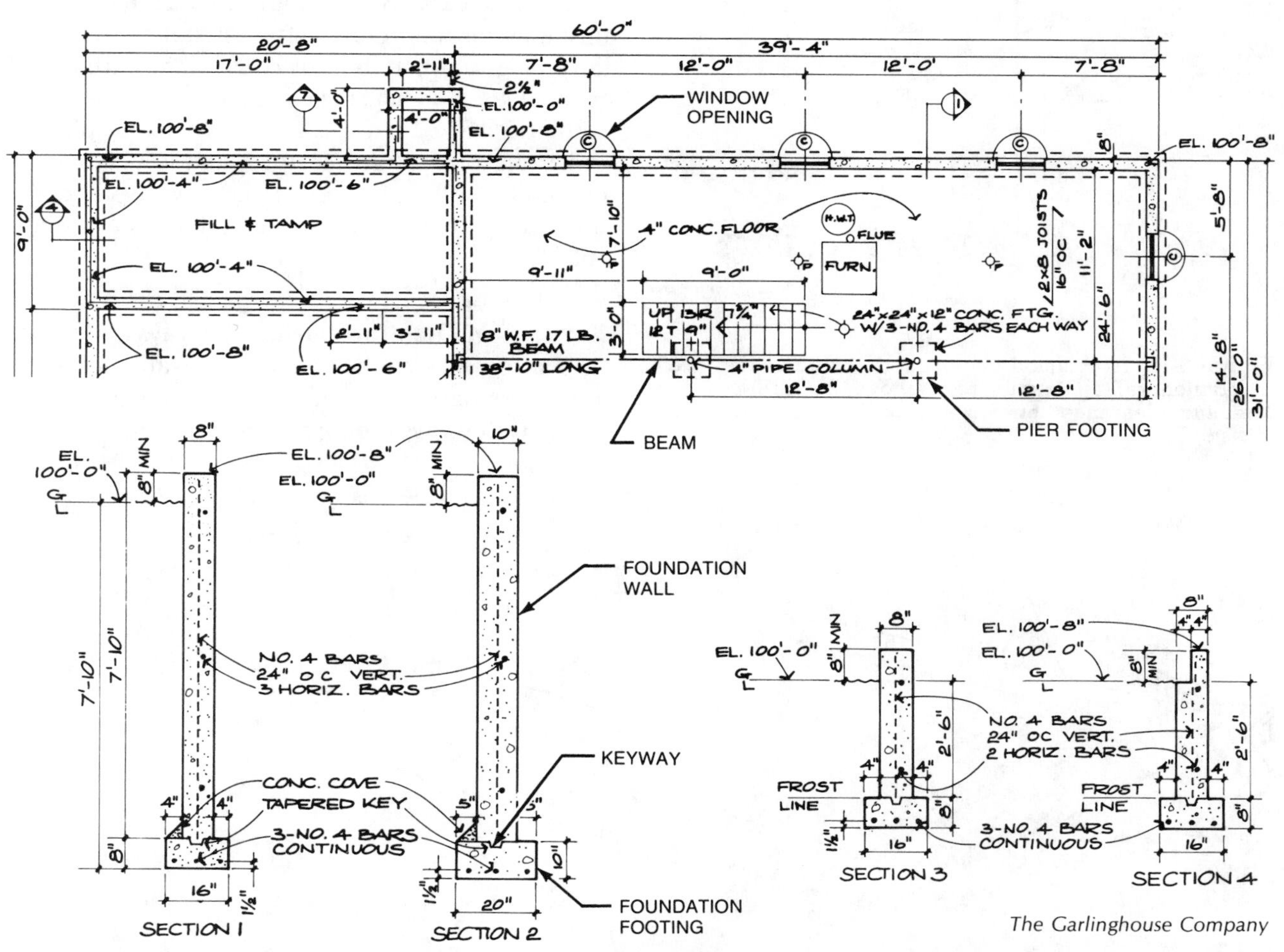

The Garlinghouse Company

Figure 3-12. A set of prints provides information for a full basement residential foundation. The details and foundation section drawings are referenced by numbers on the foundation plan.

should be well insulated to resist water and vapor penetration. A bituminous material, such as asphalt, is often applied to the outside surface of the wall for waterproofing. See Figure 3-13.

Constructing Footing Forms

Footing forms for a full basement foundation are constructed after the excavation work has been completed. Excavations for below-grade basements should extend at least 2′ outside the building lines to provide ample working space for the formwork. More than 2′ may be necessary in loose or porous soil. An excavation should extend to the bottom of the floor slab and is determined by the height of the foundation wall and how much the wall extends above the finish grade. The bottom of the excavation must be level. Therefore, the highest elevation point around the perimeter of the excavation should be used as the reference point for determining the depth of an excavation. See Figure 3-14.

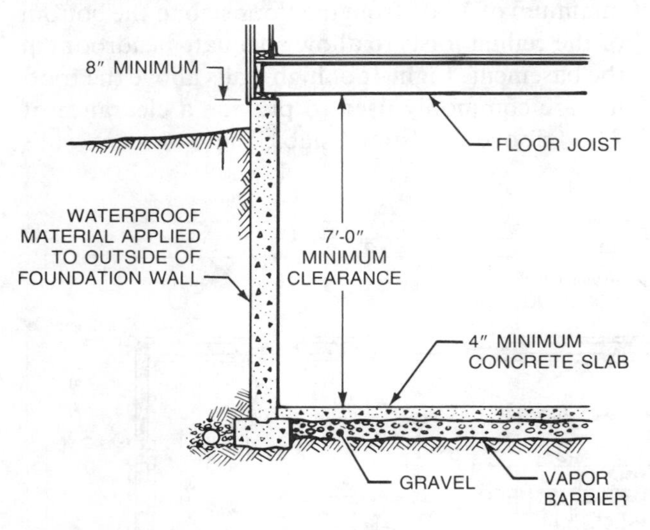

Figure 3-13. A common code practice is to extend the foundation wall a minimum of 8″ above the finish grade. Minimum clearance between the floor slab and the floor joists is 7′-0″.

Footing forms are positioned by measuring the required distance from the building lines. A transit-level or builder's level is used to establish the elevations of the footings. (See Appendix E, Leveling Instruments, for information on reading grades and elevations.) Footings are formed using an earth-formed or constructed footing method. *Earth-formed footings* can be used in firm and stable soil. Earth-formed footings are formed by digging a trench to the dimensions of the footing, and filling it with concrete. Precautions must be taken to avoid collapse of the sides of the trench.

Constructed wood forms are required in loose and unstable soil conditions. After the stakes are driven into the ground, 1″ boards or 2″ planks are nailed to the stakes and held the correct distance apart with form ties. See Figure 3-15. When using planks, stakes are placed farther apart and less bracing is required. Although wood is commonly used for stakes and bracing, many metal devices are available.

Keyways and Reinforcement. Keyways and rebars tie the footings to the foundation walls. Keyways are tapered grooves centered in the surface of the footing. Keyways are formed by pressing *key strips*

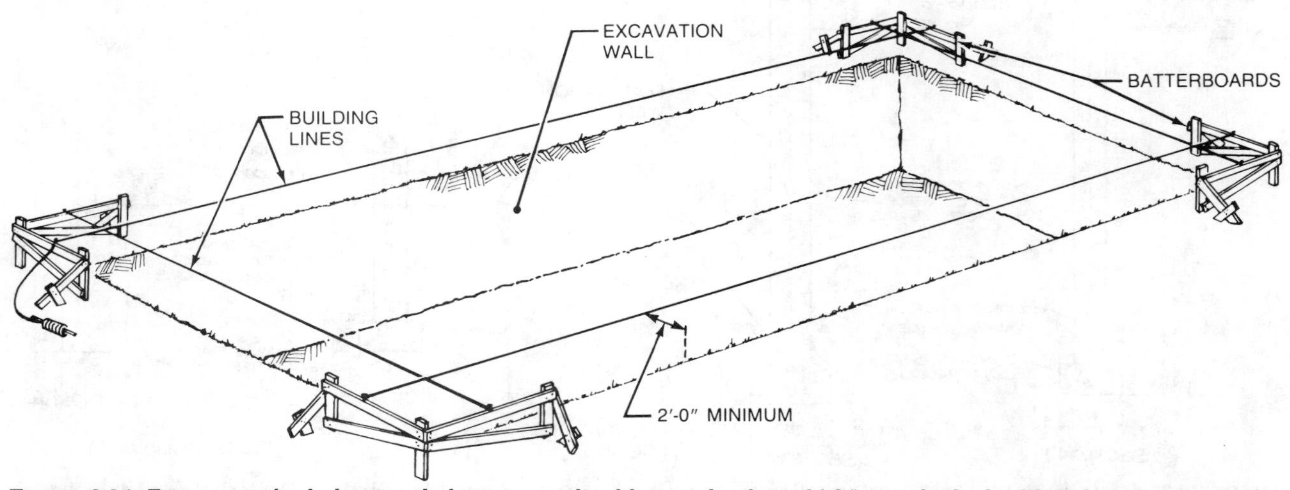

Figure 3-14. Excavation for below-grade basements should extend at least 2′-0″ outside the building lines to allow sufficient room for form construction.

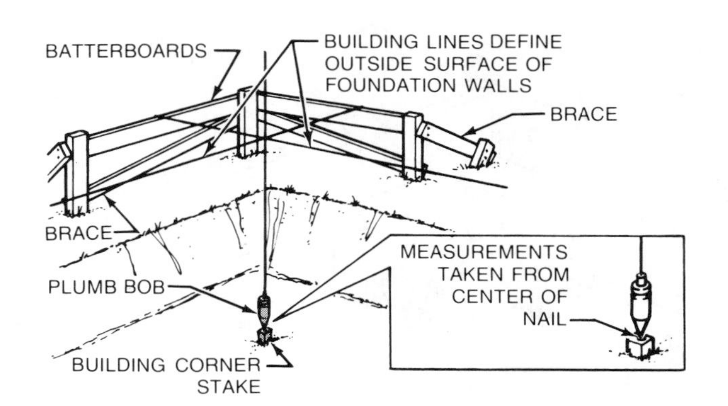

1. Stretch building lines on batterboards. Suspend a plumb bob from each building corner intersection. Drive stakes and place nails to establish building corners.

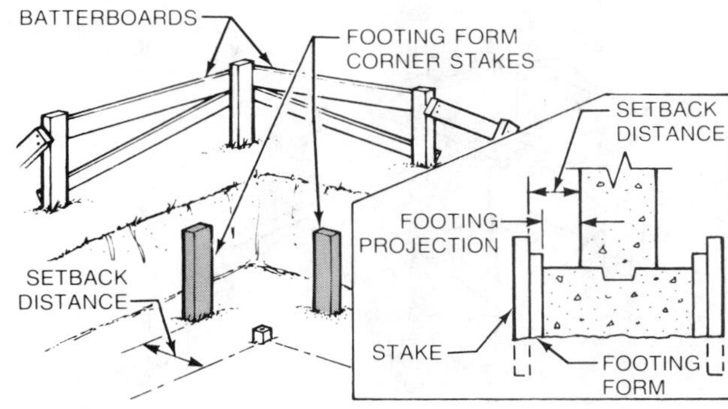

2. Measure the distance that the footing projects beyond the foundation wall plus the thickness of one form board. Drive two footing corner stakes and remove building corner stake.

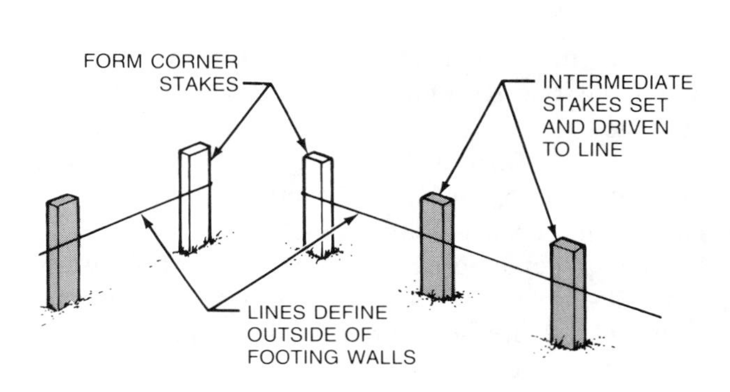

3. Stretch lines between footing corner stakes. Set intermediate stakes to lines and drive into position.

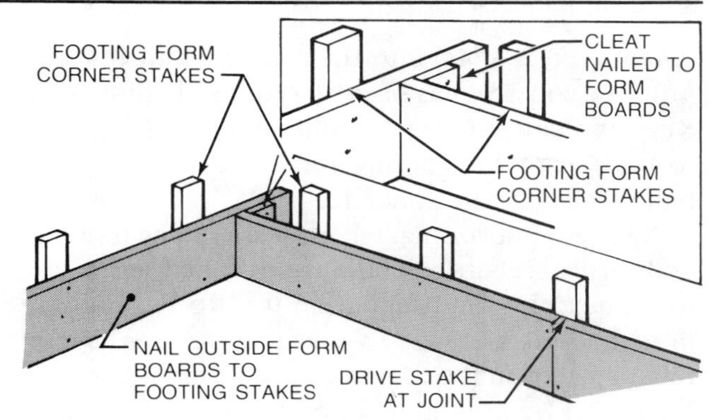

4. Mark the footing elevation on all stakes using a transit-level or builder's level. Nail form boards to stakes running one side past the adjoining side. Nail a cleat to the corner for strength.

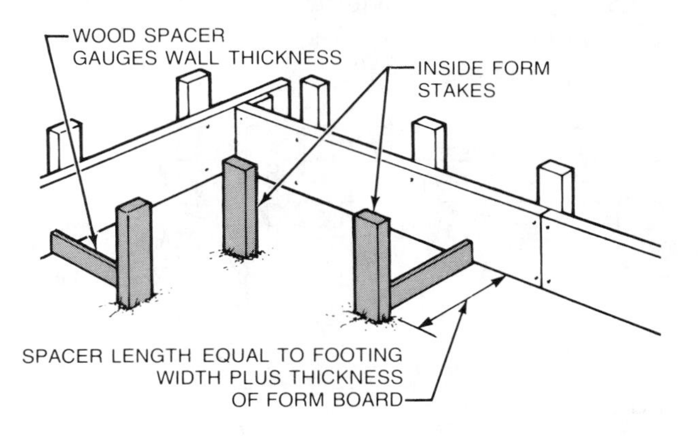

5. Cut a wood spacer the width of the footing plus the thickness of one form board. Drive stakes for inside form wall using spacer as a gauge.

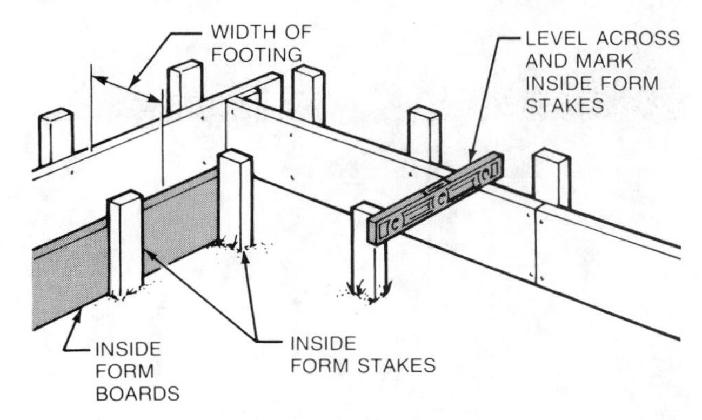

6. Level across from the outside form boards with a hand level and mark the inside form stakes. Nail inside form boards to stakes.

Figure 3-15. Wood-formed footings are constructed for loose and unstable soil conditions.

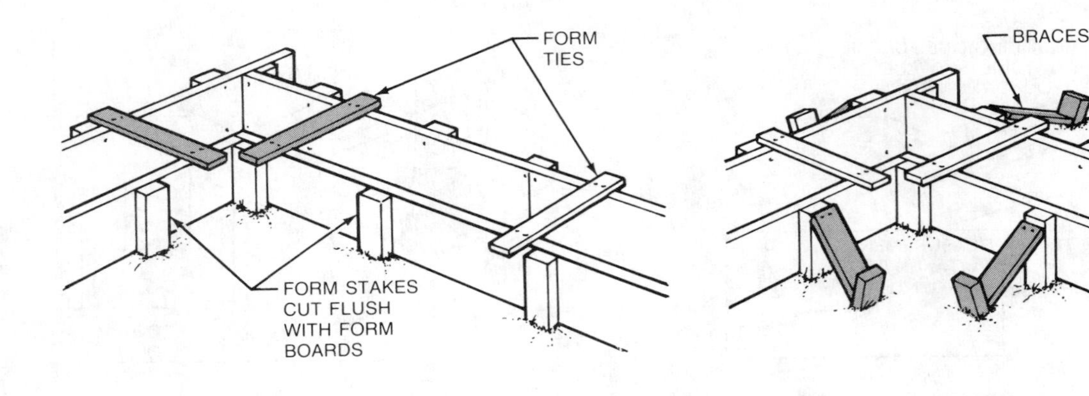

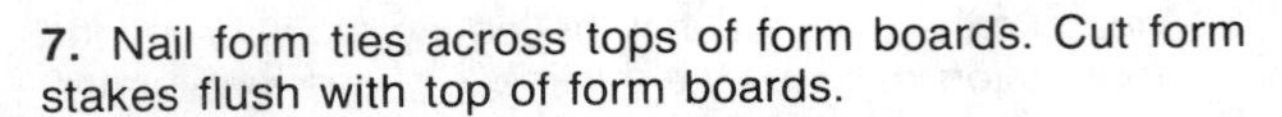

7. Nail form ties across tops of form boards. Cut form stakes flush with top of form boards.

8. Drive stakes and brace all form walls.

Figure 3-15 (continued). Wood spacers are used to establish consistent footing form width. Form ties secure the form boards in position.

(chamfered 2 × 4s) into the concrete immediately after the concrete has been placed. See Figure 3-16. Keyways may also be formed by securing a key strip to the top of the footing form with a crosspiece before placing the concrete.

A keyway, horizontal rebars placed in the footing, and vertical rebars extending above the footing are used to secure the foundation wall to the footing in seismic risk areas. The vertical rebars later tie into rebars placed in the concrete or masonry walls. See Figure 3-17. Size and number of rebars required is provided in section views of the foundation plan. The local building code should also be consulted to verify rebar requirements for the area.

Figure 3-16. The keyway secures the bottom of the foundation wall. It is formed by pressing a key strip into the soft concrete.

Constructing Wall Forms

Wall forms are constructed after the concrete has set in the footings. Outer form walls are usually constructed first to facilitate placement of reinforcement. Inner form walls, however, are constructed first when an opening must be left in the center, such as for an elevator shaft.

Base plates are used as a base for outer form walls of built-in-place or panel forming systems. After the base plates are secured, outside form walls are erected and carefully plumbed and aligned. See Figure 3-18. Rebars, electrical conduit, and other utilities are then positioned.

If window or door bucks are required, they are positioned and fastened to the outside form walls. If *beam pockets* are required, they are also formed at this time. A beam pocket is a space in the foun-

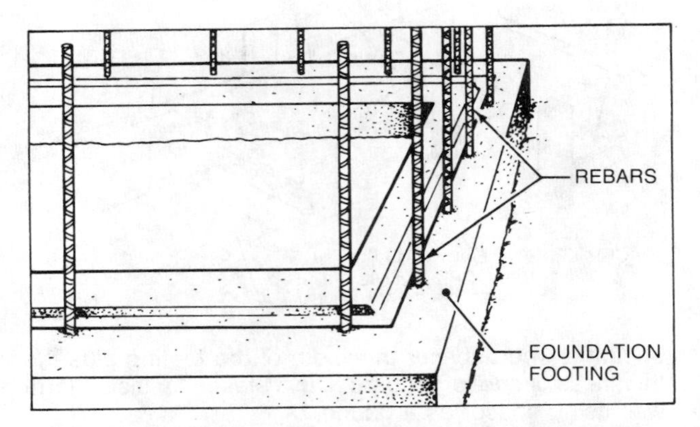

Figure 3-17. Vertical rebars extending from the footing will tie into rebars placed in the wall constructed over the footing.

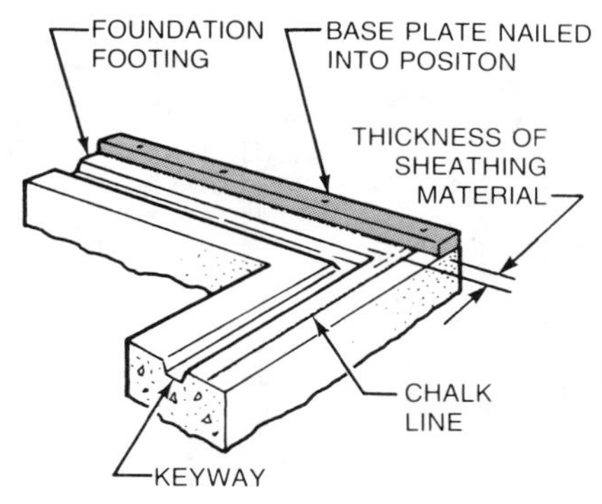

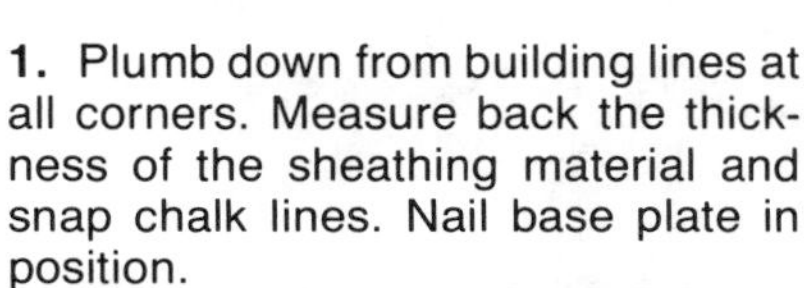

1. Plumb down from building lines at all corners. Measure back the thickness of the sheathing material and snap chalk lines. Nail base plate in position.

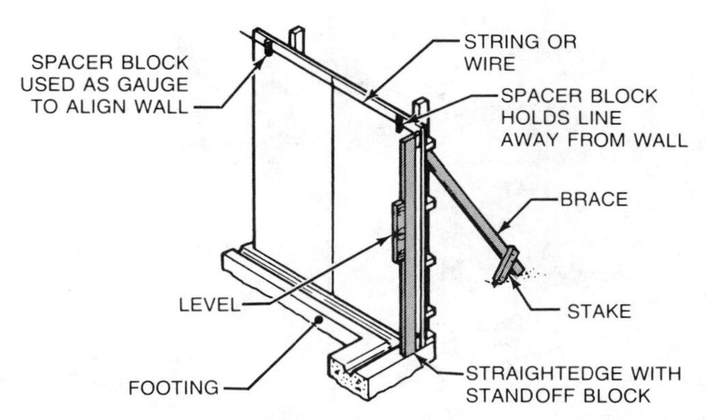

2. Construct form wall. Plumb walls with level and straightedge with 3/4″ × 3/4″ × 4″ stand-off block nailed to edge. Brace ends of form walls. Stretch a line from one end of wall forms to the other and hold away with 3/4″ spacer blocks. Align form walls by using a third spacer block as a gauge. Brace where necessary.

Figure 3-18. Form walls must be plumbed and aligned properly.

dation wall that receives a beam. The top surface of the beam is flush with the top of the foundation wall. When wood beams are used, the pockets should provide 4″ minimum bearing surface and ½″ clearance around the sides and end of the beam.

After the outer form walls are positioned, the inside form walls are *doubled up* and tied together. Doubling up is the placing of the second or opposite form wall. See Figure 3-19.

CRAWL SPACE FOUNDATIONS

A crawl space foundation is often used as an alternate foundation design in a building that does not require a full basement. In crawl space foundations, or basementless foundations, short walls are constructed over spread footings. A crawl space foundation facilitates the placement of plumbing, electrical, and other utilities. A T-foundation is commonly used and is placed monolithically. The perimeter of the superstructure is supported by the T-foundation and pier footings provide intermediate support. See Figure 3-20.

The *crawl space* is the space between the bottom of the floor joists resting on top of the foundation walls and the ground below. Information regarding the minimum crawl space distance for a building in a given area is usually provided by the local building

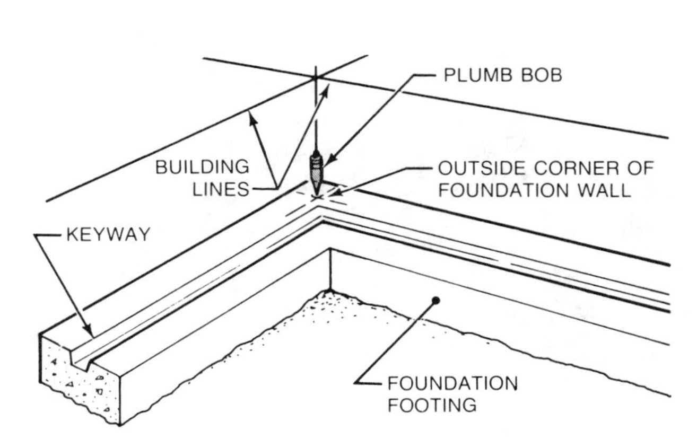

1. Plumb down from the building lines and mark the outside corners of the foundation wall on the footing.

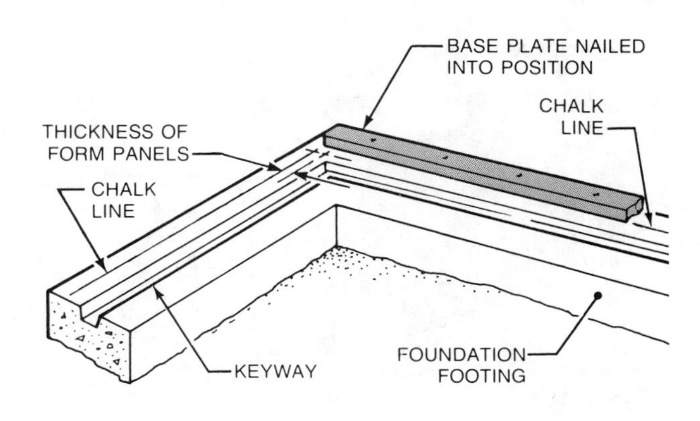

2. Measure the thickness of the form panels to the outside of the corner and snap lines. Nail base plates next to the chalk line with concrete nails.

Figure 3-19. The position of full basement wall forms is laid out on top of the foundation footing.

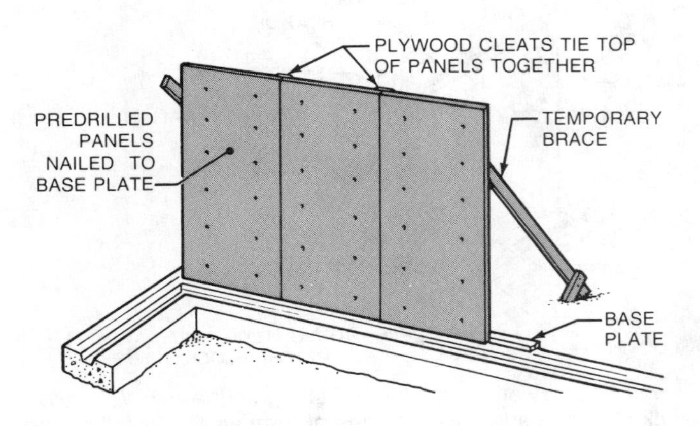

3. Nail a predrilled panel into place and support with a temporary brace. Nail additional panels into place and tie tops together with plywood cleats. Brace every third or fourth panel.

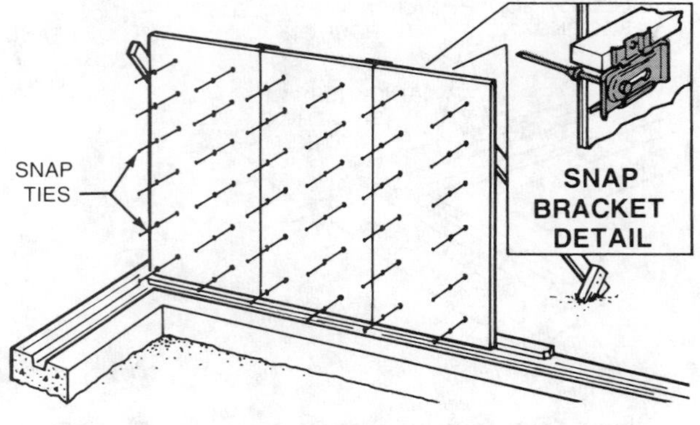

4. Place snap ties into predrilled holes and secure with snap brackets and wedges.

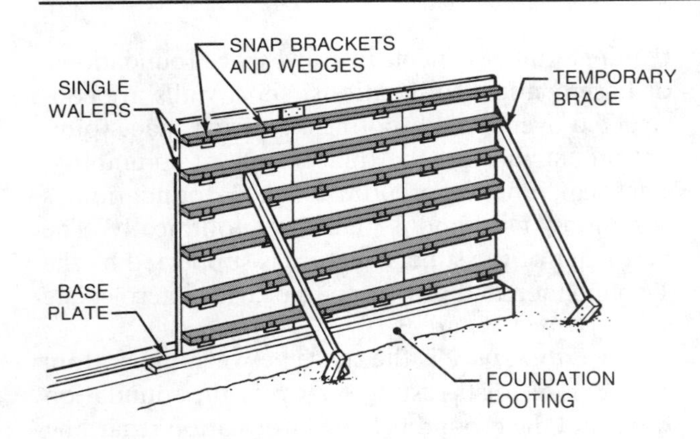

5. Place a row of walers and hand tighten wedges. Place remaining walers and tighten with a hammer.

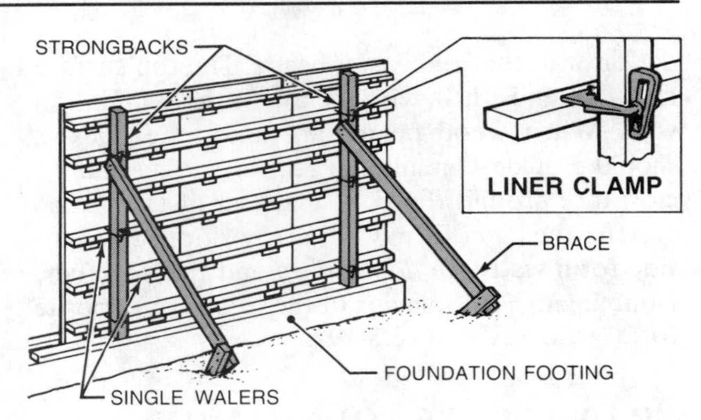

6. Attach strongbacks to walers. Align top of wall to a line and nail braces to strongbacks.

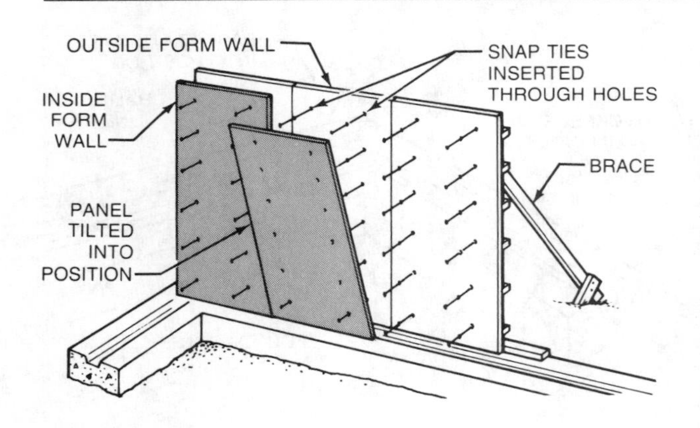

7. Tilt inside form wall into position while inserting ties through the panels.

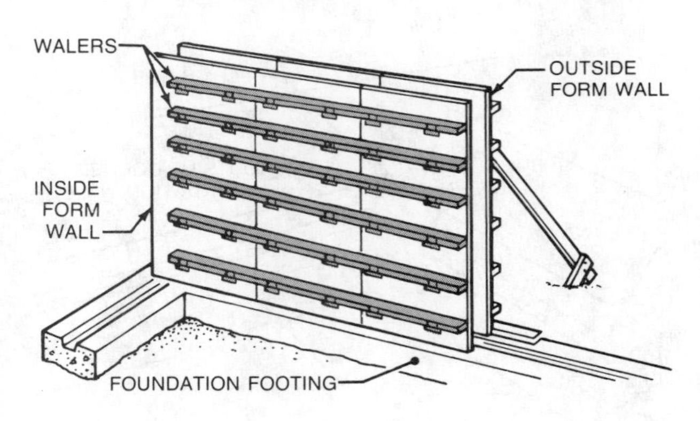

8. Place snap brackets, wedges, and walers for inside form wall.

Figure 3-19 (continued). Full basement wall forms are reinforced with strongbacks and walers.

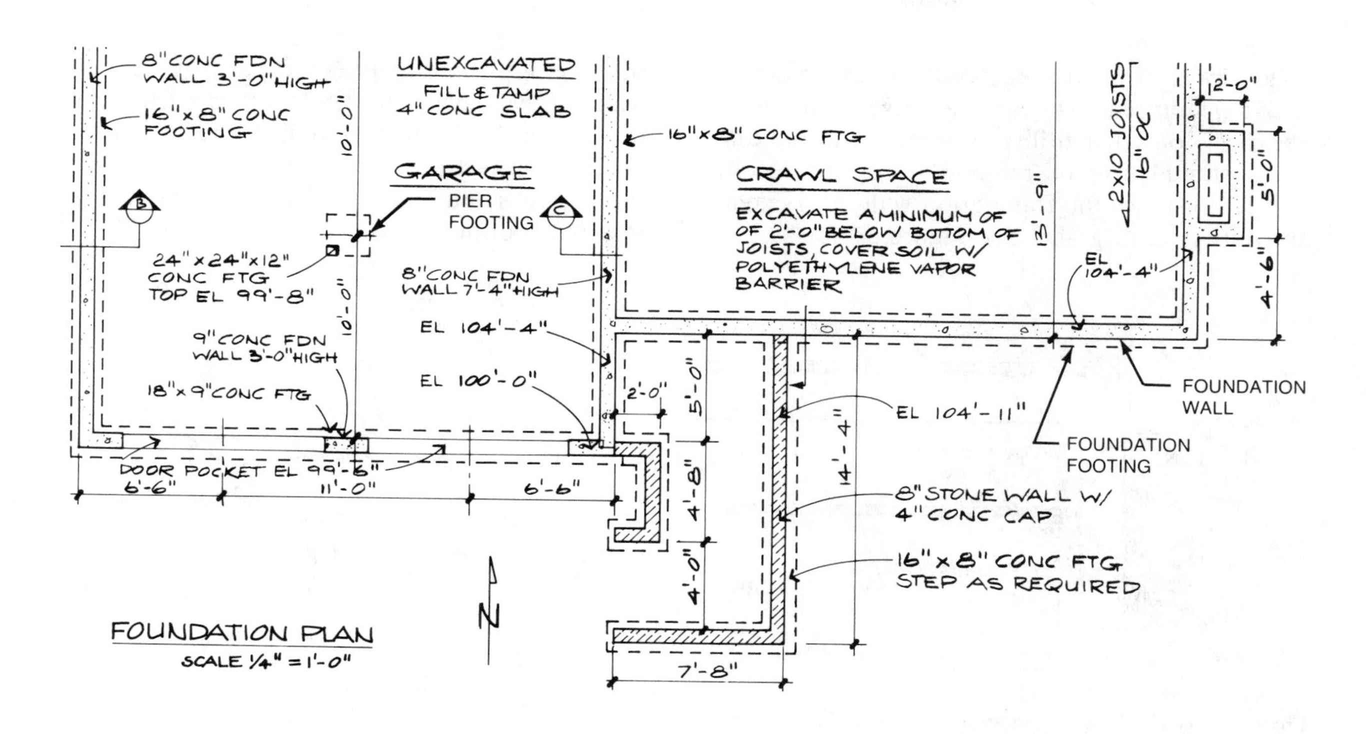

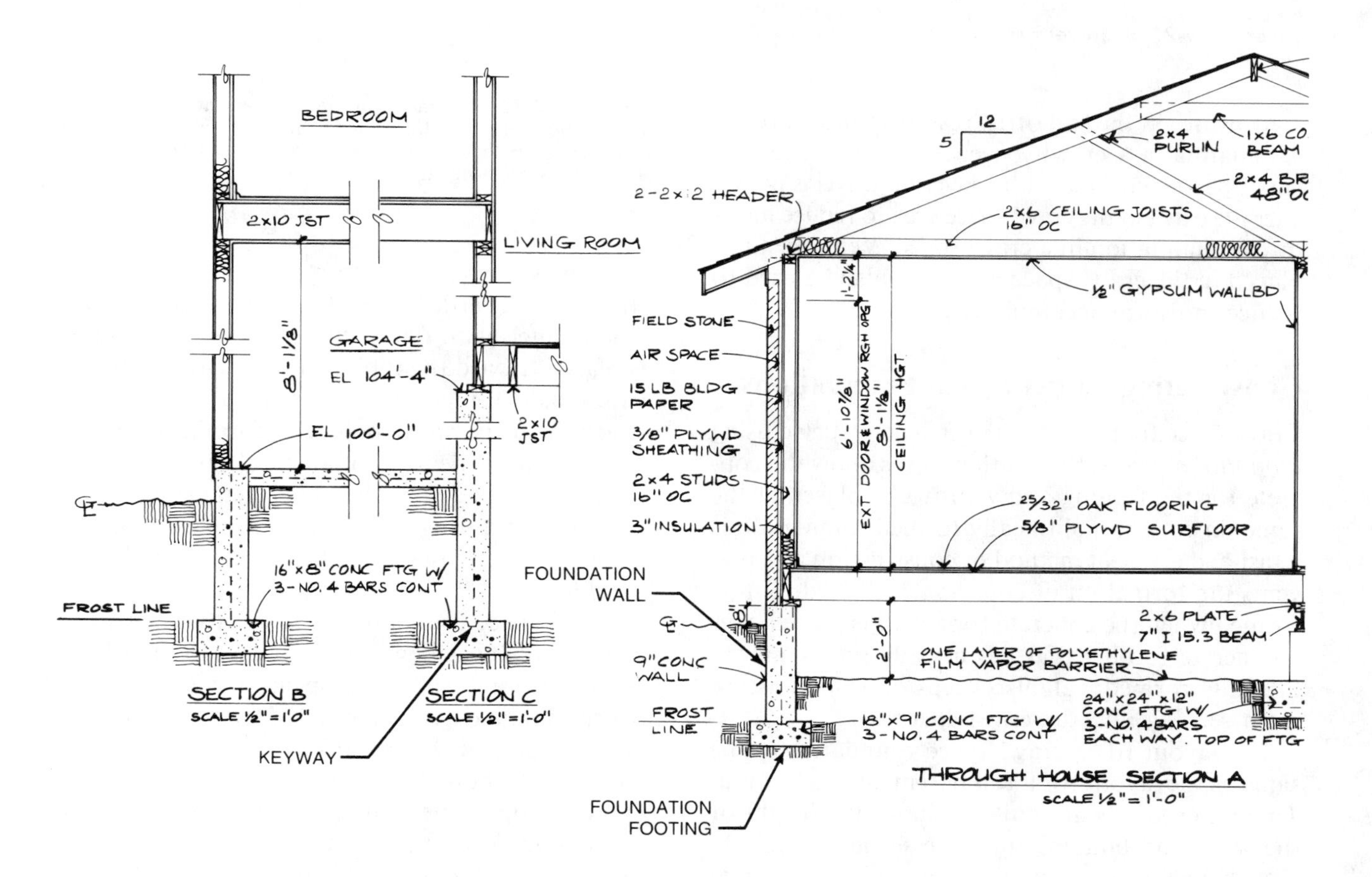

Figure 3-20. A set of prints for a crawl space foundation provides information regarding foundation walls and footings.

code. Crawl spaces range from 18″ to 24″ in height. A 24″ clearance allows easier access for plumbing, electrical, or other utility repairs. Local building codes should also be referenced for the recommended distance the foundation wall must extend above the finish grade. See Figure 3-21.

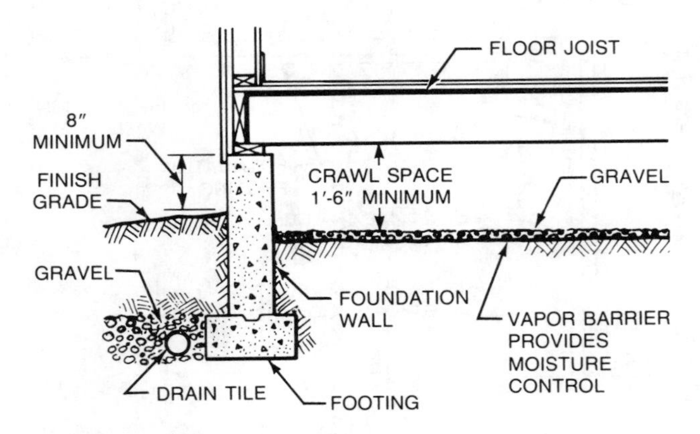

Figure 3-21. The foundation wall must extend a minimum of 8″ above the finish grade. The floor joists must be minimum of 1′- 6″ above the ground. A vapor barrier covered by 2″ of gravel provides moisture control.

Moisture in the soil often results in moisture accumulating in a crawl space. A polyethylene film vapor barrier covered with a layer of gravel is placed over the earth in the crawl space area to reduce moisture accumulation in a crawl space. Vent openings placed in the crawl space walls can also be used to reduce moisture accumulation.

Constructing Crawl Space Foundations

Crawl space foundations are often constructed using *monolithic forms*. Monolithic forms allow the concrete for the footing and walls to be placed at the same time. A monolithically formed foundation is a fast and efficient method of construction and prevents the formation of *cold joints*. A cold joint is formed when the concrete for a wall is placed over a concrete footing that has already set. A monolithically formed wall also keeps outside moisture from seeping into the crawl space.

The layout for a crawl space foundation is the same as the layout for a full basement foundation. The batterboards are built to the actual height of the walls and building lines are set in place.

Although major ground excavation is not required for a crawl space foundation, trenches are dug for the footings. Building lines serve as a guide for the width of the trench. The depth of the trench is measured vertically from the building lines. See Figure 3-22. When earth-formed footings cannot be used, the trenches must be dug wide enough for constructed footing forms.

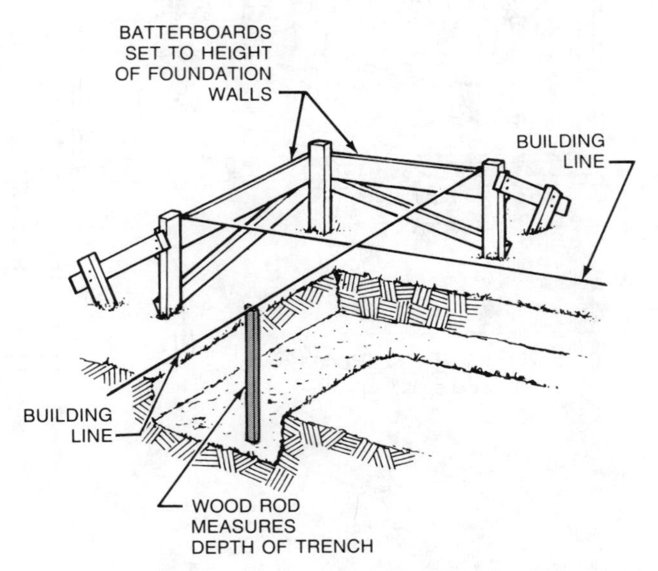

Figure 3-22. Crawl space foundation trenches are laid out according to the building lines. The depth of the trenches is measured from the building lines if the batterboards are set to the wall height.

Crawl space foundation wall forms can be constructed over earth-formed footings by securing plywood panels with flat metal stakes. A single waler is placed toward the top of the form and is braced with stakes. Temporary wood spacers are placed between the form walls, and wood form ties are nailed across the walers to hold the walls together. The temporary wood spreaders and metal stakes must be removed during the initial setting period. The panels can be reused after the forms are stripped. See Figure 3-23.

Crawl space foundation wall forms can also be constructed by nailing 2″ thick planks to short form studs. The form walls are suspended over the trench, which eliminates removing stakes from the concrete. Horizontal 2 × 4s are staked to the ground and nailed to the bottom of the form studs. Braces run from the top of the studs to the horizontal 2 × 4s. A plywood cleat is used to strengthen the tie between the horizontal 2 × 4 and the brace. A plywood template is used for spacing the forms while they are

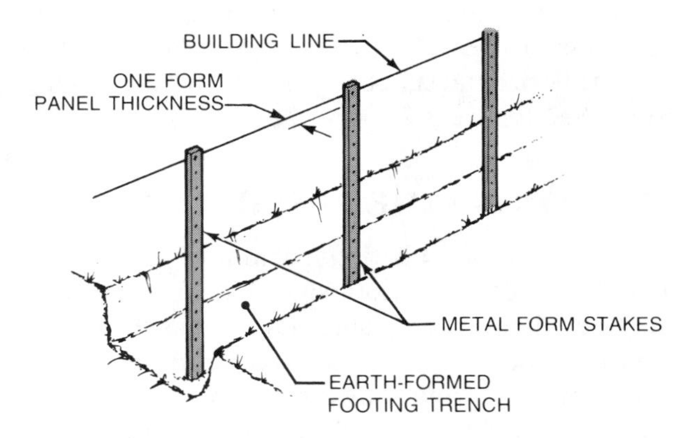

1. Plumb down from the building lines. Drive stakes one panel thickness to the outside of the building lines.

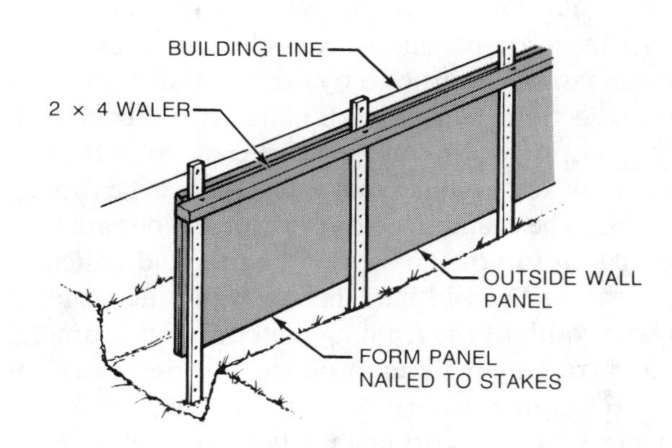

2. Mark the top elevation of the wall on the stakes. Nail the panels to the stakes and place a top waler.

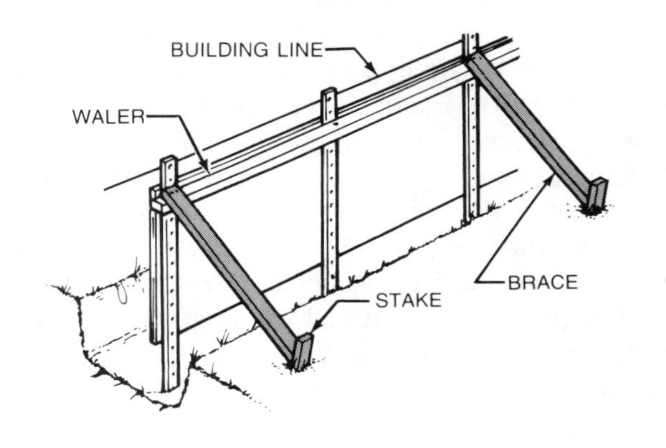

3. Plumb and brace all wall corners. Align the wall to a line and brace where necessary.

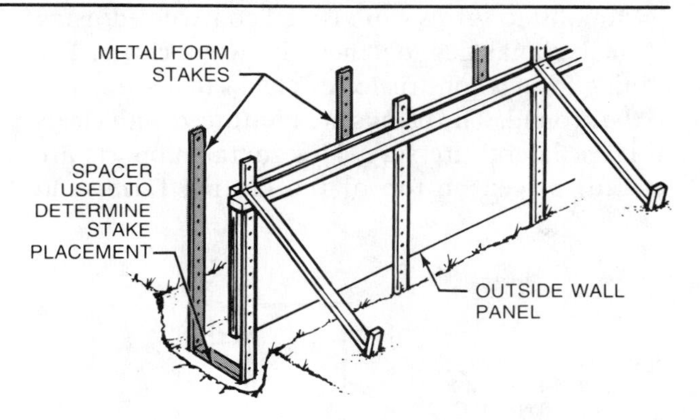

4. Cut a spacer block equal to the wall thickness plus the thickness of two panels. Drive the inside form stakes using the spacer block as a gauge.

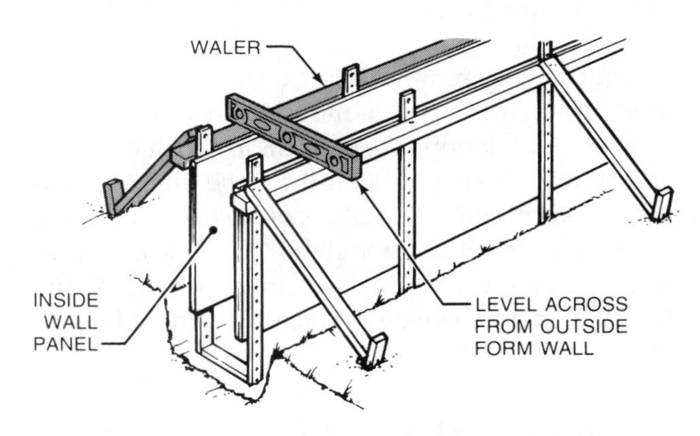

5. Level across from the outside form wall and nail the inside wall panels to the stakes. Place a top waler.

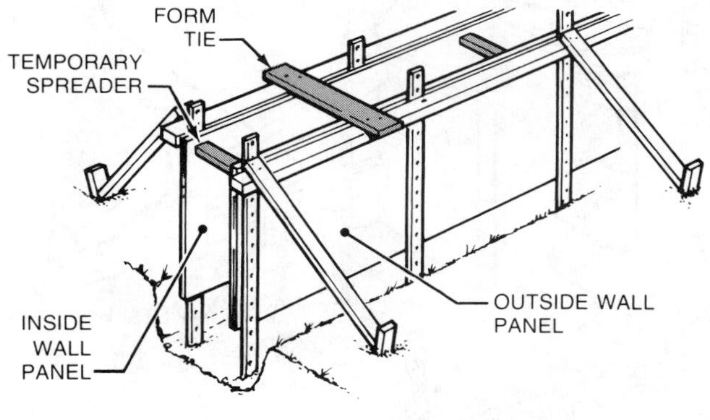

6. Place temporary spreaders (equal to the wall thickness) between the wall forms. Nail form ties across the tops of the walers. Brace inside wall form.

Figure 3-23. A low panel wall form can be constructed over earth-formed footings. Flat metal stakes are recommended to hold the form wall in position.

being secured in place. After the forms are stripped, the planks can be reused. See Figure 3-24.

In loose or porous soil conditions, monolithic forms can be constructed by using prefabricated plywood sections framed with plates and short studs. The footing forms are constructed with footing boards nailed to stakes that extend above the footing boards. The prefabricated plywood sections are then placed on top of the footing boards and nailed to the stakes. The width of the low wall is determined by the width of the framing material. For example, 2 × 4 framing material produces a wider wall than 2 × 6 framing material over the same footing. See Figure 3-25. This forming method is useful for tract home construction over level terrain where a series of similar foundations is being built. The framed plywood sections can be reused many times.

Monolithic forms can also be constructed using 2″ thick planks tied together with wood cleats. The footing form is constructed of planks that are staked to the ground. The planks are reinforced with cleats to resist lateral pressure. Horizontal supports are leveled, placed on top of the footing form, and nailed to the stakes and wall cleats. The supports suspend the wall form over the footing form. The thickness of the wall is determined by the amount that the horizontal supports project over the footing form. See Figure 3-26.

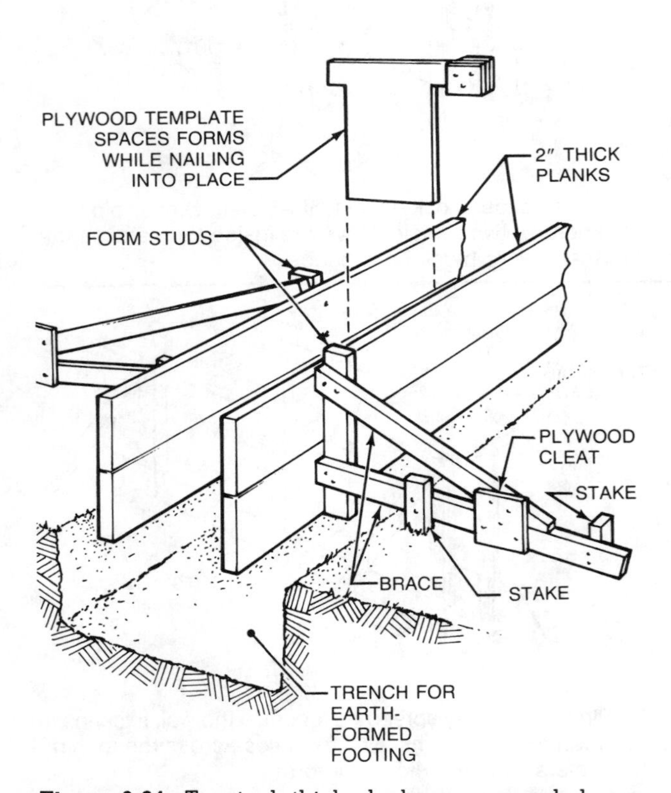

Figure 3-24. Two-inch thick planks are suspended over an earth-formed footing. This method eliminates the need to remove the stakes from the concrete.

SLAB-ON-GRADE FOUNDATIONS

Slab-on-grade foundations combine foundation walls with a concrete floor slab. The top surface of the floor slab is at the same elevation as the top of the foundation wall. The slab receives its main support from the ground directly below and is reinforced with welded wire fabric or rebars. See Figure 3-27.

Slab-on-grade foundations are commonly used in residential and other light construction. A slab-on-grade foundation is a cost-effective method that eliminates deep excavations, high foundation walls, and costly wood floor systems. Slab-on-grade foundations are not practical over steeply sloped lots, or where the water table is close to the ground surface. An example of a print for a slab-on-grade foundation system is shown in Figure 3-28.

A slab-on-grade foundation design commonly used is a floor slab constructed independently of a T-, rectangular, or battered foundation. The slab butts against the foundation wall or rests on a shoulder at the top of the wall. Because the floor slab is placed independently of the foundation, cracks around the perimeter of the slab are avoided. The independent wall and slab method is recommended in cold climates because insulation may be placed around the perimeter of the floor. See Figure 3-29.

Another commonly used design for a slab-on-grade foundation is a monolithically placed floor slab and foundation. This design, in which the walls and floor slab are placed at the same time, is common in warm climates. Rebars or welded wire fabric extending from the floor slab into the foundation walls make the floor slab an integral part of the foundation and help distribute the building load to the soil.

Constructing Slab-on-Grade Foundations

When the floor slab and foundation walls are placed independently, forms for conventional T-, rectangular, or battered foundation walls are constructed. The floor slab may either butt against the inside of

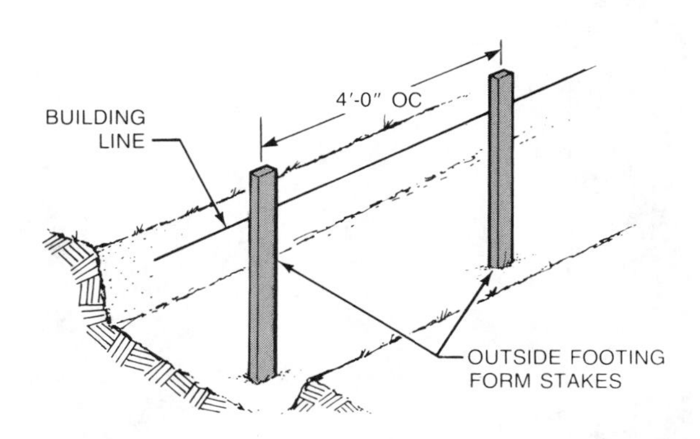

1. Stretch a line for the outside footing form stakes. Plumb and align stakes and drive into ground at 4'-0" OC.

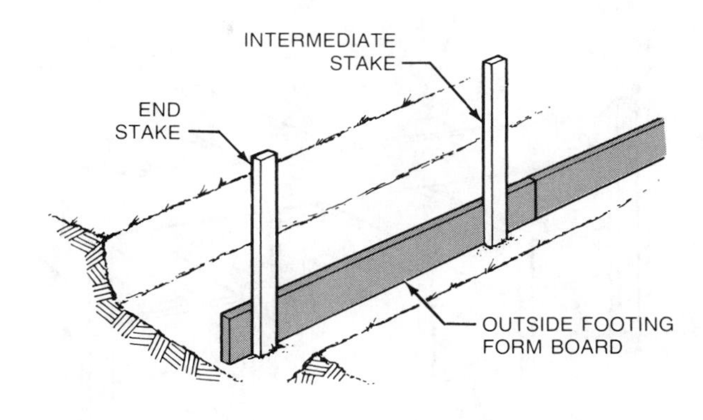

2. Establish the height of the footing on two end stakes. Snap a chalk line on intermediate stakes. Nail the outside footing form boards to the stakes.

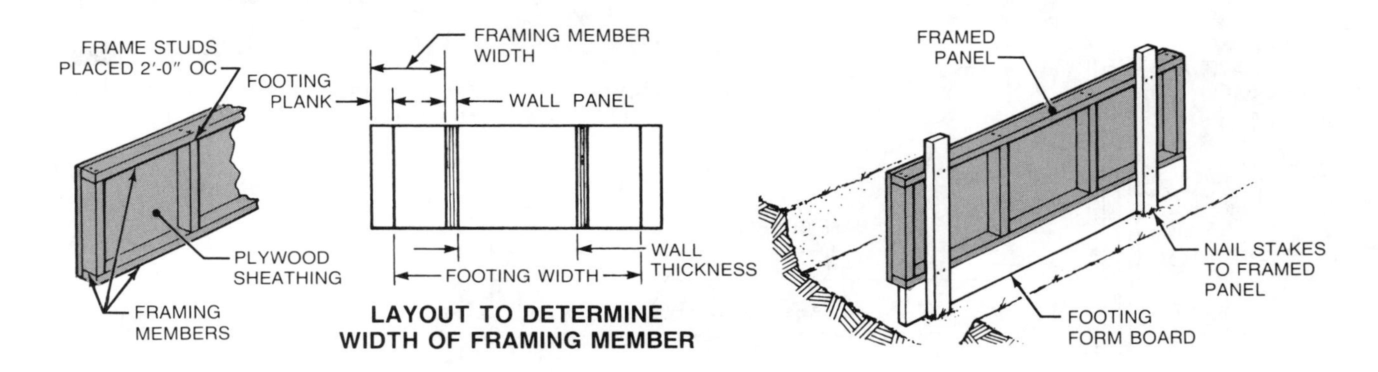

3. Construct panel sections with plywood sheathing nailed to frames. Frame width determines the wall thickness.

4. Place the framed panel on top of the footing form boards. Nail the stakes to the panel frame.

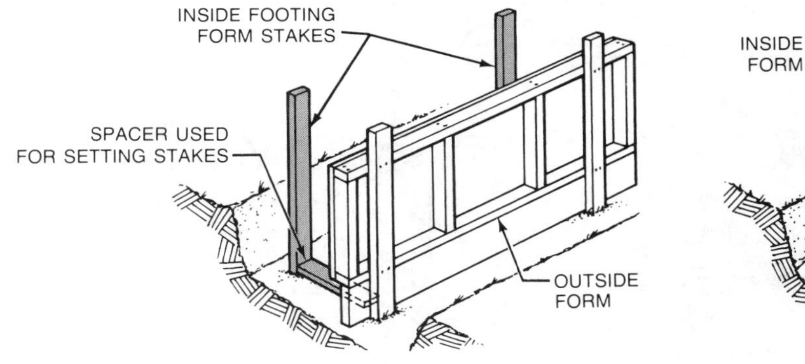

5. Cut spacers equal to the footing width plus the form board thickness and use as a gauge for setting inside form stakes. Drive stakes and nail the inside footing form boards to them.

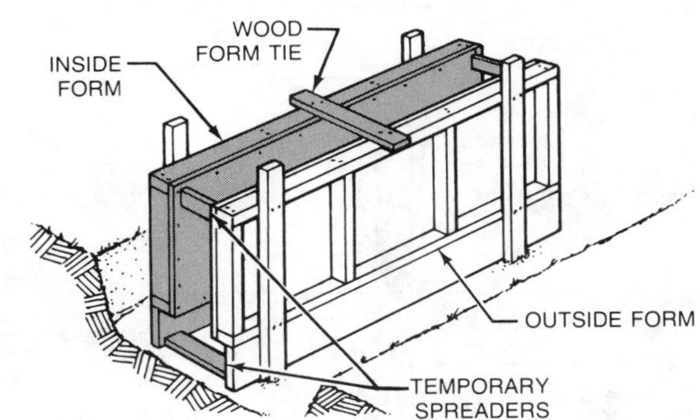

6. Place the framed panel on top of the footing form. Use temporary spreaders to position the panel. Nail the stakes to the panel frame. Nail wood form ties across the top of the form.

Figure 3-25. A framed panel wall form is used to construct a foundation. The size of the frame material will determine the width of the low wall.

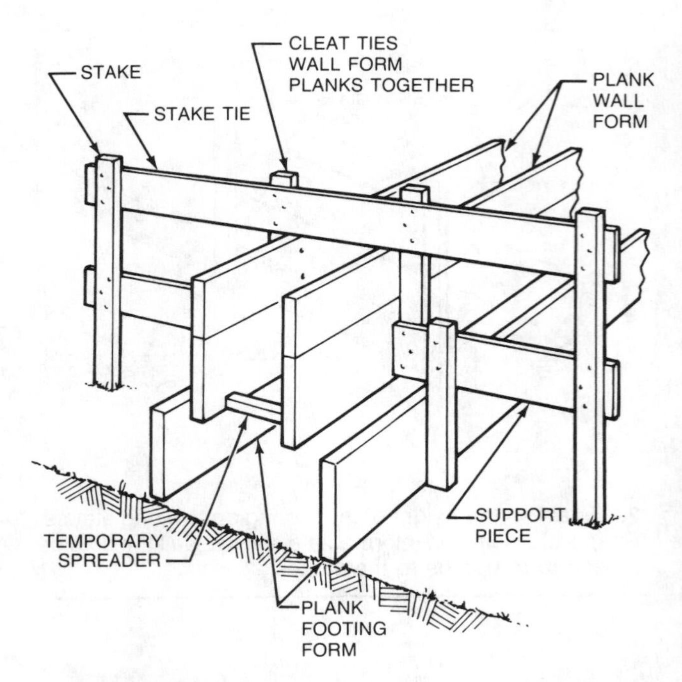

Figure 3-26. A low wall may be formed by using 2″ planks tied together with cleats.

Figure 3-27. A slab-on-grade foundation consists of the foundation walls and a floor slab.

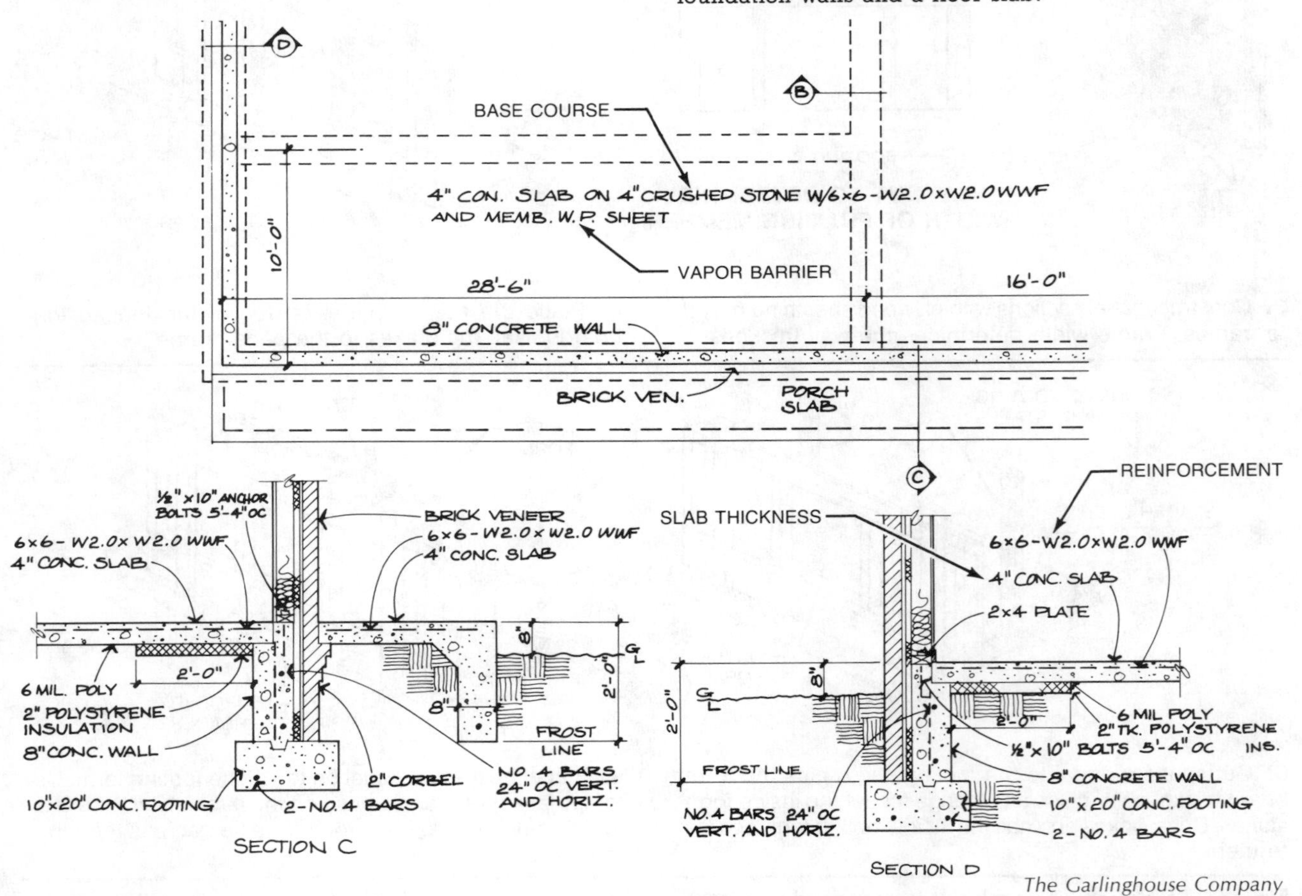

Figure 3-28. The print for a slab-on-grade foundation includes the slab thickness and type and size of reinforcement.

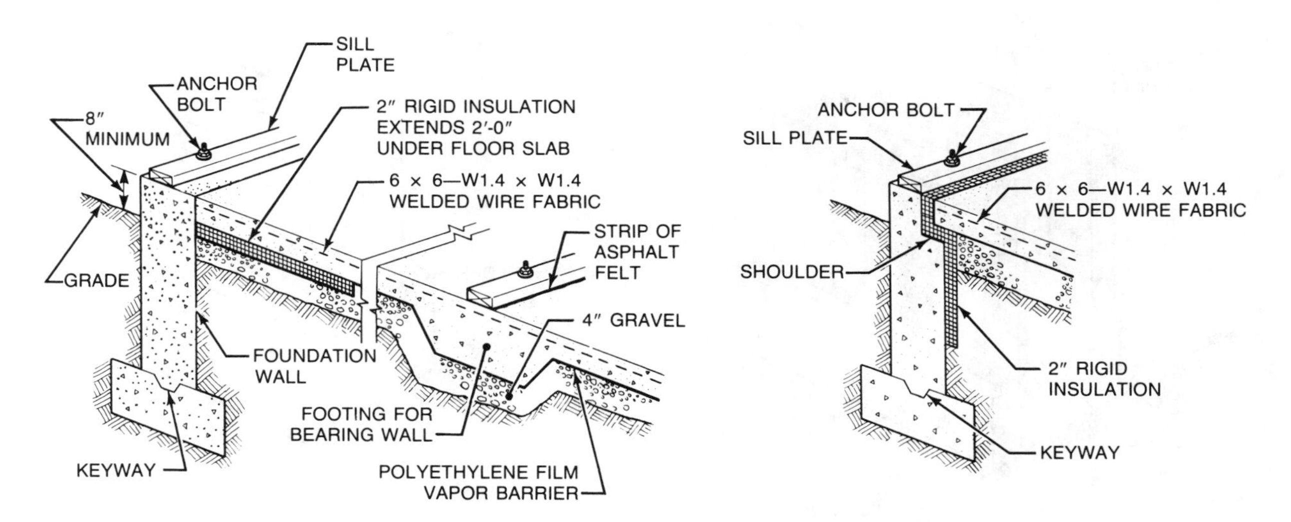

Figure 3-29. A slab-on-grade foundation is formed with the floor slab butting against the wall or resting on a shoulder. Rigid insulation is placed around the perimeter of the slab.

the foundation wall or rest on a shoulder at the top of the wall. A shoulder is formed by nailing wood members or Styrofoam equal to the depth and width of the shoulder to the top of the inside wall form.

A monolithic slab-on-grade foundation is formed by constructing an edge form along the outside edge of a trench. The edge form is constructed of plywood planks held in place with stakes and braces. A footing is formed by the concrete placed between the edge form and the opposite wall of the trench. See Figure 3-30.

GRADE BEAM FOUNDATIONS

In grade beam foundations, walls are supported by reinforced concrete piers that extend deep into the ground. A grade beam foundation is commonly used with stepped or ramped foundations erected on hillside lots where soil conditions do not provide adequate support for conventional footings. The design of grade beams and piers should conform with accepted structural engineering practices and local code requirements. The *Minimum Property Standards, One and Two Family Dwellings,* published by the U.S. Department of Housing and Urban Development (HUD), states that grade beams must extend a minimum of 8″ above the finish grade when supporting wood-frame construction over average soil conditions.

The bottoms of grade beams should extend below the frost line. The soil directly under the grade beams should be removed and replaced with coarse rock or gravel. Coarse rock or gravel reduces the chance of the ground freezing, which could cause beam movement. Other materials that drain water away from the bottom of the beam may also be used. See Figure 3-31. The grade beam should be at least 6″ thick and 14″ deep. However, in a crawl space foundation, the beam must be deep enough to provide 18″ clearance between the ground and the bottom of the floor joists. Grade beams should be reinforced with four #4 horizontal rebars.

The piers beneath the grade beams should have a minimum diameter of 10″ and may be flared at their base to cover a wider soil area. They should be spaced no more than 8′-0″ OC and extend below the frost line into firm soil. A #5 rebar should run the full length of the pier and extend into the grade beam. In seismic risk areas, additional vertical rebars extend from the piers and tie into horizontal rebars placed in the grade beam.

Constructing Grade Beam Foundations

The design and construction of a grade beam foundation form is similar to a rectangular wall form built over a foundation footing. However, holes for supporting piers are dug before the formwork begins. In firm soil the concrete is placed directly into the hole. In soft and unstable soil, a tubular fiber form is used to form the pier. After the concrete has set, the tubular fiber forms are stripped from the piers. The wall forms are then constructed directly over the concrete piers. See Figure 3-32.

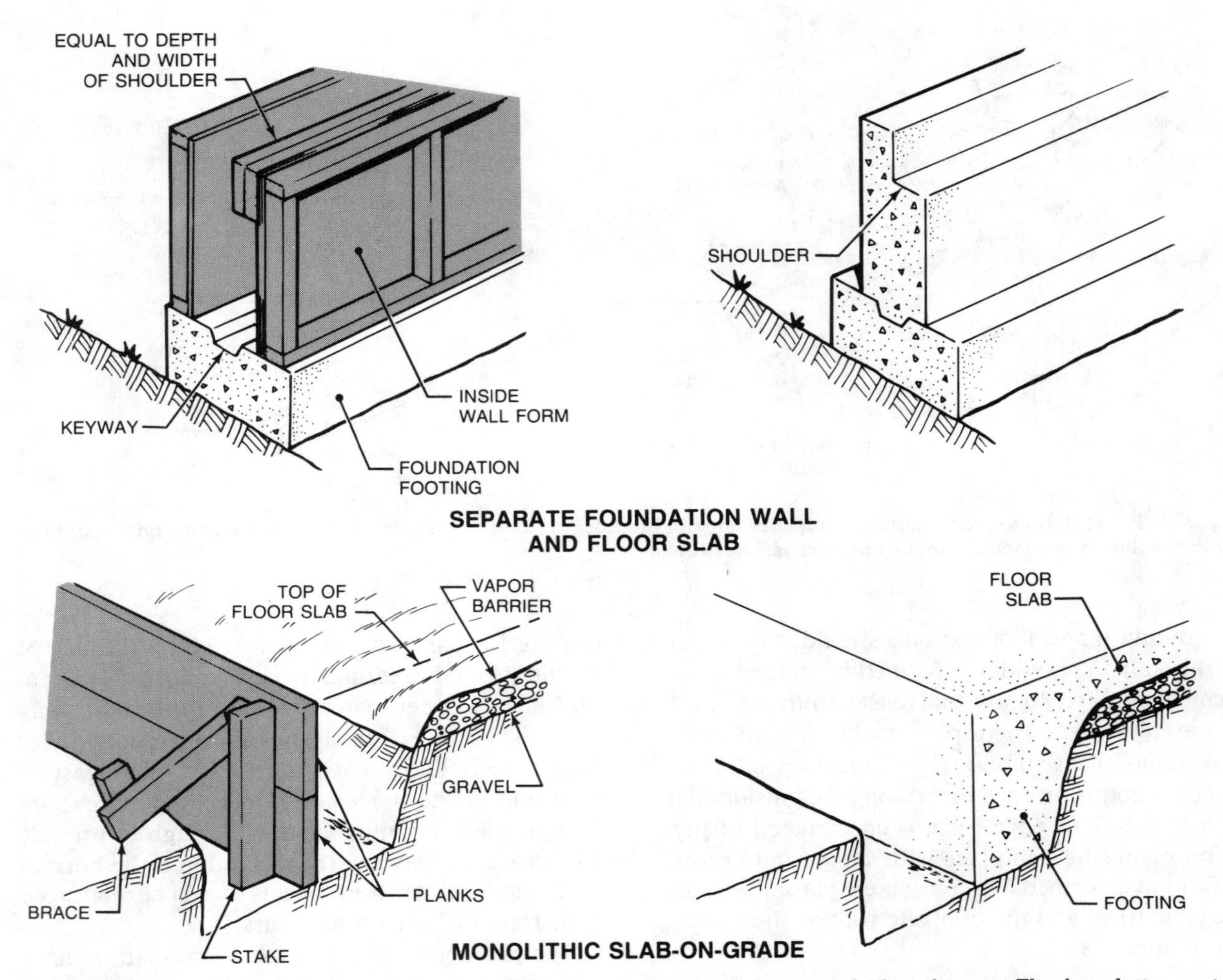

Figure 3-30. Two basic designs may be utilized when constructing slab-on-grade foundations. The foundation system may be formed separately or monolithically.

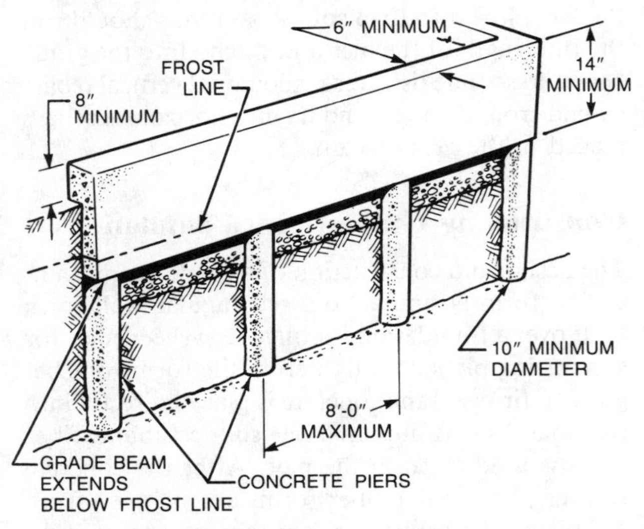

Figure 3-31. Grade beams are supported by concrete piers. A gravel base drains water away from the beam.

STEPPED FOUNDATIONS

Stepped foundations are commonly constructed on steeply sloped lots and used with crawl space or full basement foundations. A stepped foundation is shaped like a series of long steps. A stepped foundation requires less labor, material, and excavation than a level foundation on a sloped lot.

The footings of a stepped foundation must be level. Many building codes require a minimum distance of 2′-0″ between horizontal steps. The thickness of the vertical portion of the footings must be at least 6″ and no higher than three-fourths the distance between horizontal steps. See Figure 3-33.

Wall forming methods used for stepped foundations are similar to constructing rectangular wall forms. Forms for high walls are constructed on previously placed footings. Plywood or planks form

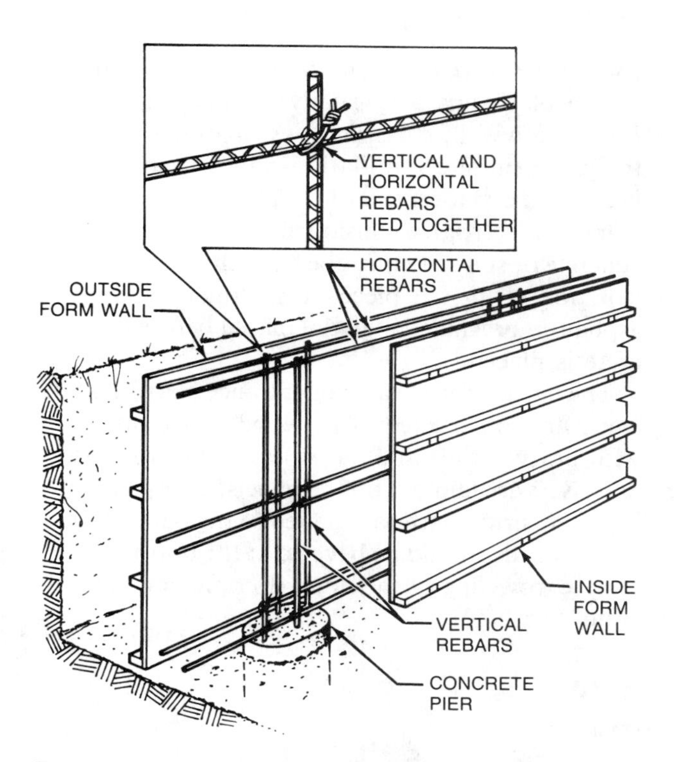

Figure 3-32. Grade beam forms are constructed over completed piers. Horizontal and vertical rebars reinforce the foundation.

the walls and footings. A *shut-off* or *bulkhead* is placed at the end of each step to hold the concrete. Shut-offs are secured in place with cleats and must be able to withstand a great amount of pressure when the concrete is being placed. The cleats are nailed to the shut-off and the whole unit is positioned to a line established for the end of the step. The cleats are then nailed from the outside of the form with duplex nails. See Figure 3-34.

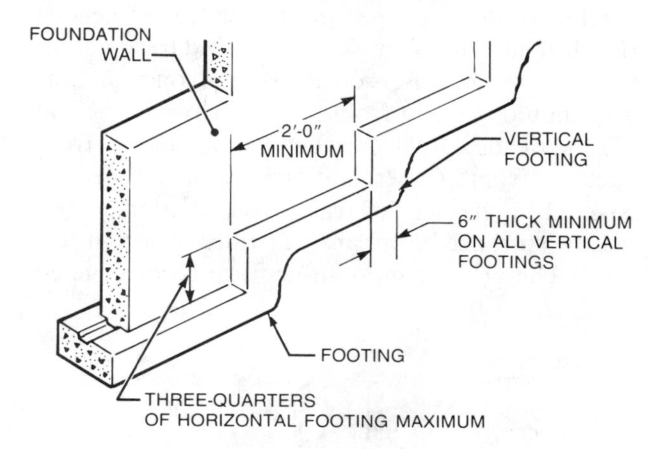

Figure 3-33. Stepped footings are constructed on sloped lots. Consult the local building code for stepped footing requirements.

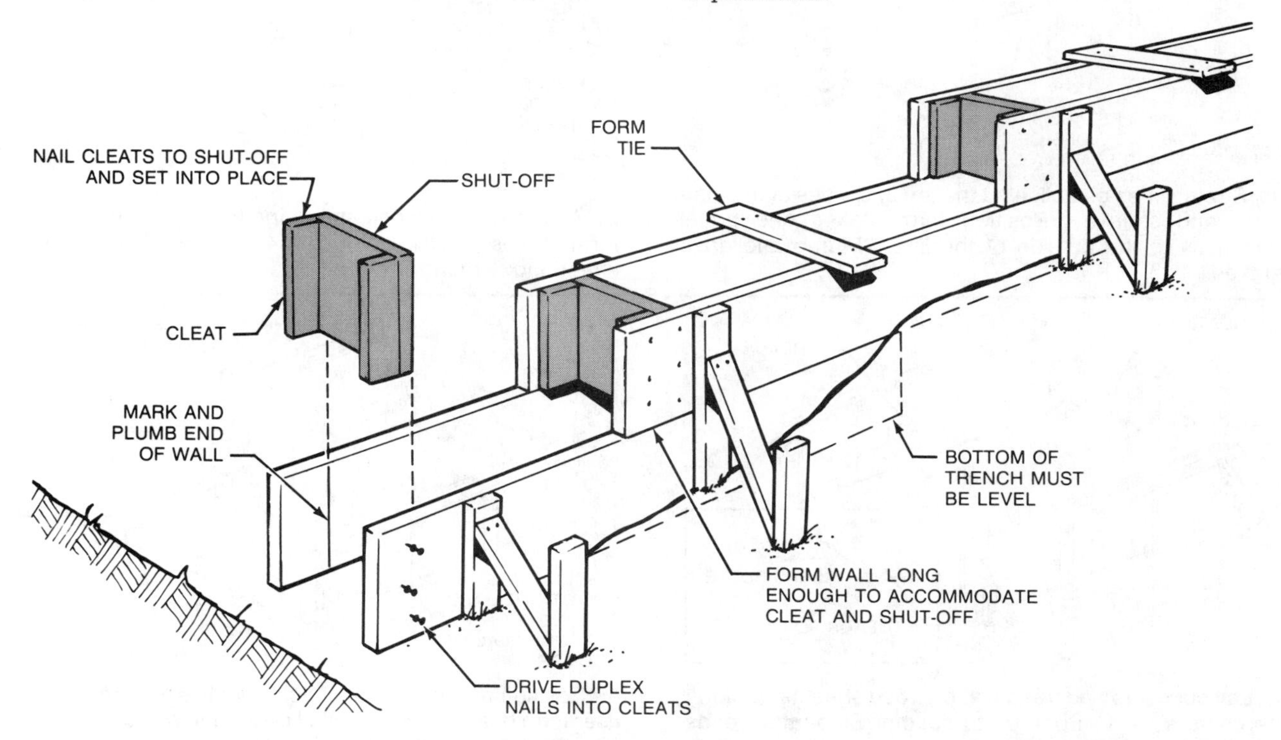

Figure 3-34. Shut-offs must be firmly secured to withstand the pressure of the concrete during placement. Cleats are nailed to the shut-off and duplex nails secure the assembly in place.

STAIRWAYS AND ENTRANCE PLATFORMS

A low entrance platform or stairway may be required to gain access to residential buildings or other light commercial structures. Concrete is commonly used to construct entrance platforms and stairways because of its durability and ability to withstand damp and wet conditions.

The risers in a stairway are all the same height, and the treads are all the same depth. Recommended riser heights range from 7″ to 7½″, and tread depths from 10″ to 12″. Risers on an exterior concrete stairway should slope in from the top between ¾″ and 1″, and treads should slope between ⅛″ and ¼″ from back to front. (Riser and tread calculations are covered in chapter 5, Heavy Construction.)

Low entrance platforms generally require a few steps. Therefore, a monolithic form is constructed for low entrance platforms and stairways. Low entrance platforms are usually formed against the foundation wall. An *isolation strip* placed between the platform and foundation wall prevents cracking caused by movement of the platform. The movement results from expansion and contraction of the concrete or soil settlement beneath the platform. An isolation strip is a piece of ½″ thick premolded asphalt-impregnated material placed before the concrete is placed.

The outside form walls for entrance platforms and stairs are constructed of plywood panels stiffened with braced studs and/or walers. After the treads and risers are laid out on the panels, cleats supporting riser form boards are nailed to the panels. Riser form boards should be beveled at the bottom to facilitate troweling of the steps after placing the concrete. See Figure 3-35.

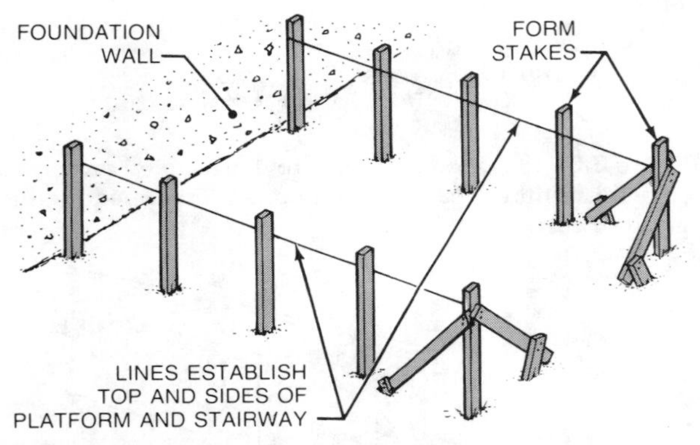

1. Stretch lines establishing the top and sides of the platform and stairway. Position form stakes one panel thickness to the outside of the lines. Plumb and drive stakes.

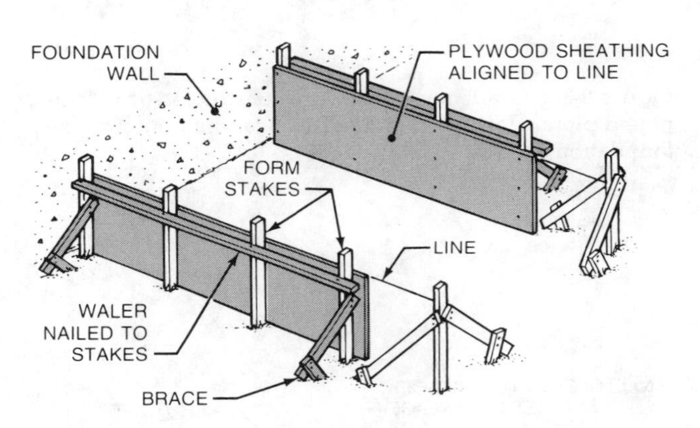

2. Align top of plywood sheathing to line and nail to the form stakes. Nail a waler 3″ – 4″ below the top of the sheathing. Brace the form.

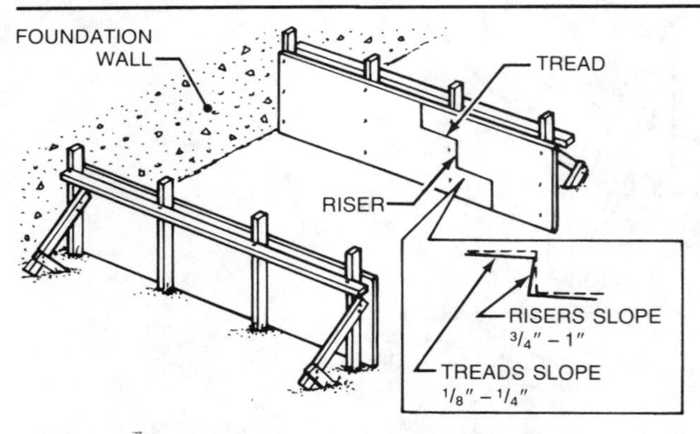

3. Lay out risers and treads on plywood sheathing. Slope the risers ¾″ – 1″ from top to bottom. Slope the treads ⅛″ – ¼″ from front to back.

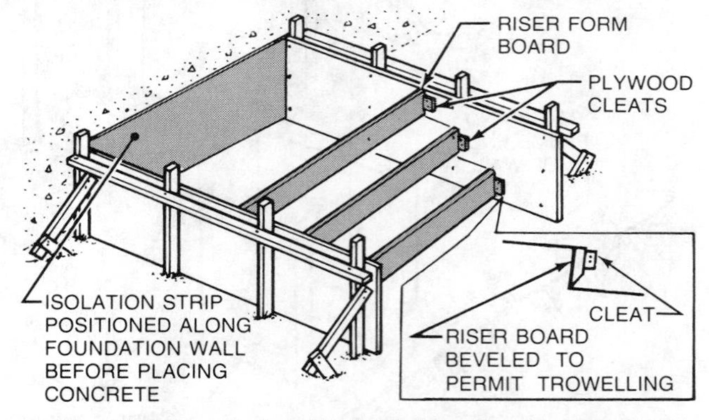

4. Position plywood cleats back from the riser mark one riser form board thickness. Nail cleats into place. Nail riser form boards to the cleats.

Figure 3-35. A low entrance platform and stairway are formed monolithically.

PIER FOOTINGS

Pier footings are a base for wood posts, steel columns, and masonry or concrete piers supporting wood beams or steel girders. The beams and girders provide intermediate support for the framed floor above. See Figure 3-36. Pier footings are used with all types of foundation construction, including crawl space, stepped, and grade beam foundations.

Most pier footings are independent structures. However, pier footings may be joined to foundation footings that support chimneys, fireplaces, or pilasters. *Pier foundations* may be used to support the superstructure. In a pier foundation, the exterior and interior walls of the building are supported by beams, posts, and pier footings.

Pier footings distribute loads over a larger soil area than other footings. The size of a pier footing is determined by the weight of the live and dead loads and the bearing capacity of the soil. Various pier designs are used, including rectangular, stepped, tapered, and circular piers. See Figure 3-37. Information regarding the size, shape, and reinforcement of pier footings is included in the section view drawings of the print. A local building code or structural engineer should be consulted if this information is not included in the prints.

Constructing Pier Footing Forms

Pier boxes, which are forms for pier footings, are fabricated and set to lines that establish the exact

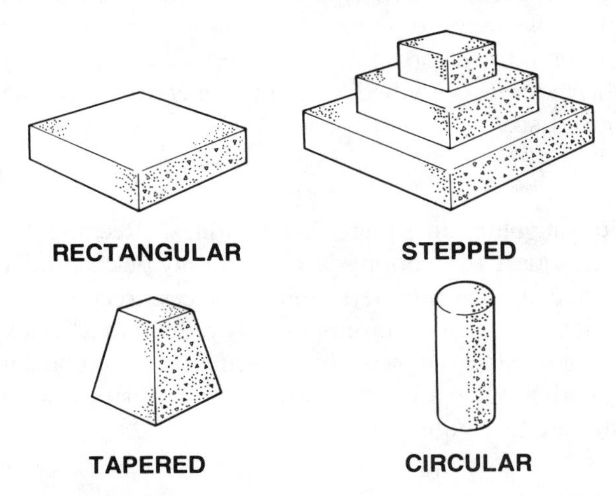

Figure 3-37. The imposed load of the building and the bearing capacity of the soil must be considered when determining pier design.

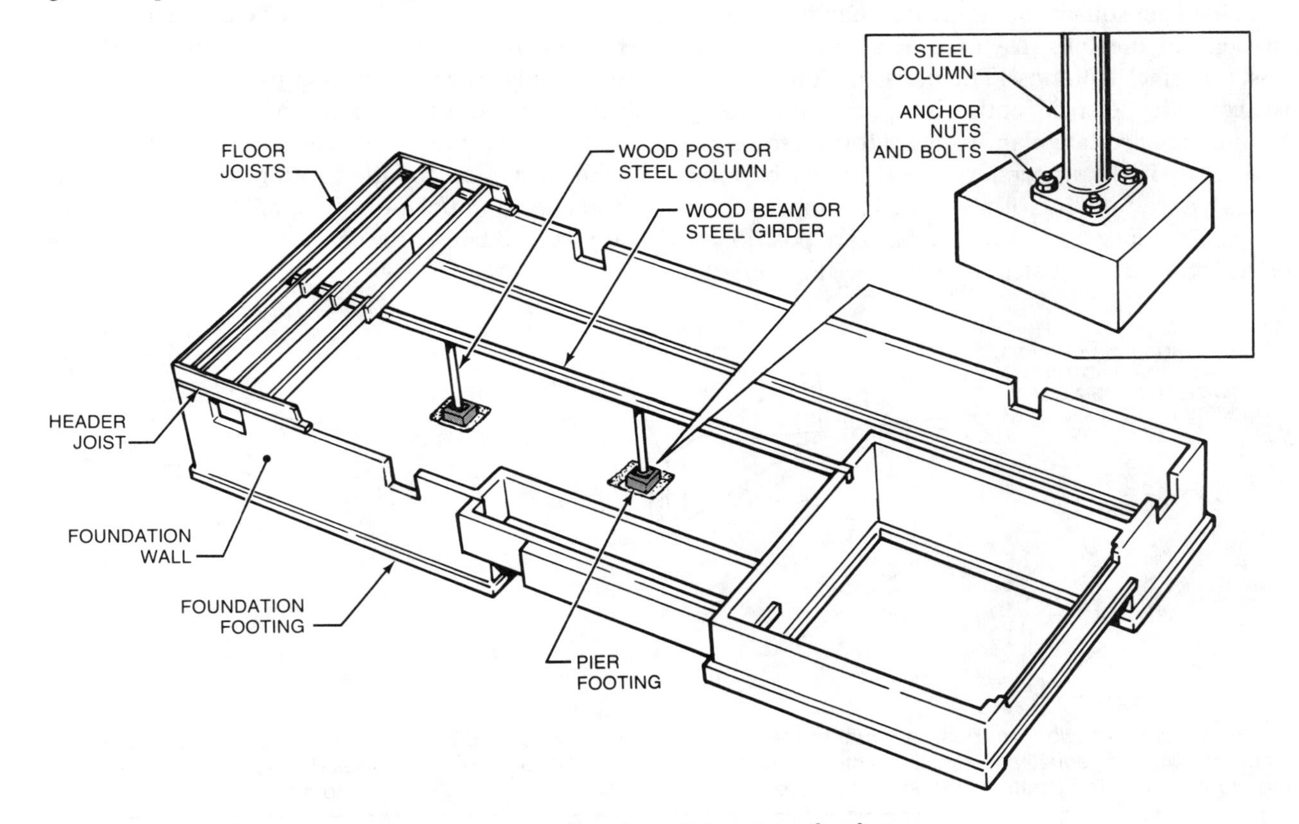

Figure 3-36. Concrete pier footings are a supporting base for posts and columns.

positions of the piers. Forms for square, battered, and stepped pier footings are made of planks or plywood. Forms for circular pier footings are often constructed of fibrous material, such as treated waterproof cardboard, or circular metal forms. The bottoms of all pier footings rest on firm soil and extend below the frost line. The forms are held in place by stakes to prevent uplift or movement. Post or column anchors may be positioned before placing the concrete or embedded in the concrete during the initial set.

Rectangular and Square Pier Footings. Rectangular and square pier footings are commonly placed under steel columns, chimneys, and fireplaces. Rectangular and square pier forms are usually built with 2″ thick planks. After the pieces have been cleated and nailed together, they are placed and held in position with stakes. See Figure 3-38.

Stepped Pier Footings. A stepped pier footing is designed for conditions where the imposed structural load per square foot is greater than the bearing capacity of the soil. It is used as a base for wood posts or steel columns. The construction of each level of a stepped pier footing is the same as building an individual rectangular or square form. However, two sides of the upper level forms must be long enough to rest on the form below. The upper forms are held in place with cleats. Rebars are positioned in the forms to tie the steps together. See Figure 3-39.

Tapered Pier Footings. Tapered pier footings have a wide base that distributes the load over a large area of soil. Tapered pier footings require less concrete than rectangular piers with the same size base. A 60° taper angle from the horizontal should be maintained to provide a safety margin based on a 45° shear stress angle.

Tapered pier forms are commonly constructed of plywood. Two sides of the form are cut to the exact width and height of the pier footing and the other two sides are cut wider to accommodate the plywood thickness and cleats. After the form has been assembled and set in place, it is staked securely to the ground to prevent uplift. Tapered pier forms are subject to greater uplift when placing concrete than rectangular or square pier forms. See Figure 3-40.

Circular Pier Footings. Deep circular pier footings are commonly used to support residential and light grade beam foundations. Shallow circular pier footings are placed beneath wood posts supporting floor beams. Circular pier footings are also used as part of the supporting structure beneath porches, decks, and stair landings.

Circular pier footings may be formed with fiber or metal forms. Fiber forms are cut from tubes made of spirally constructed fiber plies. Metal forms are often made of one-piece spring steel that clamps together at the ends. After the forms are placed in the area excavated for the footing, they are plumbed and staked in position. Post anchors or post bases are secured before the concrete is placed. See Figure 3-41.

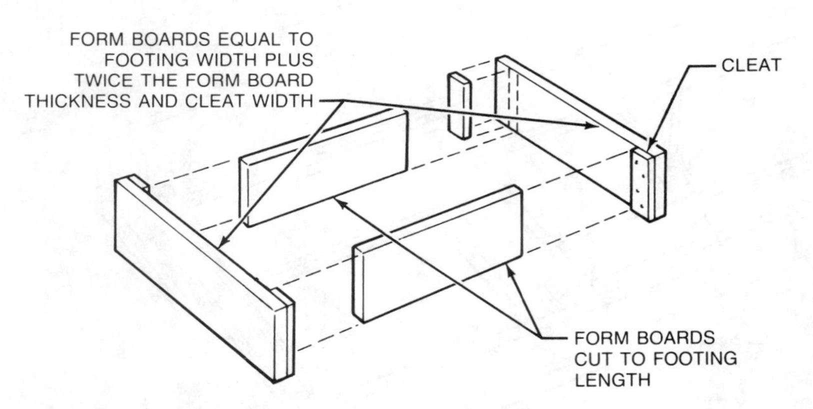

1. Lay out and cut two form boards equal to the footing length. Lay out and cut two form boards equal to the footing width plus twice the form board thickness and cleat width. Nail cleats to the longer pieces.

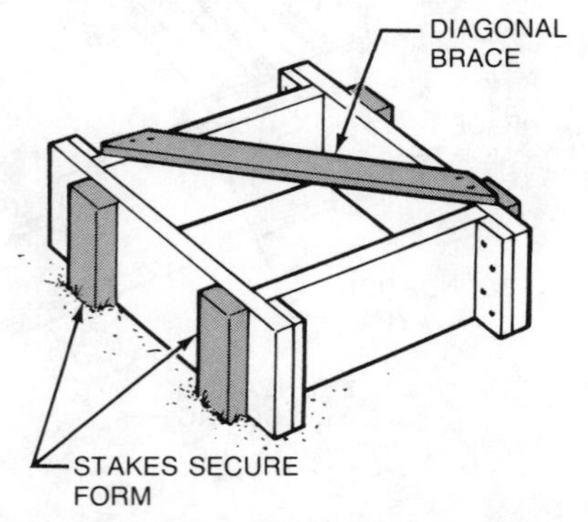

2. Nail the sides together. Square the form and nail a diagonal brace across the top. Position the form and drive stakes to secure it in place.

Figure 3-38. Rectangular pier footings are commonly used to support Lally columns or fireplaces.

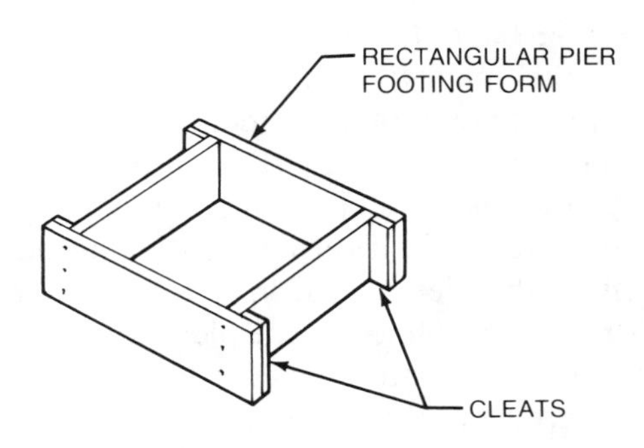

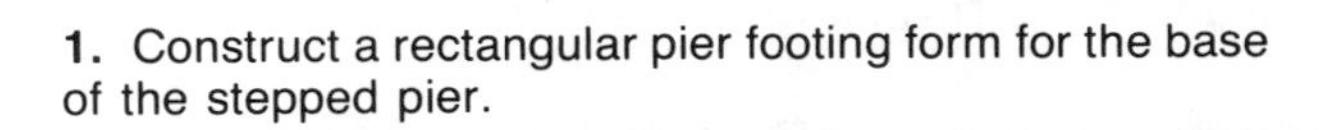

1. Construct a rectangular pier footing form for the base of the stepped pier.

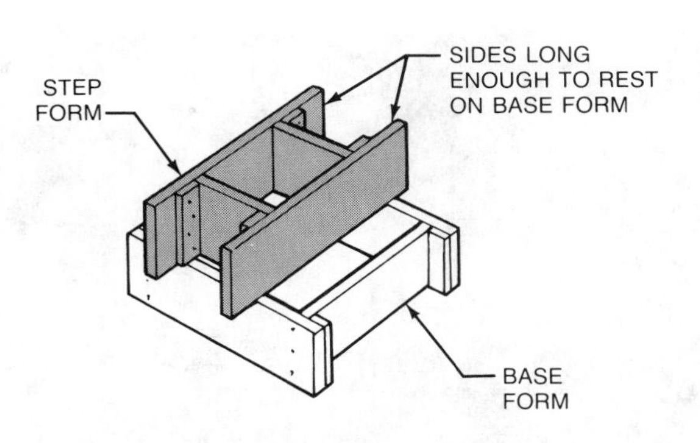

2. Construct a smaller rectangular pier footing form for the step with two sides long enough to rest on the base form.

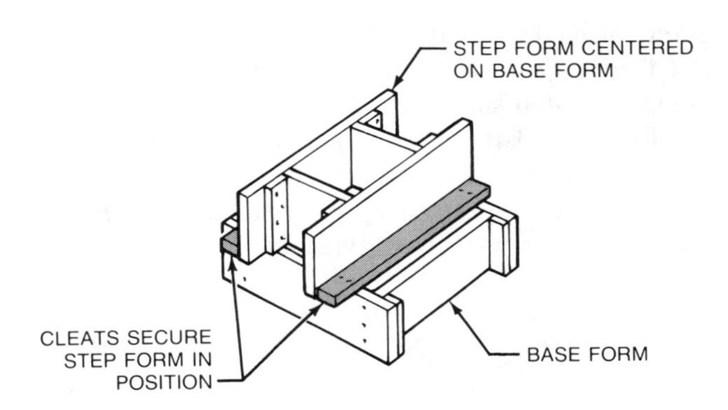

3. Center the step form on the base form. Nail cleats along the sides to secure the step form.

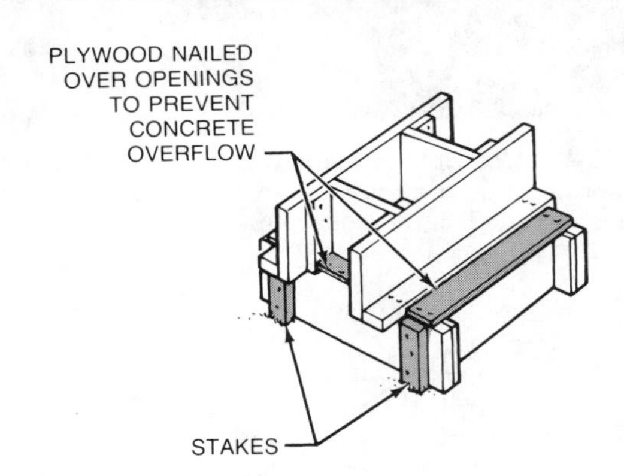

4. Nail plywood over open sections of the base form. Drive stakes to secure the stepped footing form in position.

Figure 3-39. Stepped pier footings are designed to support a structural load that is greater than the bearing capacity of the soil.

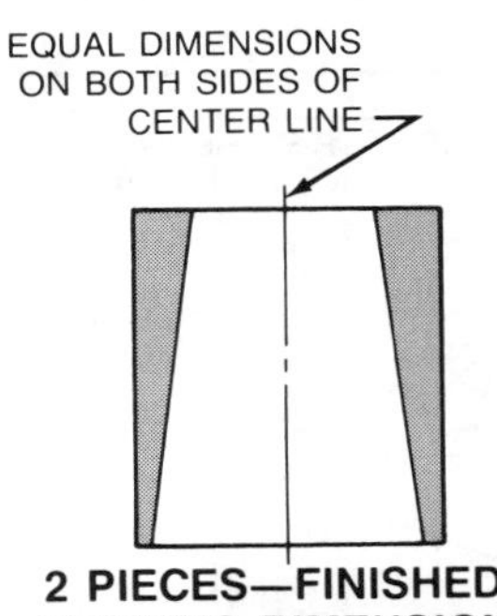

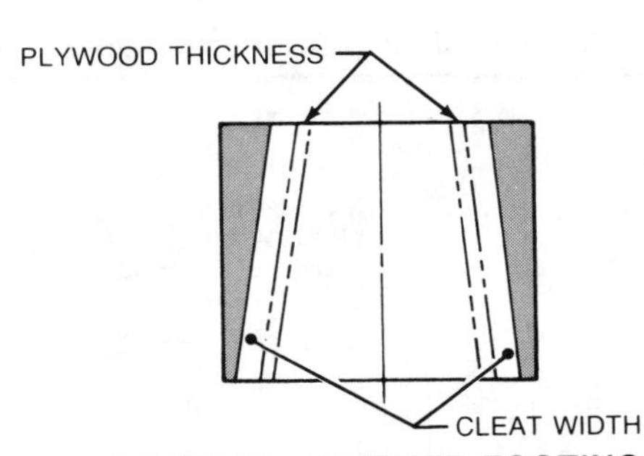

1. Lay out and cut two pieces of plywood to the finished footing dimensions. Lay out and cut two pieces of plywood to the finished footing dimensions plus twice the plywood thickness and cleat width.

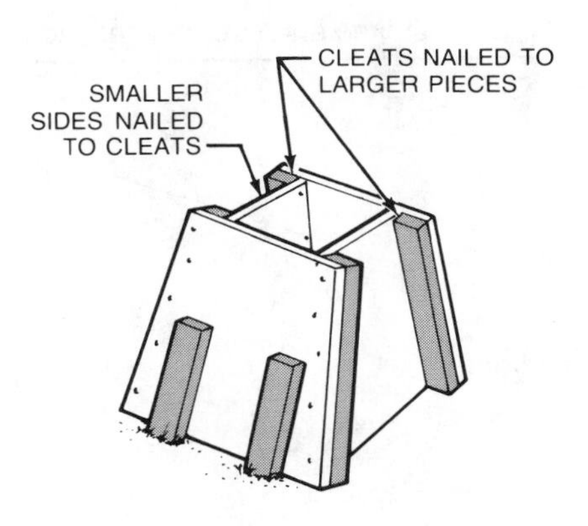

2. Nail cleats to the larger pieces. Nail the smaller sides to the cleats. Secure the form with stakes.

Figure 3-40. Tapered pier footings provide a wide base to withstand heavy structural loads.

Figure 3-41. Concrete is placed in a round fiber form when constructing a circular pier footing. A metal post base is embedded in the concrete.

Pier Footing Layout. The layout of pier footings is shown in the plan view of the foundation prints. The location of pier footings is indicated with dimensions from the foundation wall to the center of the closest pier. Other pier footings are located with center-to-center dimensions. When laying out the positions of the pier forms, lines are stretched to establish the center lines of the pier forms and are fastened to stakes, batterboards, or wall forms. The pier forms are set to the lines and leveled. The forms are then held in place with stakes and/or soil thrown against the forms. See Figure 3-42.

ANCHORING DEVICES

Anchoring devices are embedded in the top surfaces of foundation walls and pier footings. Anchoring devices fasten sill plates to the tops of foundation walls. They also are used to anchor the bottoms of wood posts or steel columns to pier footings.

Sill plates, or *mudsills,* are fastened to the tops of foundation walls to provide a nailing surface for

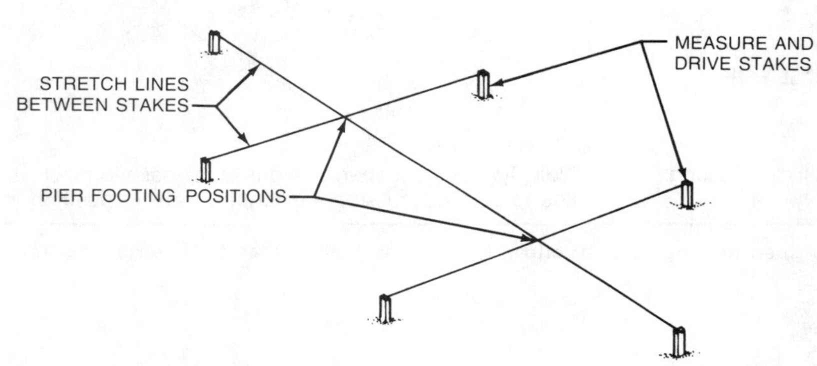

1. Drive stakes and stretch lines indicating the center lines of the pier forms.

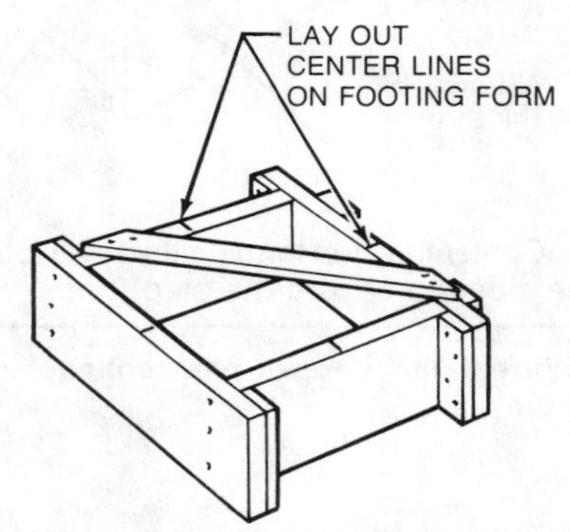

2. Lay out center lines on all four sides of the pier form.

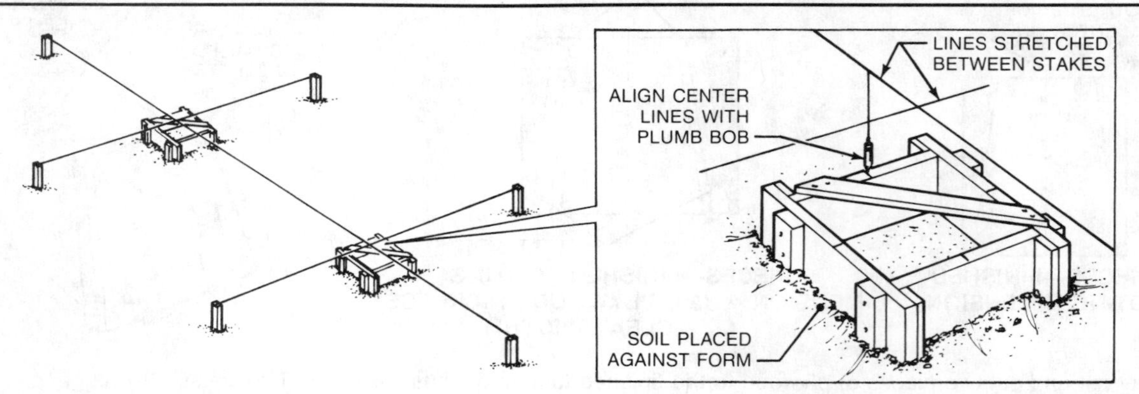

3. Align the center lines of the pier form with the stretched lines using a plumb bob or hand level. Drive stakes and place soil against the pier form.

Figure 3-42. Lines are stretched between layout stakes used to locate the position of pier footing forms.

floor joists or wall studs. In residential or other light construction, sill plates are usually 2 × 4s or 2 × 6s. See Figure 3-43. Larger and heavier buildings may require 4 × 6s. Foundation grade redwood or *pressure-treated lumber* is recommended for sill plates because of superior resistance to decay and insect attack. Redwood should be surface-treated with a wood preservative before it is bolted down to the top of the foundation wall. Pressure-treating is a process in which chemical preservatives are forced into the wood under intense pressure.

Figure 3-43. Sill plates are positioned over anchor bolts. They are later tightened down with washers and nuts.

Anchor bolts or *anchor clips* are used to fasten sill plates to foundation walls. One end of the anchoring device is embedded in the concrete while the concrete is being placed. See Figure 3-44. In some cases sill plates are fastened with *powder-actuated studs*. Powder-actuated studs are special concrete nails driven with powder-actuated fastening tools.

Anchor Bolts

Anchor bolts are the most effective anchoring device used to fasten sill plates to concrete. One end of the anchor bolt is threaded and extends above the sill plate to receive a nut and washer. The end embedded in the concrete is usually bent into an L or J shape for greater holding ability.

The depth and spacing of anchor bolts in foundation walls is shown in foundation section view drawings of the prints. If the information is not available in the prints, consult the local building code. The Uniform Building Code (UBC) recommends the following regarding anchor bolts.

1. Steel anchor bolts should be at least ½" diameter.
2. The bent end of the anchor bolt should be embedded a minimum of 7" into the concrete of reinforced masonry, or 15" into unreinforced grouted masonry.
3. Anchor bolts must be spaced not more than 6' apart. There should be a minimum of two bolts per sill plate with one bolt located within 12" of each end.

Setting Anchor Bolts and Sill Plates. Anchor bolts may be set before or after concrete is placed. When setting anchor bolts before concrete is placed, the anchor bolts are suspended in the wall form by securing them to wood templates. See Figure 3-45. The position of the anchor bolt on the wood template is determined by laying out a center line along the grain of the template. The center line across the

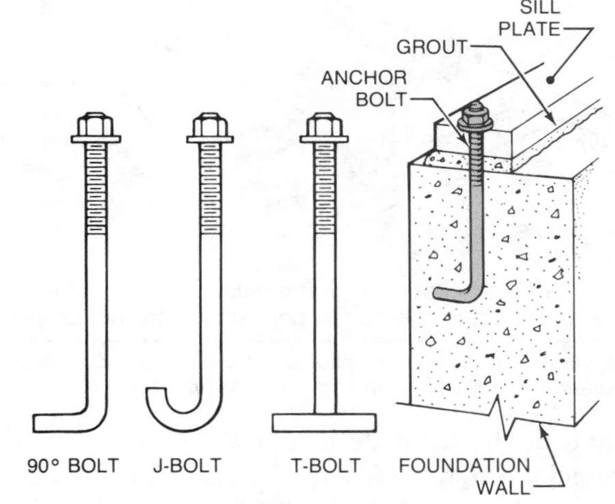

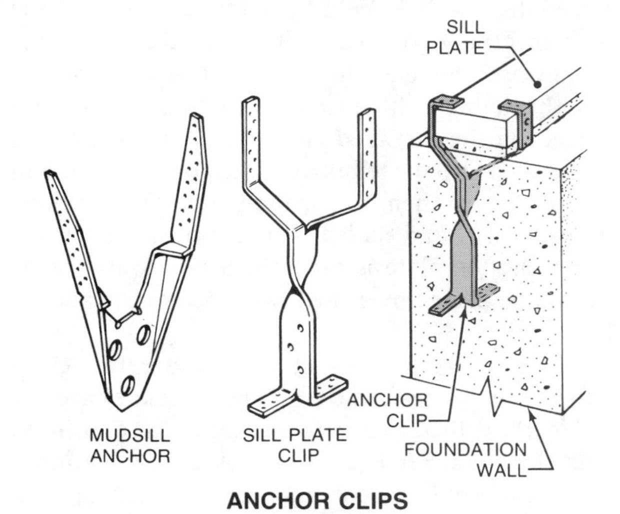

Figure 3-44. Anchoring devices secure the sill plate to the foundation wall.

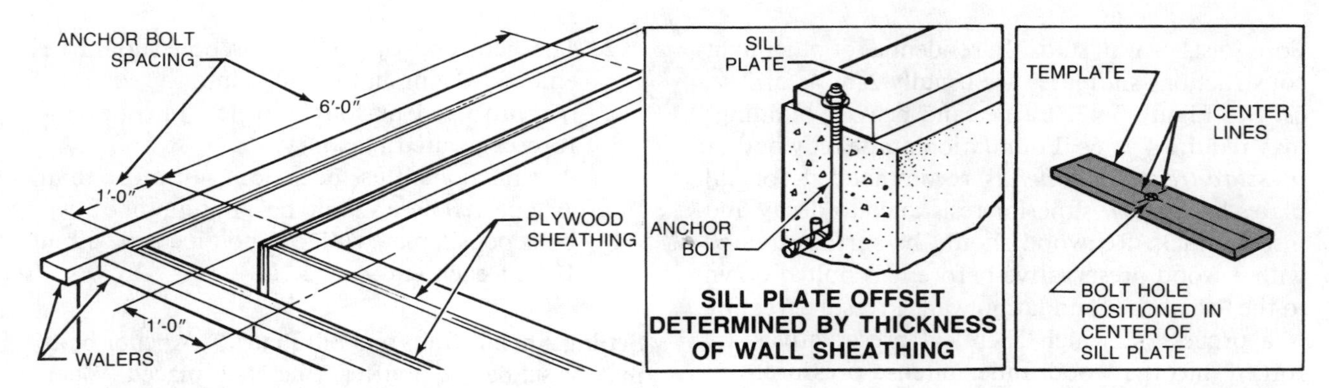

1. Mark the spacing of the anchor bolts on top of the wall form. Cut wood templates and determine the amount of offset required. Lay out the center lines and drill a snug hole for the anchor bolts.

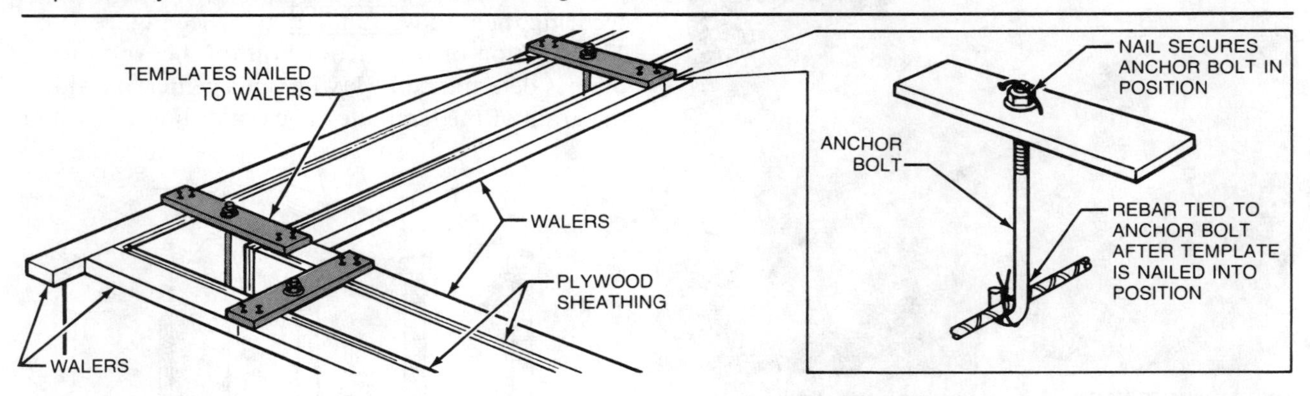

2. Insert the anchor bolt in the hole and place a nut and washer on the threaded end of the bolt. Drive a nail and bend it over to secure the bolt in position. Align the templates to the center lines on top of the form and nail into position.

Figure 3-45. Anchor bolts are set before the concrete is placed. The anchor bolts are centered in the width of the sill plates and are offset in the foundation wall.

grain of the template is laid out so the bolt will extend through the center of the sill plate. The edge of the sill plate may be flush with the outside of the foundation wall or held back the thickness of the wall sheathing. A hole is then drilled through the point where the center lines intersect. An anchor bolt fits through the hole, and a nut and washer are placed on the threaded end of the anchor bolt. A nail driven into the template is bent over the nut and bolt to secure them to the template. The template is positioned and nailed to the top of the forms. After the template is secured, a horizontal rebar should be placed over the lower ends of the anchor bolts and tied.

Once the concrete has set, the nuts are unscrewed and the templates removed. The sill plates are cut to length, drilled to accept anchor bolts, and fitted over the bolts. See Figure 3-46. A layer of *grout* (a mixture of sand, cement, and water) is often applied to the foundation wall to provide an even base for the sill plates. Local codes may require a strip of water-resistant material or metal termite shield beneath the sill plates.

When setting anchor bolts and sill plates after concrete placement, the anchor bolts and sill plates are prepared by cutting the sill plates to length and drilling holes for the anchor bolts. The anchor bolts are placed through the holes. Nuts and washers are screwed on the threaded end of the anchor bolts. When the concrete reaches the top of the wall form, the sill plates and anchor bolts are pressed in the concrete. The nuts are tightened after the concrete has set. See Figure 3-47.

Anchor bolts are also used to fasten sill plates for interior walls to concrete slabs. Lines are stretched immediately after the concrete is placed, and the anchor bolts are set to the lines. Sill plates for interior walls, however, are commonly pinned down with powder-actuated studs.

Setting Column Anchor Bolts. Anchor bolts in pier footings must be set precisely to align with holes in

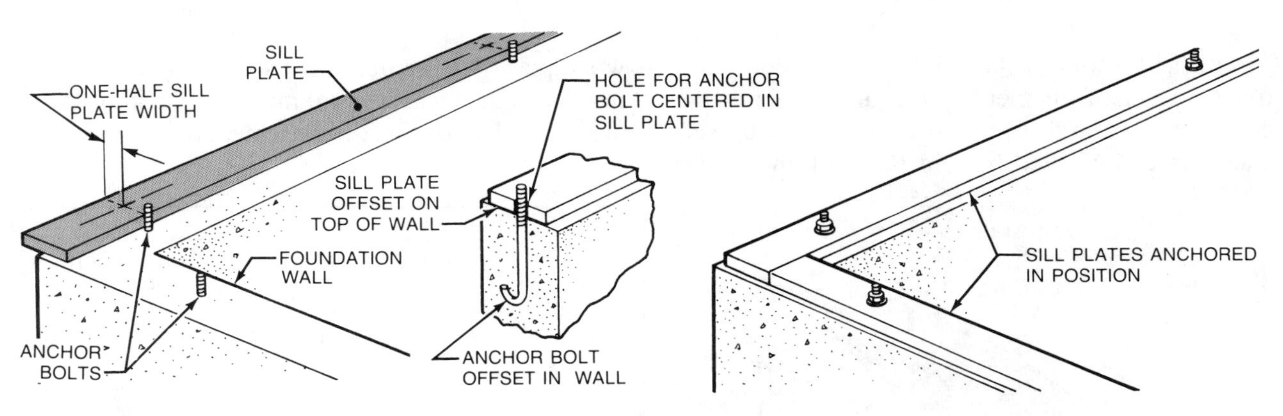

1. Cut the sill plates to length and place against anchor bolts. Square lines across the sill plate. Lay out one-half the sill plate width on the lines.

2. Drill holes in the sill plate. Position the sill plate and tighten down with washers and nuts.

Figure 3-46. Sill plates are placed over the anchor bolts and tightened down.

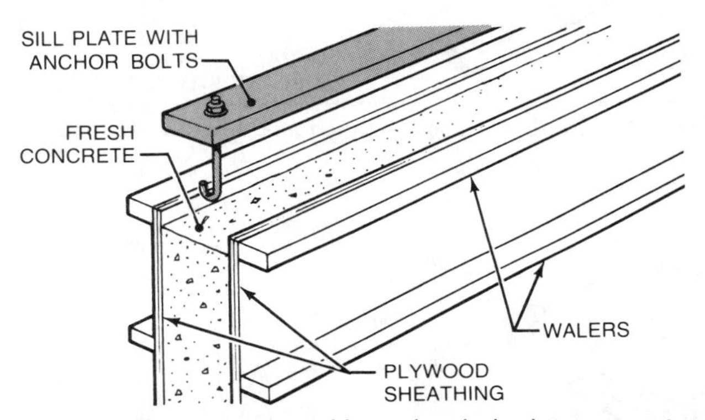

1. Press the sill plates with anchor bolts into concrete when it reaches the top of the form. Offset sill plate from outside of wall to compensate for wall sheathing. Place a wood piece over the bolts when tapping them into the concrete.

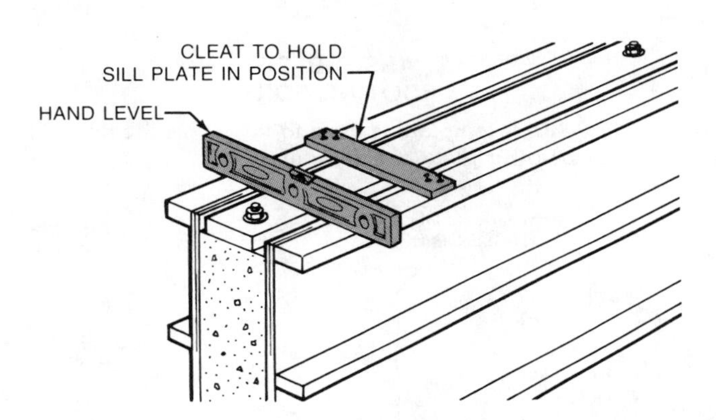

2. Level the sill plate with a hand level. Nail cleats across the form to hold the sill plate in position while the concrete sets.

Figure 3-47. Sill plates may be set in place when the concrete is placed.

the bearing plate at the base of a steel column. Holes are laid out and drilled in a template to hold the anchor bolts. The template is centered and nailed to the top of the footing form. See Figure 3-48.

Post Anchors

Post anchors secure the bottoms of wood posts or steel columns to pier footings. Post anchors are positioned in the piers while concrete is placed in the pier forms.

A *metal post base* is commonly used to fasten a wood post to a pier footing. The prongs of the metal post base are set in the concrete during concrete placement. The upper section of the metal post base is nailed to the post when the post is positioned. The metal post base must be at the correct elevation and position to ensure the proper position of the wood post.

An *adjustable metal post base* is secured to a pier footing with a ½″ J-bolt that had been set during the placement of concrete. After the concrete has set, the adjustable metal post base is fastened to the anchor bolt. A standoff plate provides a flat bearing surface for the post and keeps the post off the surface of the concrete, protecting it from wood rot and termite damage. See Figure 3-49.

Pier blocks, steel dowels, or *anchor bolts* may be used to anchor posts or columns. See Figure 3-50. Pier blocks are pieces of wood cut from the same lumber used for sill plates. Nails driven at an angle into the bottom of the pier block are pressed into the fresh concrete. A wood post is set in place and is toenailed to the pier block. When a steel dowel is used, the bottom of the wood post rests directly on the surface of the pier. A steel dowel is embedded in the pier and the exposed section of the dowel penetrates a predrilled hole in the bottom of the wood post. Anchor bolts are used to fasten the bearing plates of steel Lally columns. The anchor bolts are set in place during the placement of the concrete.

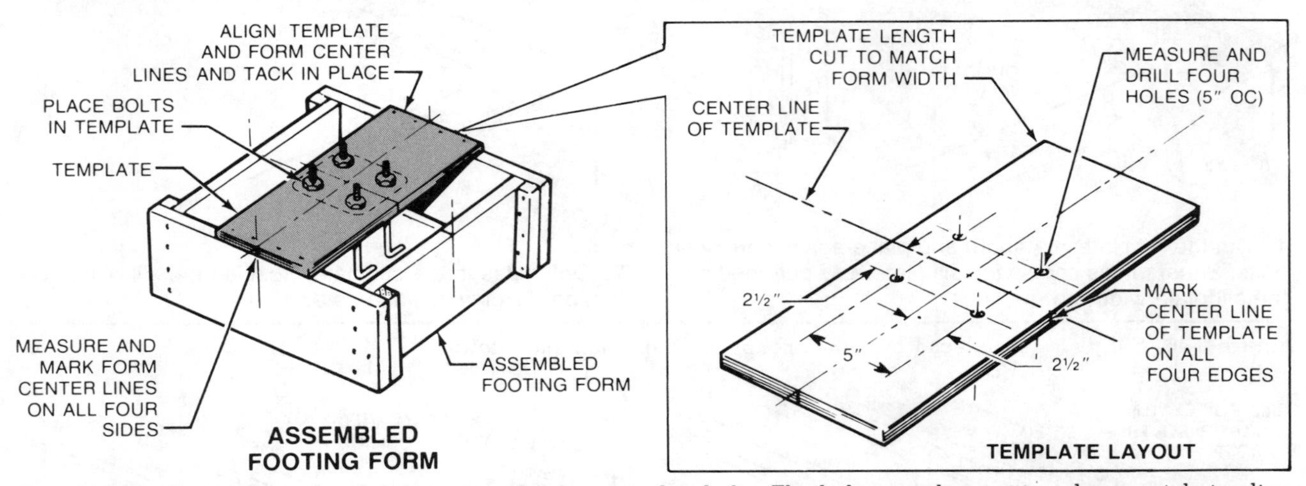

Figure 3-48. A template is used to layout steel column anchor bolts. The bolts must be positioned accurately to align with the bearing plate of the column.

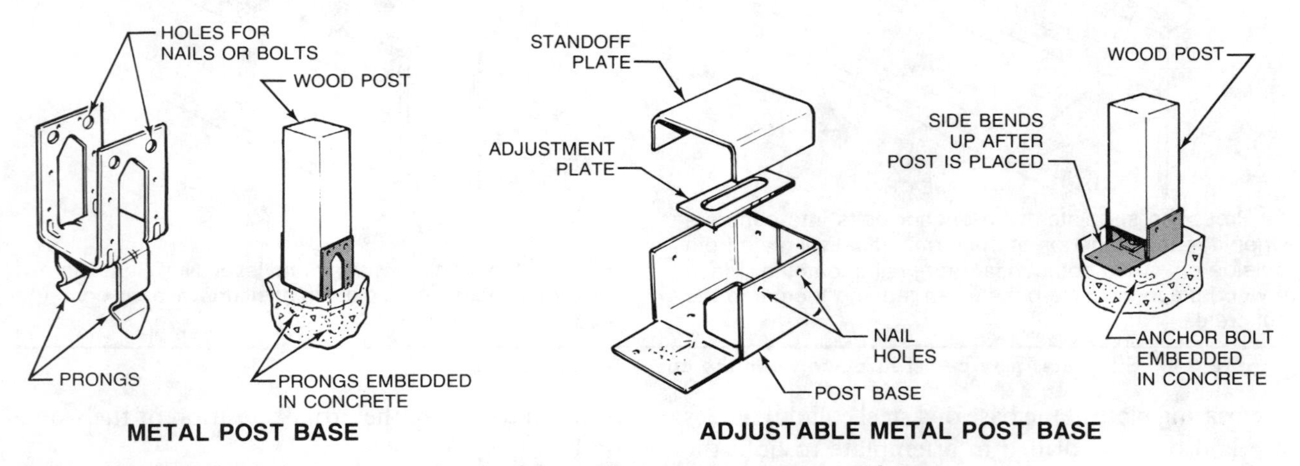

Figure 3-49. Metal post bases are used to anchor wood posts to concrete pier footings.

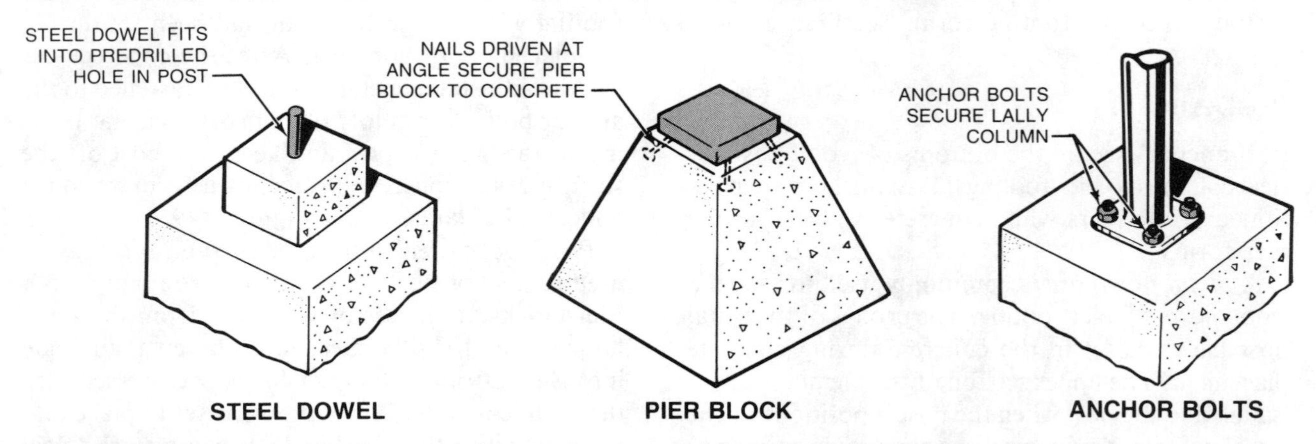

Figure 3-50. Posts or columns are secured to piers using pier blocks, steel dowels, or anchor bolts.

Chapter 3—Review Questions

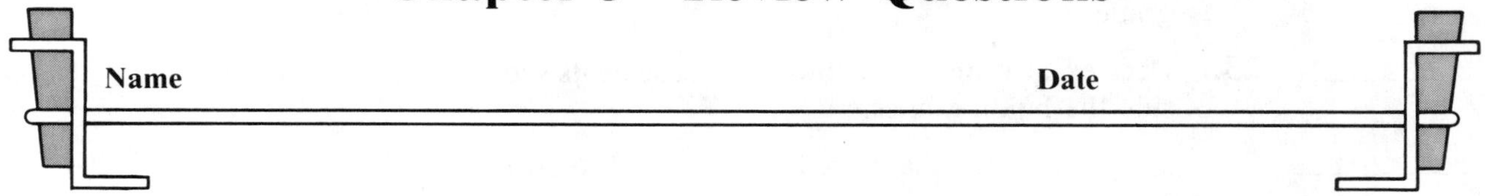

Completion

_________ **1.** A(n) _________ is the base for a foundation wall.

_________ **2.** A T-foundation has a(n) _________ footing that rests on bearing soil.

_________ **3.** The width of a footing must be _________ the thickness of the wall.

_________ **4.** Concrete pier footings are pedestals for posts or _________.

_________ **5.** The front setback of a building is measured from the front _________ line.

_________ **6.** The _________ method is used for squaring building lines without a transit-level.

_________ **7.** Batterboards hold the _________ lines in position.

_________ **8.** If the diagonal measurements of a square or rectangular shape are the same, all the lines are _________ to each other.

_________ **9.** A(n) _________ is a saw cut made in the batterboards to hold building lines.

_________ **10.** _________-formed footings are used in firm and stable soil.

_________ **11.** A(n) _________ is a tapered groove at the center of a footing.

_________ **12.** _________ views provide information regarding the size and spacing of rebars.

_________ **13.** Beam pockets should provide _________″ clearance around the sides and ends of wood beams.

_________ **14.** A(n) _________ column is a steel column commonly supported by a square or rectangular pier.

_________ **15.** A(n) _________ pier footing is constructed when the imposed structural load per square foot is greater than the bearing capacity of the soil.

_________ **16.** A(n) _________° angle should be used for the slope of a tapered pier.

_________ **17.** Crawl space foundations are also called _________ foundations.

_________ **18.** A(n) _________-foundation design is used most often for crawl space foundations.

_________ **19.** The minimum distance required from the bottom of the floor joists resting on crawl space foundations to the ground is _________″.

_________ **20.** _________ pressure against a foundation wall is caused by the force of the earth against the wall.

_________ **21.** _________ loads are the constant weight of the entire superstructure.

_________ **22.** A foundation wall should extend at least _________" above the outside finish grade.

_________ **23.** Deep excavations for below-grade basements should extend at least _________' outside the buiding lines.

_________ **24.** _________ are chamfered 2 × 4s used to form keyways.

_________ **25.** A(n) _________ joint is formed when fresh concrete is placed over a concrete surface that has set.

Multiple Choice

_________ **1.** The advantage of a monolithic form for a crawl space foundation is that _________.
- A. concrete for the walls and footing is placed at the same time
- B. a cold joint will not be formed
- C. water cannot seep through the outside
- D. all of the above

_________ **2.** When placing concrete for an entrance platform against an existing foundation wall, _________.
- A. place wire mesh first
- B. place an isolation strip between the platform and wall
- C. place a vapor barrier between the platform and wall
- D. embed steel dowels in the foundation wall.

_________ **3.** A grade beam is _________.
- A. placed under a row of piers
- B. a foundation level with the finish grade
- C. a foundation wall supported by concrete piers
- D. a low foundation

_________ **4.** In seismic risk areas, piers should be tied to grade beams with _________.
- A. angle iron
- B. rebars
- C. wire mesh
- D. all of the above

_________ **5.** Slab-on-grade foundations _________.
- A. are an integral wall and slab foundation system
- B. do not require foundation walls
- C. are more costly to construct
- D. do not require reinforcement

_________ **6.** _________ plates are wood plates bolted to the top of a foundation wall.
- A. Anchor
- B. Base
- C. Bottom
- D. Sill

_______ **7.** _______ is the recommended lumber for sill plates.
A. Pine
B. Hemlock
C. Redwood
D. Oak

_______ **8.** Steel columns are usually anchored with _______.
A. anchor bolts
B. angle iron
C. powder-actuated studs
D. all of the above

Identification

_______ **1.** Line

_______ **2.** Footing

_______ **3.** Stake

_______ **4.** Spacer block

_______ **5.** Brace

_______ **6.** Hand level

_______ **7.** Straightedge

_______ **8.** Form wall sheathing

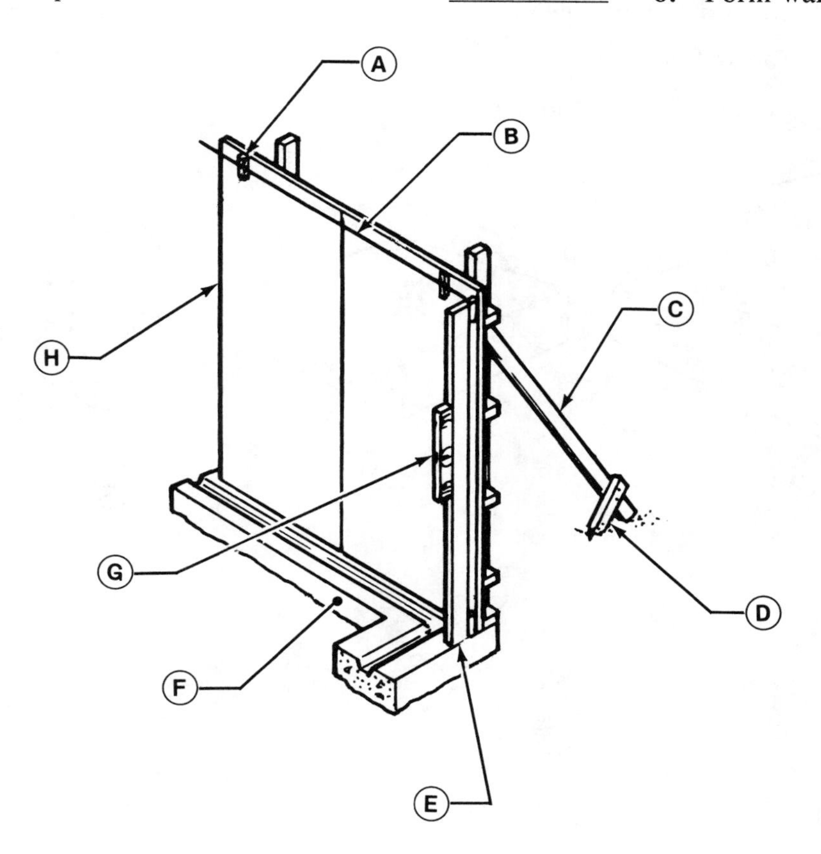

_______ 9. Battered foundation
_______ 10. L-foundation
_______ 11. Rectangular foundation
_______ 12. T-foundation

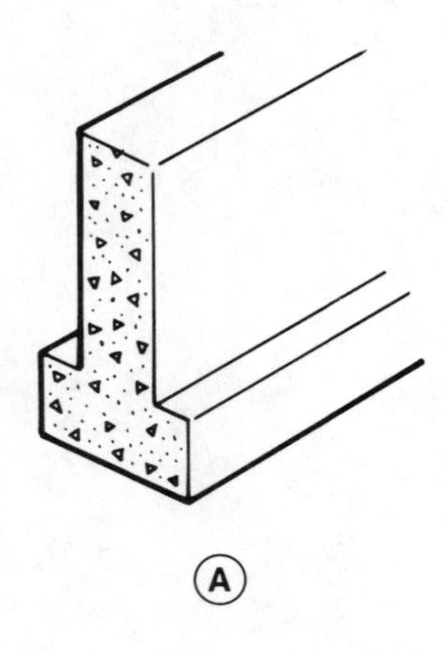

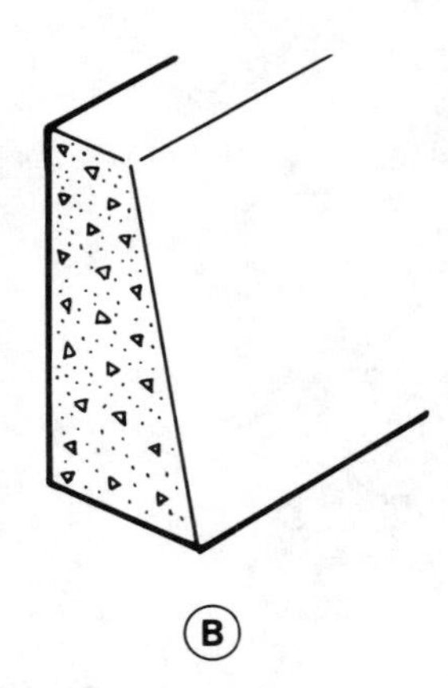

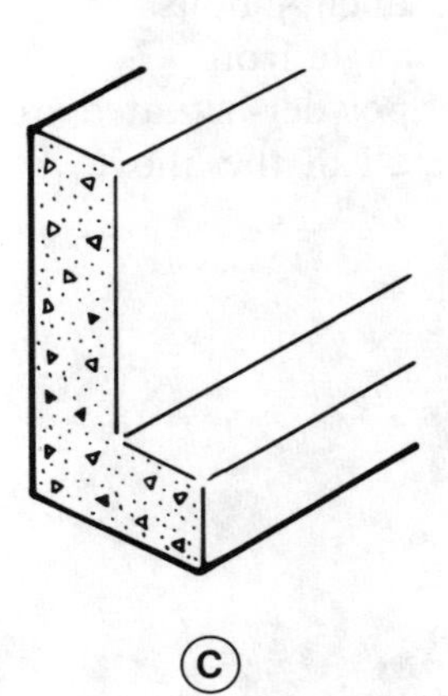

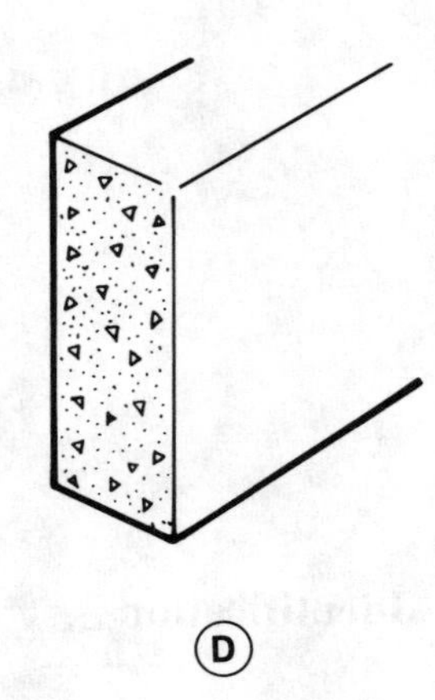

_______ 13. Circular pier footing
_______ 14. Stepped pier footing
_______ 15. Tapered pier footing
_______ 16. Rectangular pier footing

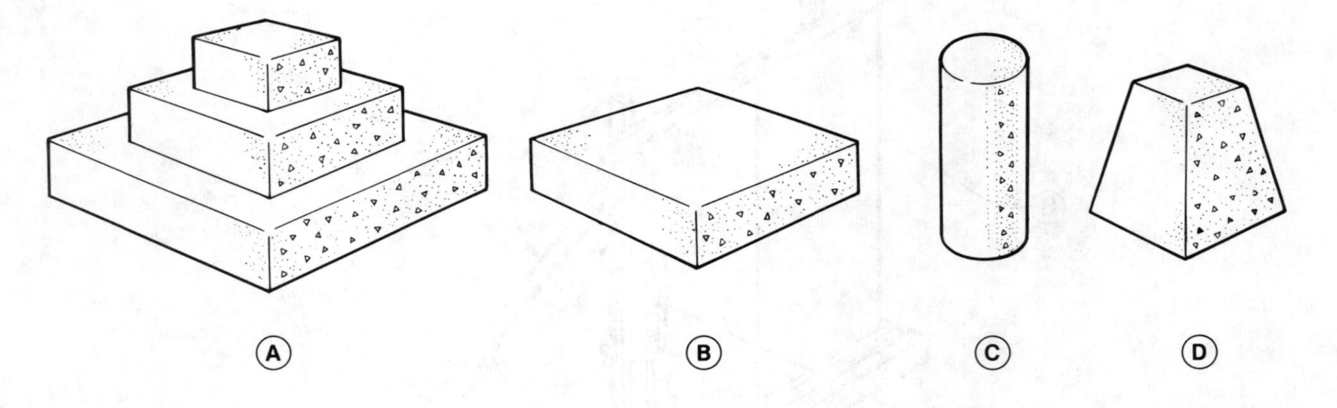

CHAPTER 4
Flatwork

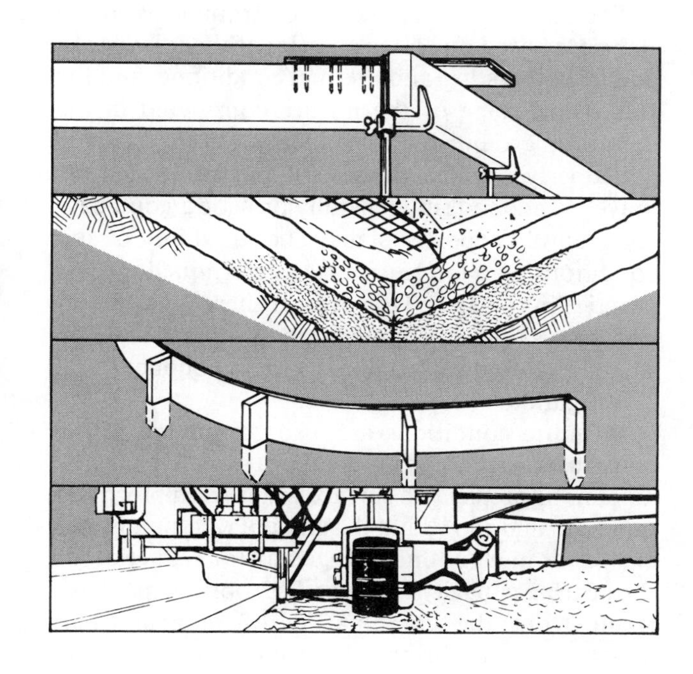

Flatwork is the preparation of the placement area, construction of forms, and placement of concrete for floors, driveways, walks, and patios. The placement area is prepared by laborers. Forms are constructed by carpenters and cement masons work and finish the concrete as it is placed.

Flatwork preparation includes accurate layout of the finished slab elevation with a builder's level or transit-level. When the elevations are established, edge forms are constructed and/or a screed system is placed. Edge forms are used in flatwork such as driveways or walks to retain the concrete in a specific area. A screed system is used to maintain proper floor elevations in the interior areas of the flatwork.

Preparations must also be made for construction, isolation, and expansion joints. Construction joints are used where fresh concrete butts against the edge of a section of concrete that is set. Isolation joints are used to prevent cracking between two adjacent sections of concrete. Expansion joints are used in large sections of flatwork when a great amount of expansion and contraction is anticipated.

Driveways and walks are constructed for vehicle and pedestrian use. Driveways are 4″ to 6″ thick and are reinforced with welded wire fabric or rebars. Public sidewalks, front walks, and service walks are commonly located around residential structures. Walks are usually 4″ thick, although thicker slabs reinforced with rebars may be required where heavy vehicles pass over the walk.

GROUND-SUPPORTED SLABS

Ground-supported slabs are commonly placed for ground-level floors of residential and commercial buildings that do not have basements. For these types of structures, ground-supported slabs are usually less costly to build than wood-framed floors. Ground-supported slabs are also used in the construction of garage floors and below-grade basement floors.

The design and construction of ground-supported slabs are based on soil properties at the job site. In addition, moisture and thermal conditions, and the shape and slope of the lot are considered in the design of ground-supported slabs.

Properly drained dense soil mixtures, such as gravel, sand, and silt, generally provide a good base for ground-supported slabs. The suitability of soil conditions on a job site is often determined by past practice in the area. However, if there is any question about the soil composition at a particular job site, a qualified soil engineer should conduct a soil investigation.

Moisture conditions are also considered in the construction of ground-supported slabs. The amount of predictable surface water from precipitation (rain and snow) and the amount of groundwater can result in a volume change and/or reduction of the bearing capacity of the soil. Problems may also result from the combination of moisture and temperature conditions. Low temperatures cause groundwater to freeze and pose the danger of frost heave below the slab.

Ground-supported slabs are placed on level ground. Steeply sloped lots are not practical for ground-supported slabs because of high excavation costs and potential water drainage problems.

Slab-on-grade Floors

Slab-on-grade floors (also called slab-on-ground) are usually integrated with a slab-on-grade foundation system. Slab-on-grade floors are placed after the foundation has been constructed, or monolithically along with the foundation walls. The floor slab is at the same elevation as the top of the foundation wall. In most slab-on-grade floors, the top surface of the slab is at least 8″ above the finish grade level at the perimeter of the foundation walls. Floor slabs for residential buildings are a minimum of 4″ thick. Thicker floor slabs may be required in commercial structures for supporting heavy loads.

When a slab-on-grade floor is placed after the foundation has been constructed, rigid insulation is recommended around the perimeter of the slab to reduce heat loss. The insulation should be at least 1″ to 2″ thick and extend 24″ vertically below grade level or 24″ horizontally under the concrete slab. See Figure 4-1.

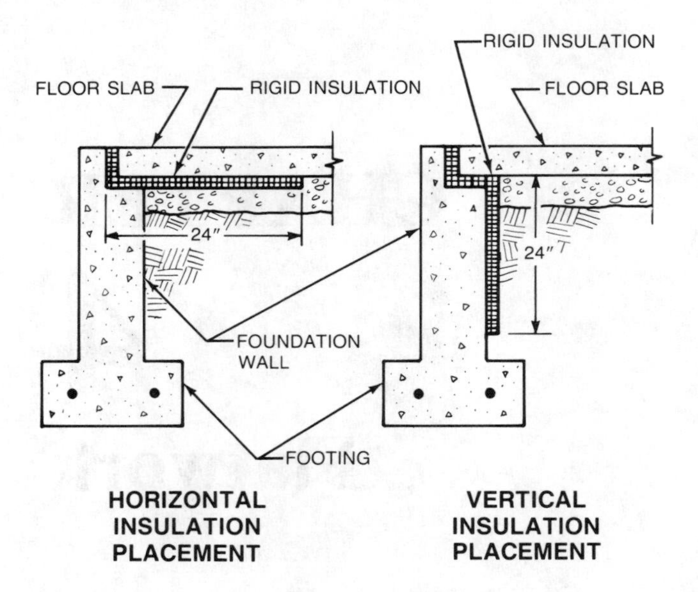

Figure 4-1. Rigid insulation extends 24″ horizontally and vertically from the perimeter of a slab-on-grade foundation.

Site Preparation. Site preparation, including all groundwork, must be completed before the concrete is placed for a slab-on-grade floor. Site preparation may only require removing the topsoil to reach undisturbed soil, or it may require excavating deep enough to place a layer of compacted fill and a gravel base course. See Figure 4-2.

Groundwork provides support for the slab and controls ground moisture. Vapor barriers are commonly used to contain ground moisture beneath the slab. All pipes and ducts to be embedded in the concrete must be set in place before the concrete slab is placed. Site preparation for a slab-on-grade floor requiring fill and a gravel base course is as follows:

1. Remove all topsoil and dig down to firm and undisturbed soil. Excavation must be deep enough to hold the layers of fill and gravel.
2. Place and compact fill material. Fill is required when the ground surface is uneven or where a gradual slope must be leveled. The fill should be compacted in 4″ to 12″ courses by using

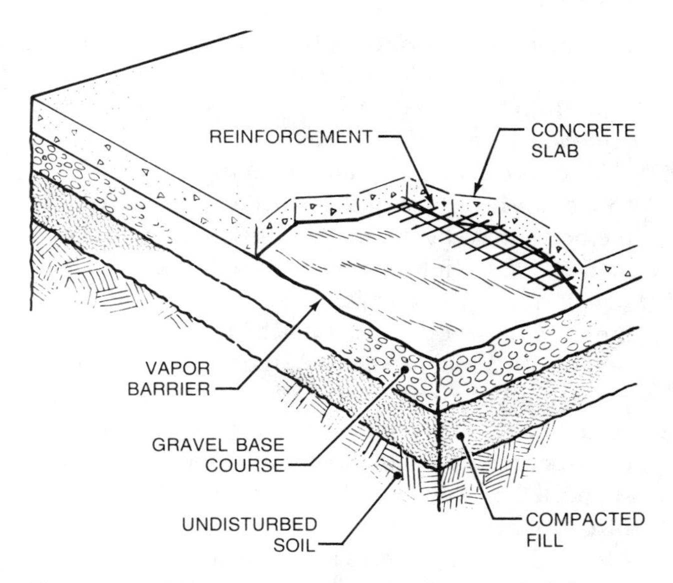

Figure 4-2. The building site for a slab-on-grade floor must be excavated to undisturbed soil. Compacted fill, a gravel base course, and a vapor barrier are placed in the excavation.

hand or power equipment. Fill material should be free of vegetation and other foreign material that might cause uneven settlement.

3. Install pipes, drains, ducts, and other utility lines.
4. Place a *gravel base course* at least 4″ thick to control the capillary rise of water through the slab bed. The base course also provides uniform structural support for the concrete slab and reduces the amount of heat lost to the ground.
5. Place a moisture-resistant vapor barrier over the base course, directly beneath the floor slab to prevent moisture from seeping through the slab. Six-mil polyethylene film is commonly used as a vapor barrier. All joints are lapped at least 6″ and should fit snugly around all projecting pipes and other utility openings. Precautions must be taken so the vapor barrier is not punctured during construction work.
6. Place perimeter insulation along the foundation walls if required.
7. Place reinforcement in the slab according to the foundation section views. The reinforcement is usually at the center of the slab or approximately 1″ to 1½″ from the top surface.

Forming Slab-on-grade Floors. When placing a floor slab independently of the foundation walls, forming the edges of the slab is not required because the perimeter of the slab butts up against the foundation walls. When the concrete slab is placed monolithically with the foundation walls, the outside foundation form boards also form the edge of the slab. This forming method consists of 2″ thick planks held in place with stakes and braces.

Construction Joints. A construction joint is formed where the fresh concrete floor section butts up against the edge of the concrete floor section that has already set. A construction joint is formed by staking down a 2″ thick *bulkhead* (plank) at the outer edges of the concrete placement area. The top of the bulkhead is positioned at the height of the floor surface and a beveled *key strip* is fastened to the bulkhead. Metal, wood, and premolded key strips are commonly used to form a keyway for the floor. The keyway secures the edge of the next floor section in position. See Figure 4-3.

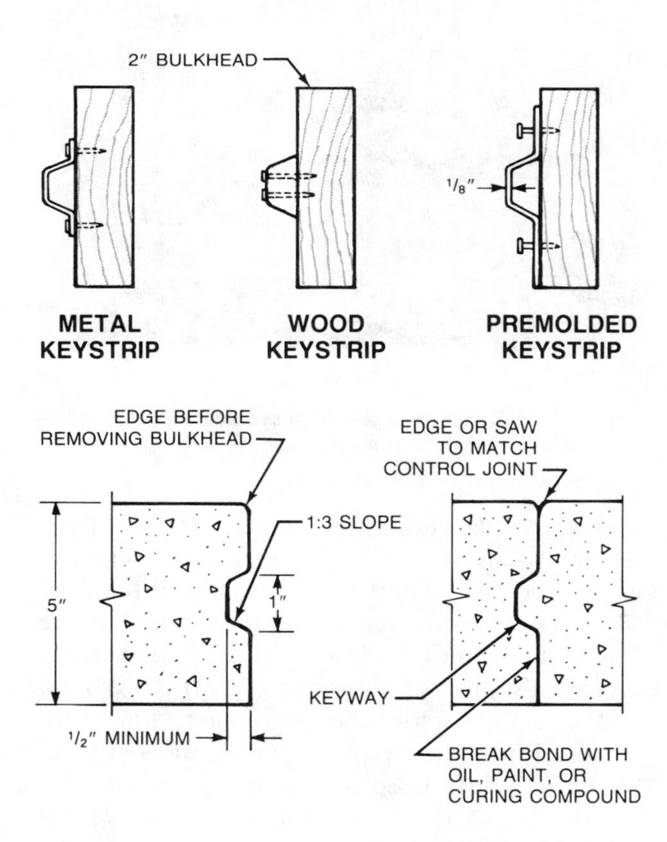

Figure 4-3. Bulkheads are used to form construction joints when a concrete slab is placed in sections. Premolded key strips are permanently embedded in the slab.

Concrete for large floor areas in commercial buildings such as warehouses, factories, and stores is placed in sections. Therefore, provisions must be made for *construction joints* when placing large floor slabs. See Figure 4-4.

Walsh Construction Company of Illinois

Figure 4-4. Concrete is placed in sections for large industrial or commercial concrete slabs.

Screeding. Placing concrete for large floor sections requires the use of a *screed system* to maintain proper floor elevations in the interior areas of the slab. The screed is positioned with its bottom edge at the finish elevation of the floor surface. Wood stakes or screed supports hold the screed off the ground and allow rebars to be positioned. Lines are stretched from the top of the outside walls or form boards to adjust the screeds to their proper height. A *strike board* acting as a straightedge to level the concrete is placed between the screed boards. Strike boards are held at the same level as the screeds by cleats nailed at opposite ends. See Figure 4-5. As the concrete is being placed and consolidated, the cement masons *strikeoff* the concrete by moving the strike board along the screeds with a saw-like motion. The screeds and their supports are then removed from the concrete.

Screeds can also be placed with the top edge flush to the finish surface of the concrete. Two screeding methods may be used with this system. In one method, a section of the floor slab is placed and struckoff to the screeds. The screeds are then removed and the concrete is placed for the next floor section. In a second method, the screeds remain in place until the entire slab has been placed. The screeds and their supports are removed and the cavities are filled with concrete. Metal pipe screeds supported by wooden stakes or adjustable chairs are often used with this method. See Figure 4-6. Mechanical equipment is also available for screeding operations and is often used when placing larger slabs.

Control Joints. Control joints, also called *relief* or *contraction joints,* confine and control cracking in concrete slabs caused by expansion and contraction. A control joint is a groove in the surface of the slab a depth of one-fourth the slab thickness. Cracks occurring in the future will be confined to the area beneath the control joints. Control joints may be formed with a special grooving tool when the concrete is being finished off. They may also be cut into the slab after the concrete has set using a power saw equipped with an abrasive blade. Recommended spacing for control joints is 15′ to 20′. See Figure 4-7.

Expansion Joints. Expansion joints are used in slabs that cover large areas of commercial buildings, and where a great amount of expansion and contraction is anticipated. Expansion joints run through the complete thickness of the slab and are filled with a piece of preformed asphalt-impregnated fiber material. The fiber material is tacked to the form board before the concrete is placed and remains in place when the form board is removed. See Figure 4-8.

Steel Reinforcement. The two types of steel reinforcement used in ground-supported floor slabs are rebars and welded wire fabric. Floor slabs that support heavy loads and moving vehicles are reinforced

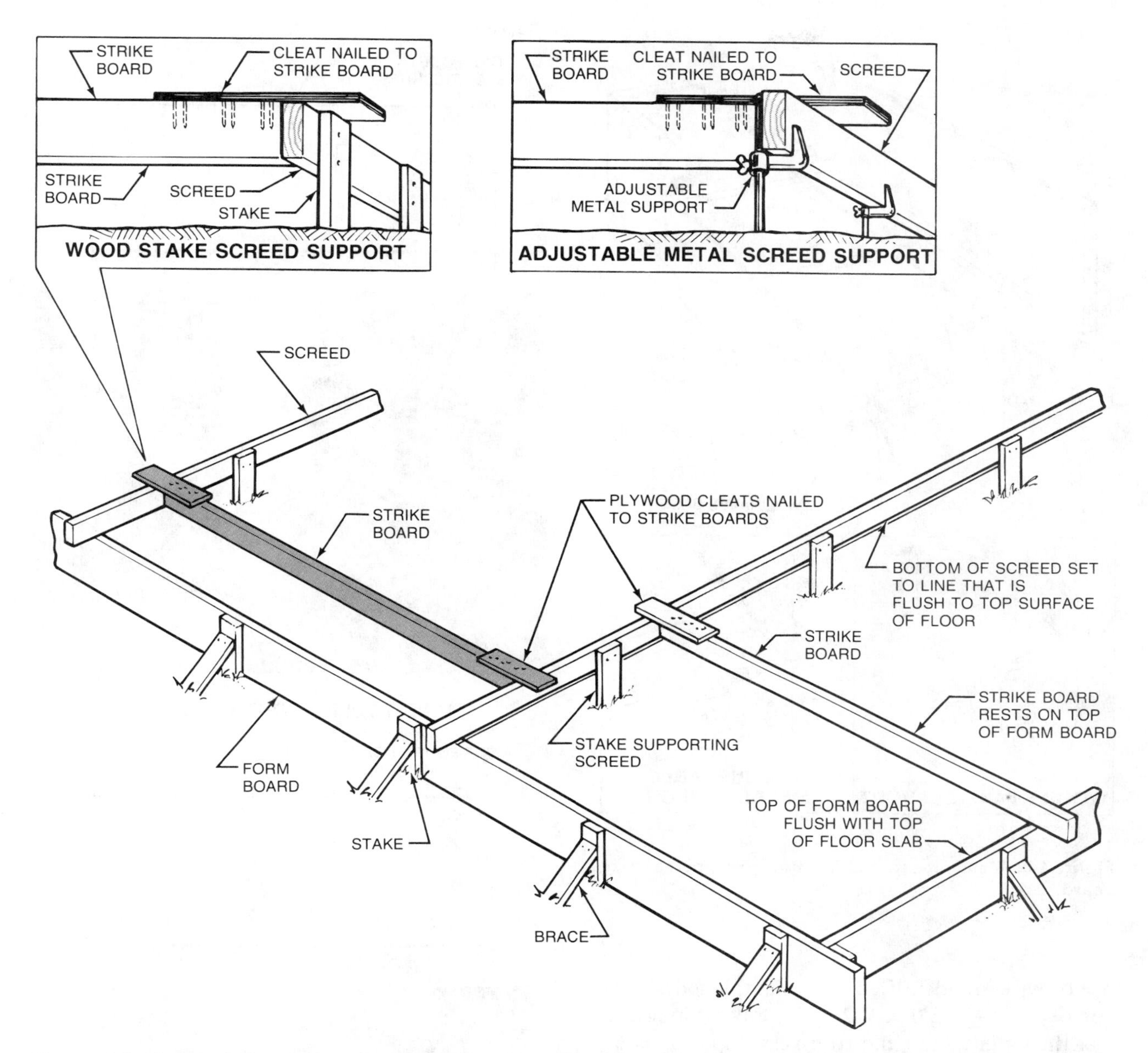

Figure 4-5. A screed system is used to strikeoff concrete placed for a concrete floor slab. The screed is supported by wood stakes or adjustable metal supports.

with rebars, which are the same type of deformed rebars used to reinforce concrete walls. Information on the size and spacing of the rebars is found in section view drawings of the foundation plans.

Floor slabs that support light loads are often reinforced with *welded wire fabric* (WWF). Welded wire fabric, also called *wire mesh*, is heavy wire arranged in a square or rectangular pattern and welded at the intersections of the wires. It is available in rolls and sheets. Two types of welded wire fabric are *smooth* and *deformed*. Smooth fabric, also called *plain fabric,* is anchored to the concrete by its welded connections. Deformed fabric is anchored by deformations in the wire as well as the welded connections. Information on the size and type of welded wire fabric to be placed in the floor slab is commonly shown in foundation section views of the prints.

Welded wire fabric is represented by long dashed lines in the section drawings of the floor slab. In the drawings, spacing of the wire, type of wire, and

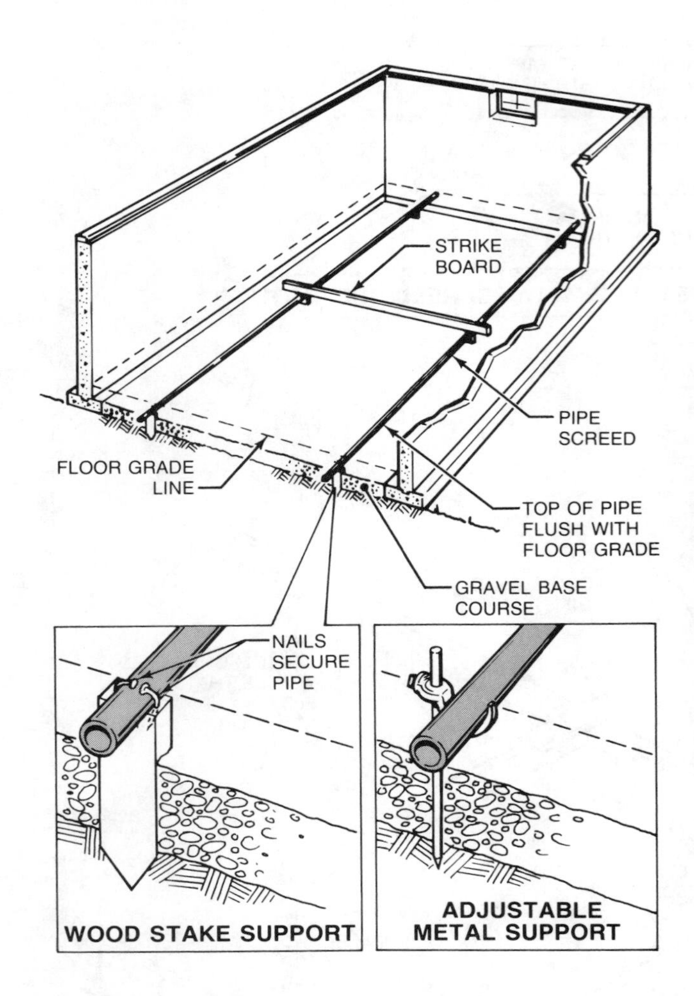

Figure 4-6. Metal pipe screeds are used to support a strike board.

size of wire are identified. For example a designation of 6 × 6—W2.0 × W2.0 in a drawing indicates that the longitudinal and transversal spacing is 6″ × 6″. The letter W indicates that the welded wire fabric is smooth wire. (Deformed wire is identified by the letter D.) The cross-sectional area of the longitudinal and transversal wires is 2.0. An older method of designating wire mesh size utilizes gauge measurements. For example, 6 × 6—8 × 8 indicates that the longitudinal and transversal wires are spaced 6″ × 6″, and the mesh is constructed of #8 gauge wire. See Figure 4-9.

Welded wire fabric is laid down shortly before the concrete arrives on the job to avoid rust and corrosion. Welded wire fabric should be placed approximately 1″–1½″ from the surface of the finished floor slab. Welded wire fabric commonly is laid on

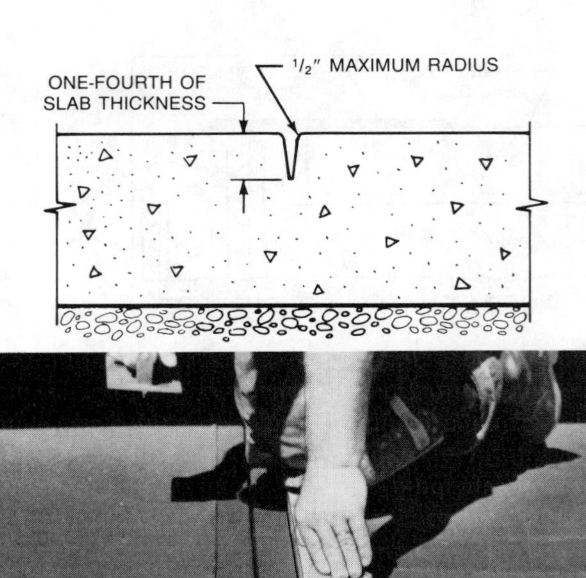

Portland Cement Association

HAND-TOOLED CONTROL JOINT

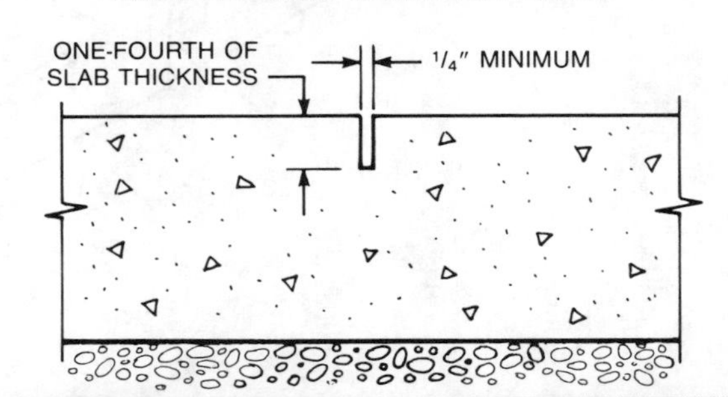

Portland Cement Association

SAWED CONTROL JOINT

Figure 4-7. Control joints in a concrete slab confine cracking resulting from expansion and contraction of the slab. Control joints are hand tooled or cut into the slab.

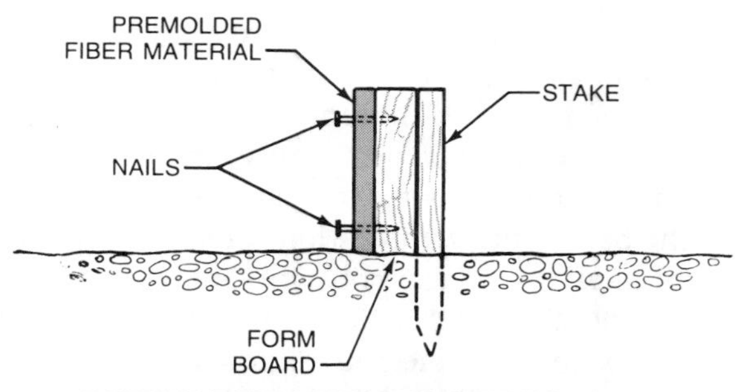

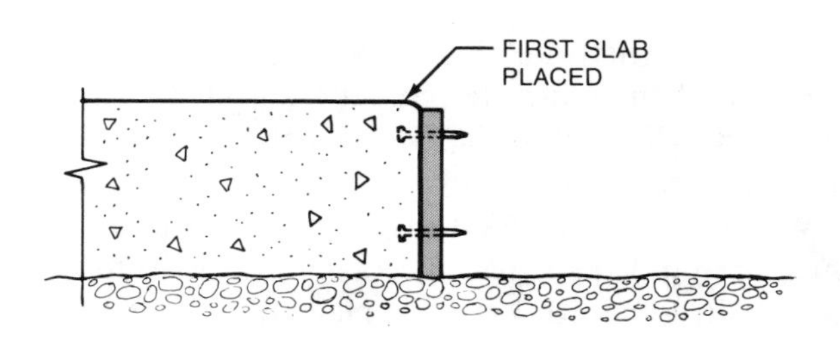

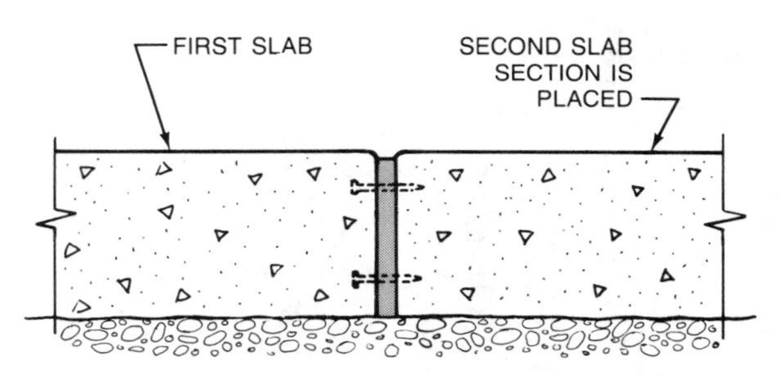

Figure 4-8. An expansion joint is used when a great degree of expansion and contraction is anticipated.

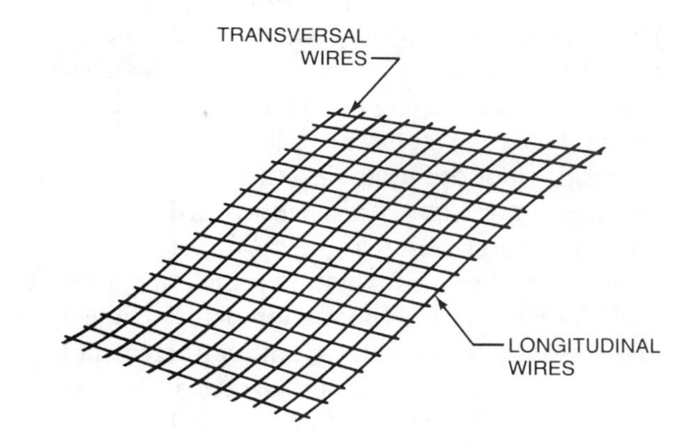

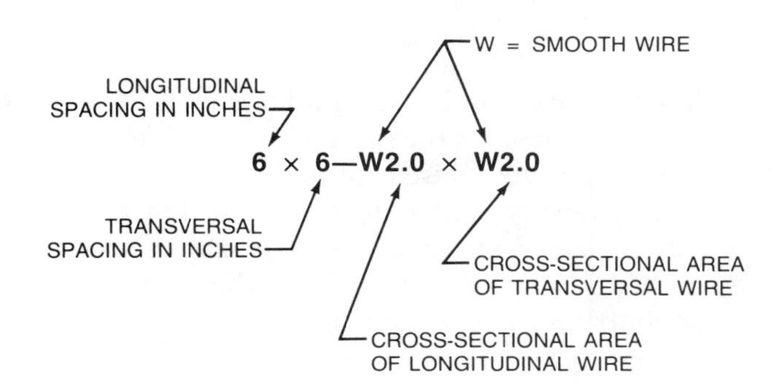

COMMON STOCK SIZES OF WELDED WIRE FABRIC

Style Designation		Steel Area sq. in. per ft.		Weight Approx. lbs. per 100 sq. ft.
New Designation (by W-Number)	Old Designation (by Steel Wire Gauge)	Longit.	Trans.	
ROLLS				
6 × 6—W1.4 × W1.4	6 × 6—10 × 10	.028	.028	21
6 × 6—W2.0 × W2.0	6 × 6—8 × 8*	.040	.040	29
6 × 6—W2.9 × W2.9	6 × 6—6 × 6	.058	.058	42
SHEETS				
6 × 6—W2.9 × W2.9	6 × 6—6 × 6	.058	.058	42
6 × 6—W4.0 × W4.0	6 × 6—4 × 4	.080	.080	58
6 × 6—W5.5 × W5.5	6 × 6—2 × 2**	.110	.110	80

*Exact W-number size for 8 gauge is W2.1.
**Exact W-number size for 2 gauge is W5.4.

Wire Reinforcement Institute

Figure 4-9. Welded wire fabric is used to reinforce concrete floor slabs.

the ground and pulled into position with a hook while the concrete is being placed.

Basement and Garage Floors

Basement and garage floors may be placed after the foundations have been constructed and are often placed after the building has been framed. This protects the slab from weather damage and other possible damage from construction work.

In basement and garage floors, the perimeter of the floors butt against the foundation walls. The ground area below the basement and garage slabs is excavated to firm soil. It is filled and compacted if necessary, and a base course of gravel is laid down. A vapor barrier is placed over the gravel and should extend over the foundation footings. Rebars and welded wire fabric are often used to reinforce basement and garage slabs.

Basement Floors. Most residential basement slabs are 4″ thick and are often reinforced with welded wire fabric. However, basement slabs that support heavy loads have thicker slabs reinforced with rebars. If floor drains are installed, the areas around the drains are slightly pitched toward the drain to facilitate water removal.

Basement floor information is included in section view drawings of the foundation plans. These drawings show slab thickness, type of reinforcement used, and the distance between the surface of the slab and the ceiling joists above. See Figure 4-10.

Isolation Joints. An isolation joint is used to prevent cracking between the slab and foundation walls when constructing a basement floor slab. An isolation joint is very similar to the expansion joint; therefore, the term expansion joint may also be used to describe the gap created where concrete floors butt up against foundation walls.

Without an isolation joint, there is a weak bond between the floor slab and foundation wall since the slab is not placed monolithically with the wall. Ground heaving may also cause cracks in and around the slab, allowing ground moisture to seep into the basement. An isolation joint usually consists of caulking or asphalt-impregnated material placed around the perimeter of the slab to isolate the floor slab from the foundation walls.

When using caulking for an isolation joint, an oiled, wedge-shaped strip is placed against the foundation wall before the slab is placed. The strip should be ½″ thick with the width equal to the thickness of the floor slab. After the slab has been placed and the concrete has set, the oiled strip is removed. The area between the slab and the wall is then filled with a caulking material.

When using an asphalt-impregnated strip, a ½″ thick premolded strip is placed against the foundation wall. After the slab is placed, the asphalt-impregnated strip remains in the concrete. See Figure 4-11.

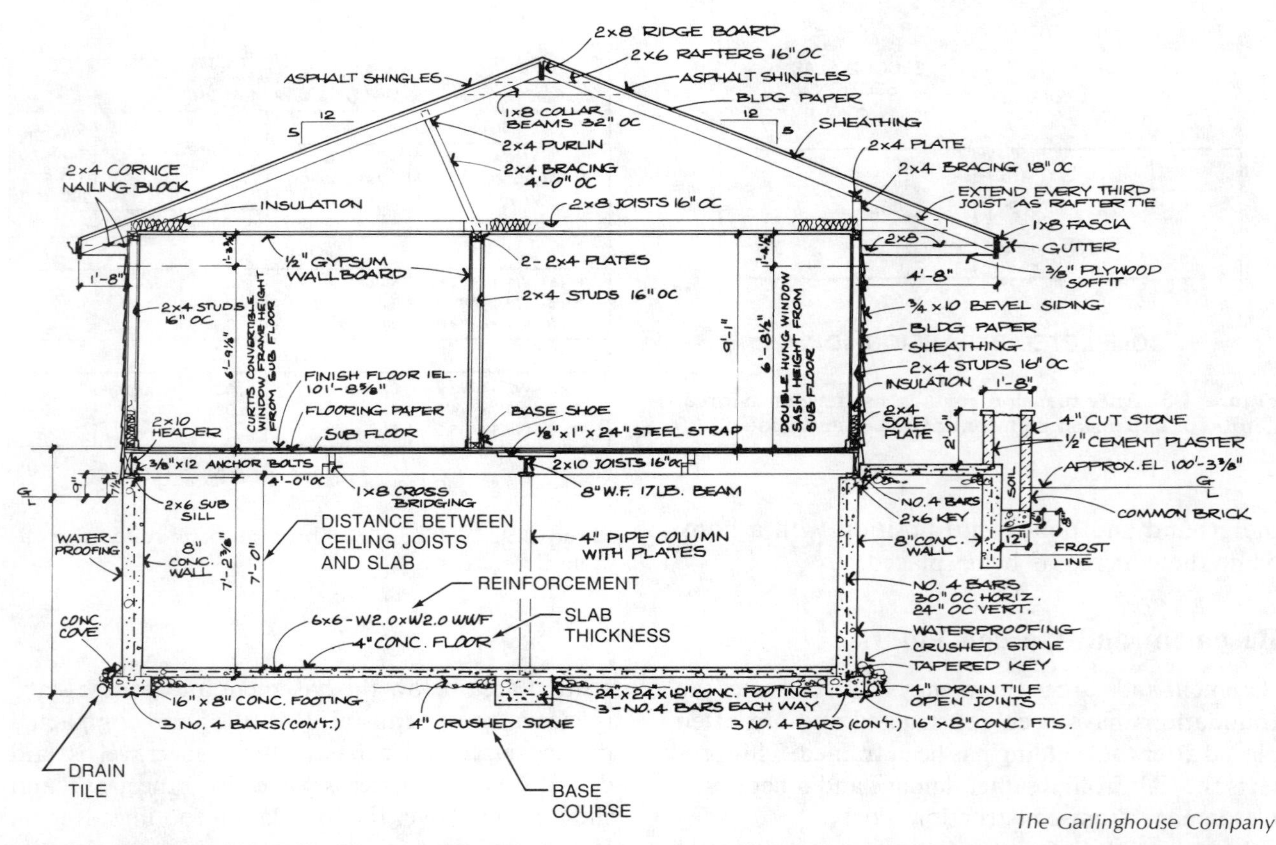

The Garlinghouse Company

Figure 4-10. The section view of a print provides information regarding the location, thickness, and reinforcement of the basement floor slab.

Before placing the concrete slab, floor elevations must be established by measuring from the bottom of the ceiling joists. Screeds are set up close to the foundation walls and at intervals throughout the slab area. See Figure 4-12.

Garage Floors. Specifications for a garage floor elevation are included in a section view of the garage.

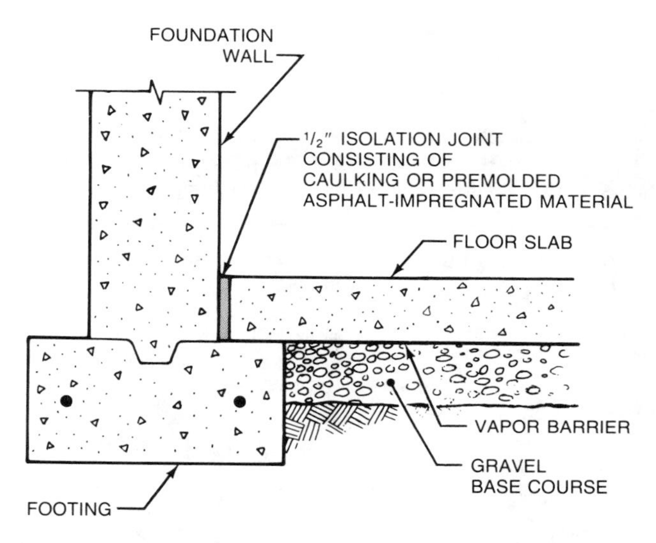

Figure 4-11. An isolation joint contains caulking or a premolded asphalt-impregnated strip.

In a residential structure with an attached garage, the section of the garage floor adjacent to the building is lower than the finish floor elevation of the building. The garage floor is sloped $\frac{1}{8}$" to $\frac{1}{4}$" per foot toward the front of the garage for proper drainage.

Once points for the floor elevation at the rear of the garage are established, the slope from back to front is calculated by multiplying the length of the floor by the slope per foot. This amount is subtracted from the elevation at the rear of the garage and marked at the front wall. For example, a garage floor measuring 16' from front to back with a $\frac{1}{8}$" per foot slope has a total floor slope of 2" (16' × $\frac{1}{8}$" = 2"). After garage floor elevations are established, chalk lines are snapped along the back and side walls. A line is then stretched across the garage opening and a form is constructed to this line. No other forms are required because the slab is formed on three sides by the foundation walls. See Figure 4-13.

EXTERIOR FLATWORK

Exterior flatwork includes driveways, walks, patios, concrete curbs and gutters, and other exterior con-

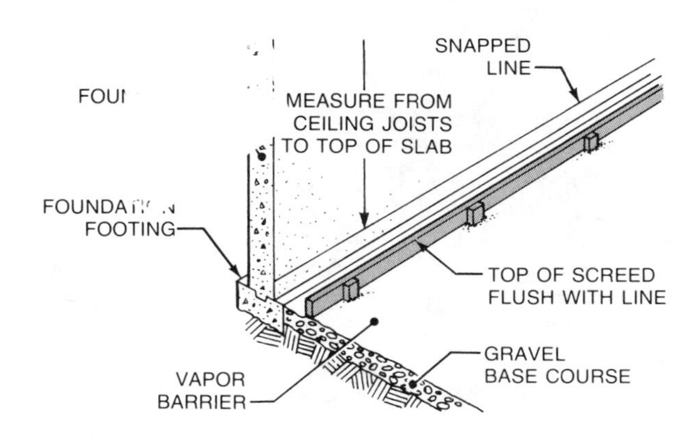

1. Measure the distance from the ceiling joists to the top of the slab and snap a line. Set a screed close to the wall and stake in position with the top of the screed flush with the line.

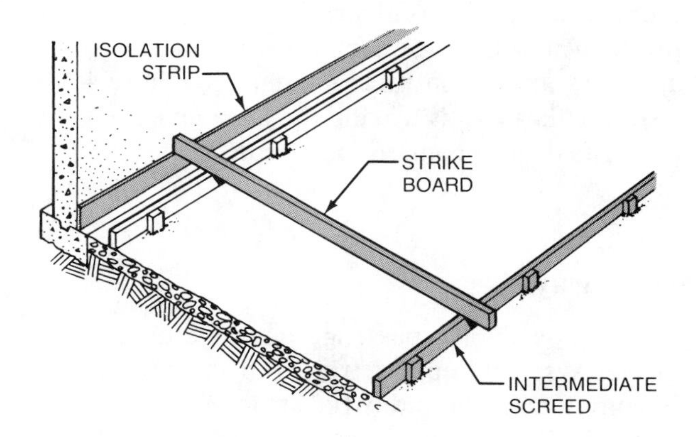

2. Set and stake intermediate screeds. Cut the strike board to length and place across the screeds. Construct isolation joints along the walls.

Figure 4-12. Screeds are set up close to the foundation walls when concrete is placed for a basement floor slab. An isolation joint is placed along the foundation wall.

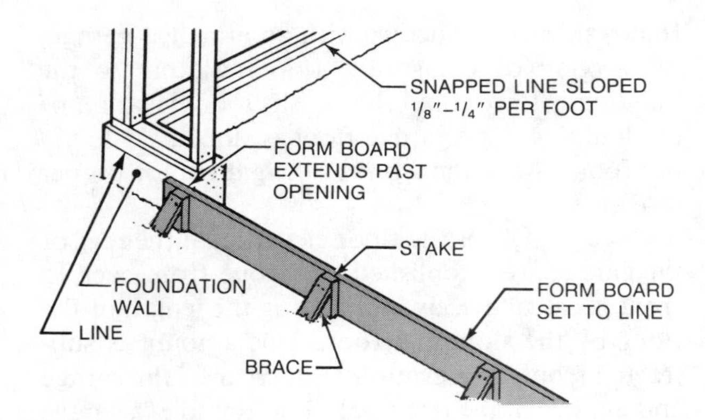

1. Snap a line along the foundation wall sloped toward the front of the garage. Stretch a line across the garage opening. Set form boards to the line. Secure with stakes and braces.

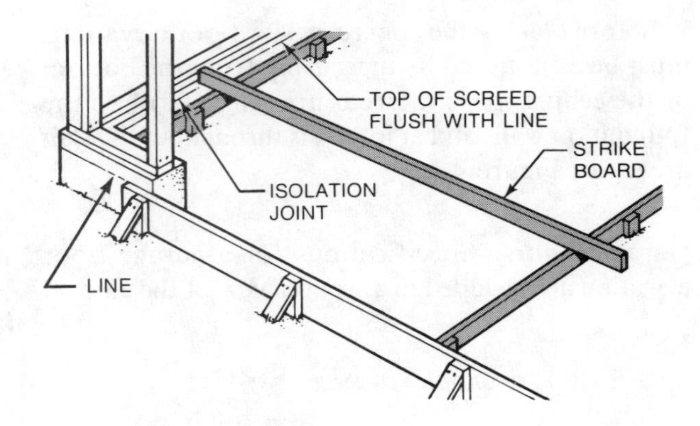

2. Set and stake screeds with the tops flush with the lines. Cut the strike board to length and place across screeds. Construct isolation joints along the walls.

Figure 4-13. Garage floors are sloped 1/8" to 1/4" per foot toward the front of the garage. The garage floor is formed on three sides by the foundation walls and on one side by a form board.

crete slabs placed around a building. Forms for driveway, walk, and patio slabs are laid out and placed so that the top edges of the form are set to the finish surface of the slab. Excavation may be required for fill and a gravel base course beneath the slab. The required depth for ground excavation below the slab is measured down from the top edge of the form.

Rebars or welded wire fabric are used where the slabs are subjected to great pressures such as in driveways or other areas supporting moving vehicles. Sloping of exterior flatwork is very important for proper water drainage. Isolation and control joints are necessary to control cracking. See Figure 4-14. Exterior flatwork is usually the last concrete work performed on a construction project.

Driveways

Driveway slabs for passenger cars are usually 4". Driveways that support truck movement, such as in commercial and industrial structures, should be 6" thick. Reinforcement consisting of welded wire fabric or rebars is normally required for both. The finished surface of the garage (or carport) end of the driveway should be ½" below the surface of the garage floor for proper water drainage. See Figure 4-15.

Most single-car driveways are 8' to 12' wide, and double-car driveways are 15' to 20' wide. Consult local building codes for the minimum widths of driveways. Driveways should be flared and widened at the curbs because the back wheels of a vehicle turn in a smaller radius than the front wheels. The

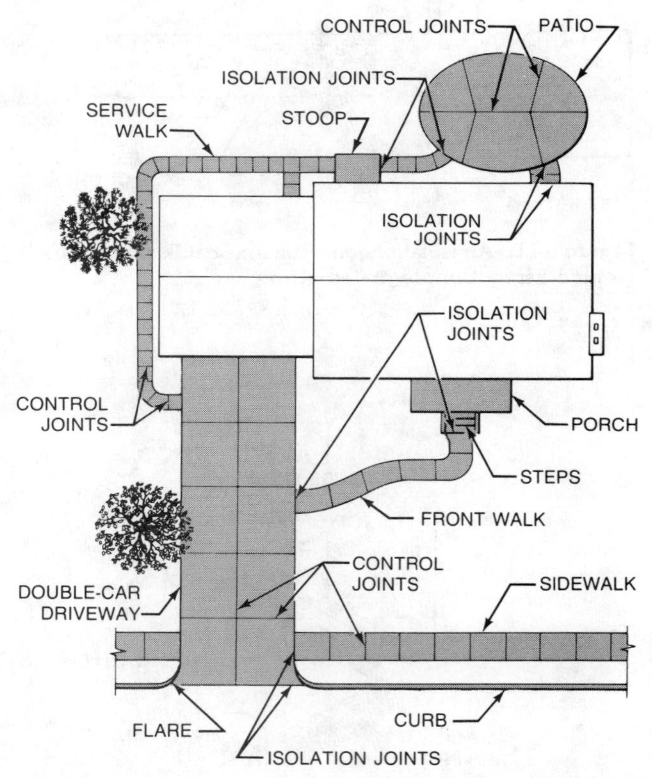

Portland Cement Association

Figure 4-14. Exterior flatwork includes driveways, walks, and patios. Isolation and control joints are used to control cracking.

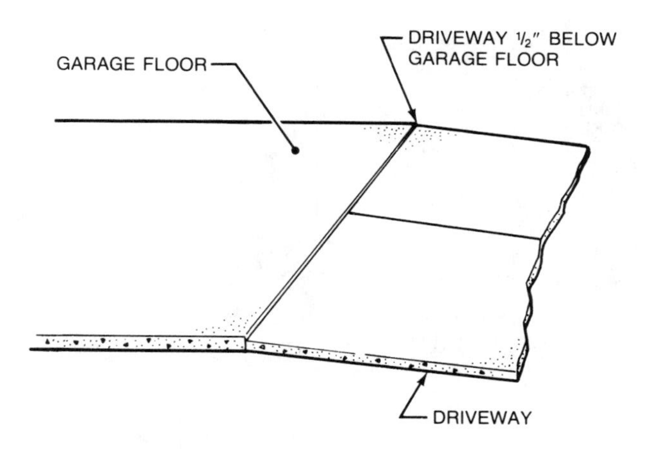

Figure 4-15. The finished surface of the garage or carport end of a driveway should be 1/2" below the surface of the garage floor. The driveway is sloped away from the garage to facilitate drainage.

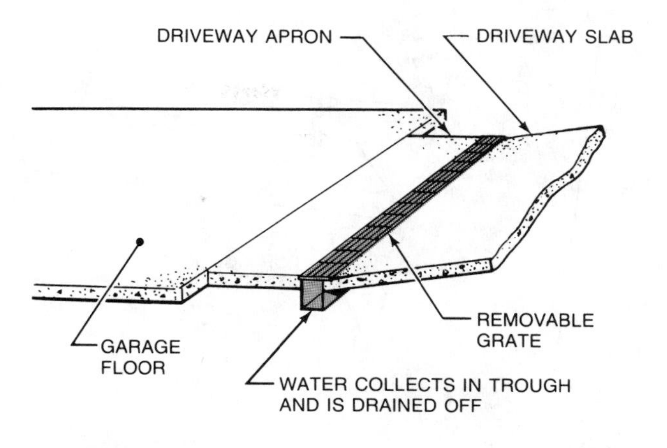

Figure 4-16. A concrete trough with a removable grate diverts water away from a garage that has a driveway sloping toward the garage.

width of the driveway is sloped 1/8" or 1/4" per foot from side to side. Wide driveways are pitched from the center in both directions. This requires placing a screed at the center of the driveway area before the concrete is placed.

A gradual slope from the garage area to the street curb is recommended. However, a garage considerably above or below the street level may have a steep slope. In conditions where a steep slope exists, an abrupt grade change should be avoided to prevent scraping of bumpers and undersides of cars on the driveway or sidewalk.

Driveways that slope down to the garage area should have a drain directly in front of the garage entrance. A recommended drain system involves installing a removable grate over a concrete trough that runs the full width of the garage. See Figure 4-16. Water collected in the trough is discharged by plastic pipe or drain tile into the surrounding soil. The water can be also be discharged through a culvert or storm sewer, if allowed by local code.

Control joints are placed in the driveway to control cracking. Control joints that run the width of the driveway should be spaced no more than 10' apart. A driveway 10' or more in width should have a control joint running down the center of the drive.

Constructing Driveway Forms. Driveway forms usually consist of staked 2 × 4s or 2 × 6s placed on edge running the length of the driveway. Form boards are not required at the garage end or at the opposite end if the garage and front sidewalk have been placed. However, provisions should be made for isolation joints where the driveway butts up against the garage floor or sidewalk. See Figure 4-17.

Walks

The types of walks generally found around buildings are *public sidewalks, front walks,* and *service walks.* Public sidewalks run along the street that border the building lot. Front walks extend from a driveway or public sidewalk to the front entrance of a building. Service walks extend from a driveway or sidewalk to a rear entrance.

Public sidewalks are usually 4' to 5' wide. They are placed next to the street curbs or separated from the curbs by a planter strip. Front walks are usually 3' wide. Service walks are usually 2'-6" wide and run along the side of a building. They are usually 2' away from the foundation.

A walk that butts against an entrance should be 5" to 6" below the surface of the stoop or door sill. If the walk butts against a stairway, the distance from the surface of the walk to the first tread should be the same as the individual riser height.

Isolation joints should be placed where walks butt up against driveways, stoops, or steps. Control joints should be spaced not less than 40" apart in walks 2' wide, and every 5' in walks 3' or wider.

A common thickness for walks is 4", although thick slabs and rebars may be required where heavy vehicles, such as trucks, cross over the walks. Walks

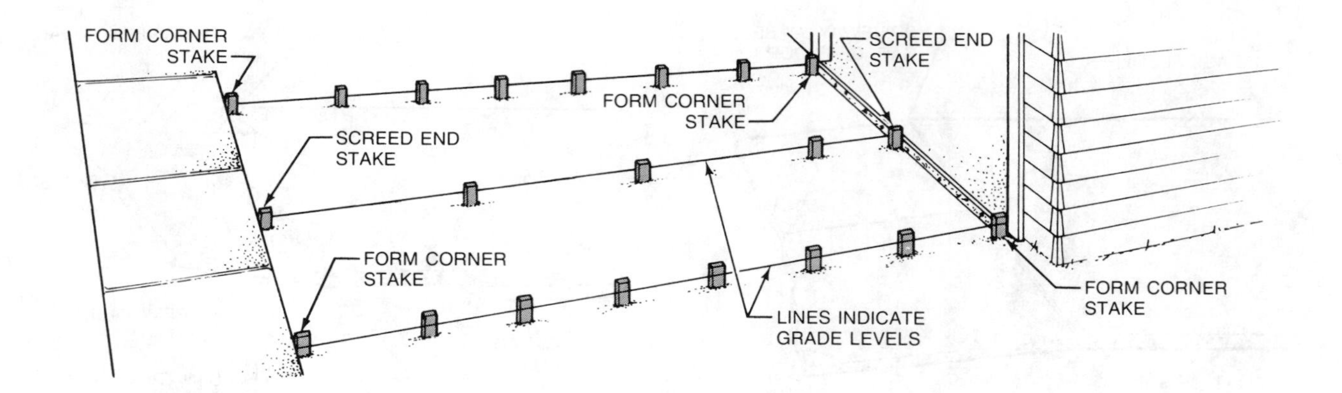

1. Drive form corner stakes. Drive screed end stakes. Lay out grade mark on the stakes and stretch lines. Drive intermediate form and screed stakes at 4′-0″ OC.

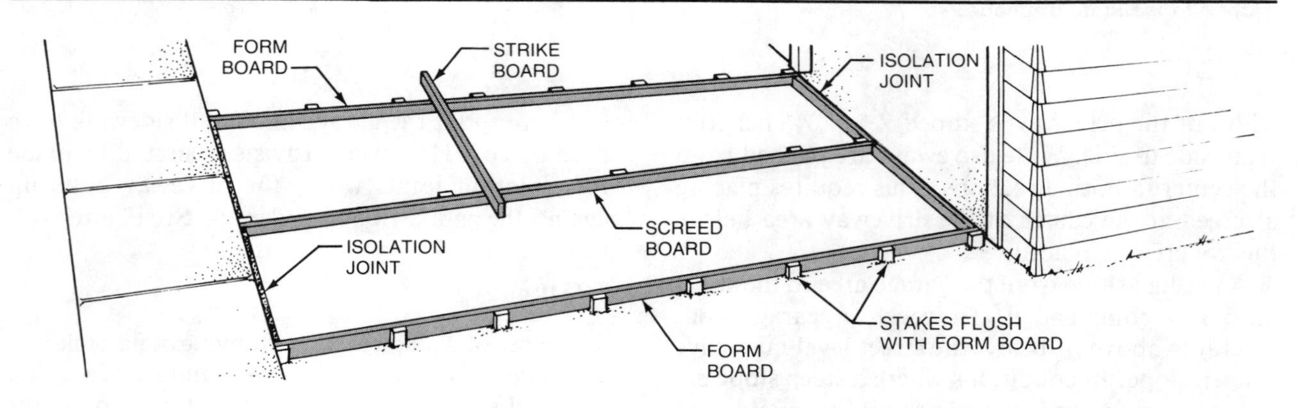

2. Nail form and screed boards to stakes. Cut the form stakes flush with the form boards. Cut a strike board to length and place on the screeds. Construct isolation joints along adjoining concrete surfaces.

Figure 4-17. Two edge forms are required to form a driveway. The garage floor and sidewalk are used to form the concrete on both ends.

should be sloped from side to side $\frac{1}{8}$″ to $\frac{1}{4}$″ per foot to allow for water drainage. Specific dimensions for walks are found in local building codes.

Constructing Walk Forms. Walk forms are usually constructed of 2 × 4s or 2 × 6s held in place by wood or metal stakes. See Figure 4-18. Reusable metal forms are also available. Control joints are tooled or cut in the surface of the walk to control cracking of the concrete.

Laying Out and Forming Curves

A curve is laid out using a line that is equal in length to the curve's radius. A round stake is driven at the center point of the curve, and one end of the line is tied loosely around the stake. The other end of the line is tied to an individual flat stake that is driven along the edge of the curve. The line is then released from the flat stake and the procedure is repeated for the remaining stakes in the curve. See Figure 4-19.

The material used to form the curve is determined by the radius of the curve. A $\frac{3}{4}$″ piece of plywood can bend sufficiently for long radius curves. Hardboard or $\frac{1}{4}$″ plywood is used for short radius curves. Saw-kerfing $\frac{3}{4}$″ material is another method used to form short radius curves. See Figure 4-20.

Patios

A concrete patio is an exterior slab constructed next to a building. It is primarily used for recreational purposes and can be designed in a variety of shapes. See Figure 4-21. Ground preparation is basically the same as for other slab work.

Patio slabs should be at least 4″ thick and pitched 1/8″ to 1/4″ per foot in one direction. Patio slabs are often reinforced with welded wire fabric. Isolation joints are constructed where any part of the patio meets adjacent concrete walls or walks. Control joints should be provided at a maximum of 10′ intervals in both directions.

Permanent wood divider strips may be used in place of control joints in the patio. Redwood, cedar, cypress, or pressure-treated 2 × 4s are often used for this purpose because of their attractive appearance and resistance to decay and insect attack. Divider strips are set to lines extending from the edge forms of the patio. They are secured by stakes driven below the surface of the dividers. The top of the wood strips should be taped to protect them from damage when the concrete is being placed. Divider strips also serve as screeds when striking off the concrete.

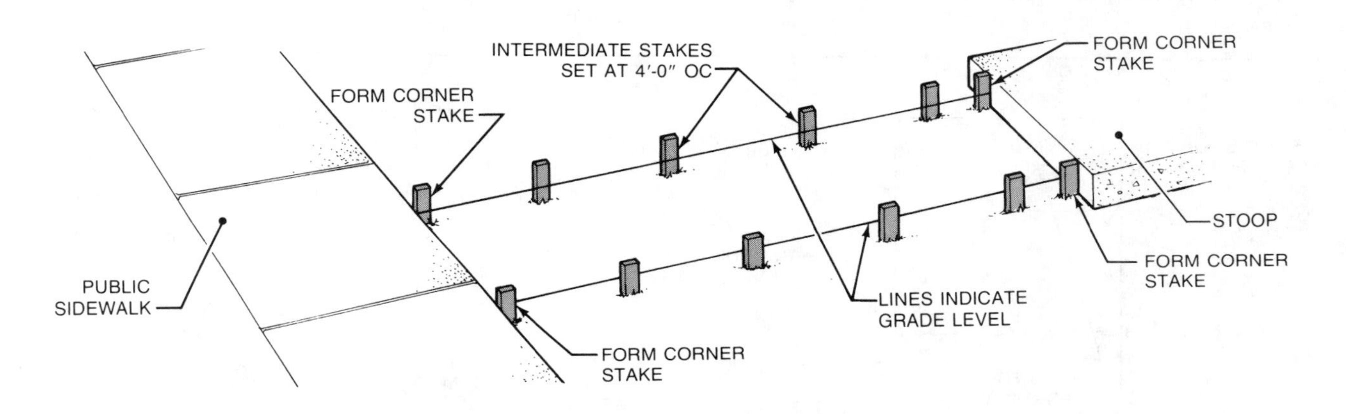

1. Drive stakes at the four corners of the walk. Mark the grade levels and stretch lines. Drive intermediate stakes at 4′-0″ OC.

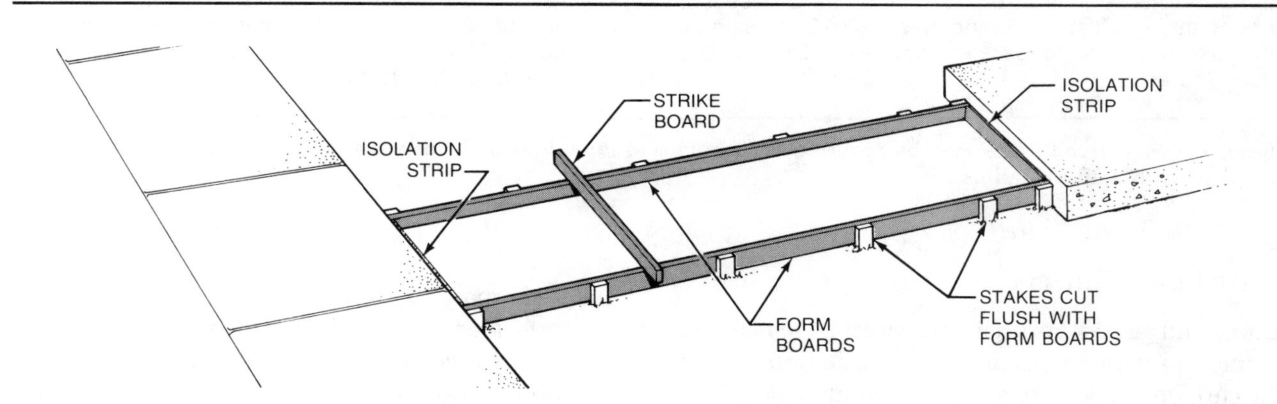

2. Snap grade levels on the stakes and nail form boards to the stakes. Cut the stakes flush with the form boards. Cut a strike board to length and place across form boards. Construct isolation joints at adjoining concrete surfaces.

Figure 4-18. Two-inch thick planks are used to form the sides of walks. A public sidewalk and house stoop are used to form the concrete on both ends.

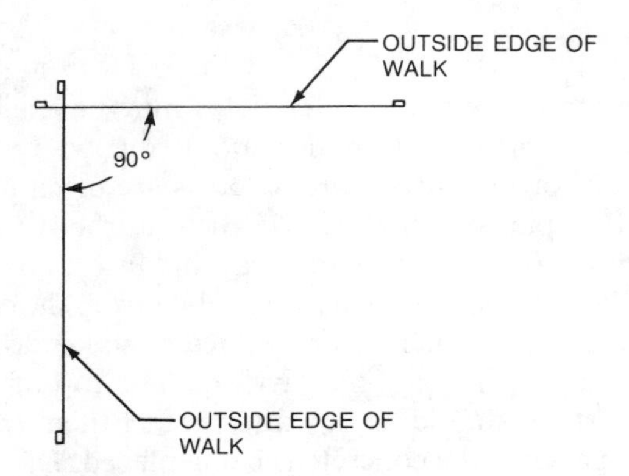

1. Drive stakes and stretch lines at a 90° angle to each other, representing the stake line for the outer edge of the walk.

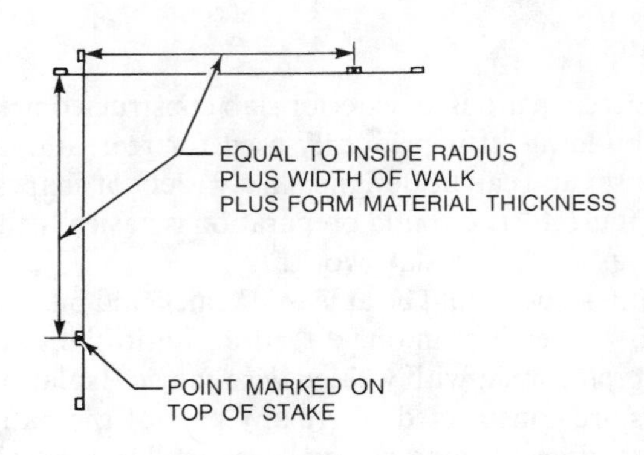

2. Measure the outside radius of the curve. This equals the curve radius (15′-0″) plus the width of the walk or driveway (5′-0″) plus the form material thickness. Drive stakes and mark the outside radius (20′-0$^1/_4$″) on top of the stakes.

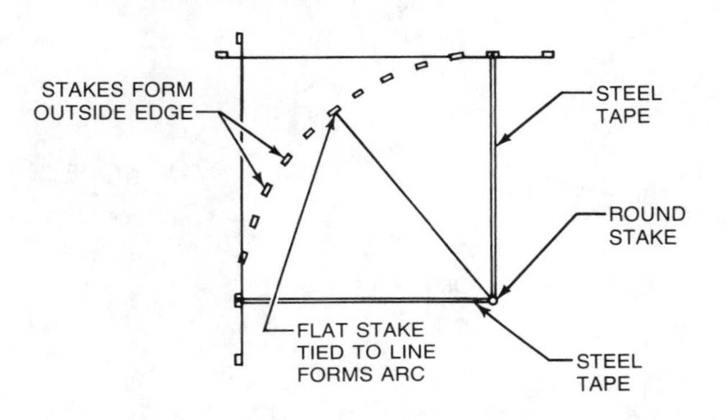

3. Measure the outside radius from both corner stakes using two steel tapes. Drive a round stake at the intersection. Tie a loose knot in one end of a line and secure around the round stake. Swing an arc (20′-0$^1/_4$″) using a flat stake tied to the opposite end of the line. Drive stakes along the arc.

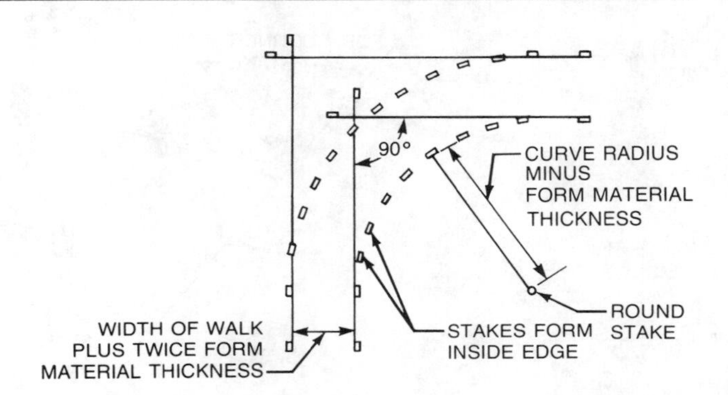

4. Measure in from the outside edge the width of the walk or driveway plus twice the form board thickness (5′-0$^1/_2$″). Drive stakes and set lines to form a 90° angle. Swing an arc measuring the curve radius minus the form material thickness (15′-0″ − $^1/_4$″ = 14′-11$^3/_4$″). Drive stakes for the inside edge of the walk or driveway.

Figure 4-19. Curved forms may be required to form curves for sidewalks or driveways. The example is a 5′-0″ wide sidewalk with a 15′-0″ radius.

Curbs and Gutters

Curbs and gutters bordering the street pavement are formed in various designs. Two basic designs are the curb only, and curb and gutter combination. The top of a curb is usually flush with the sidewalk and runs along the street edge except where driveways are located. Expansion joints are provided in the curbs to control cracking and are spaced according to local code requirements.

Wood forms are usually built with 2″ thick planks. Stakes are driven to established lines. After the grade marks have been marked on the stakes, the form planks are nailed into place.

Prefabricated metal curb and gutter forms are commonly utilized to form curbs and gutters for large construction projects. See Figure 4-22. The edge forms are staked to the ground with metal stakes. The division plates are positioned and the curb face form is suspended from the division plates. After the concrete has initially set, the curb face form and division plates are removed and the curb and gutter is finished.

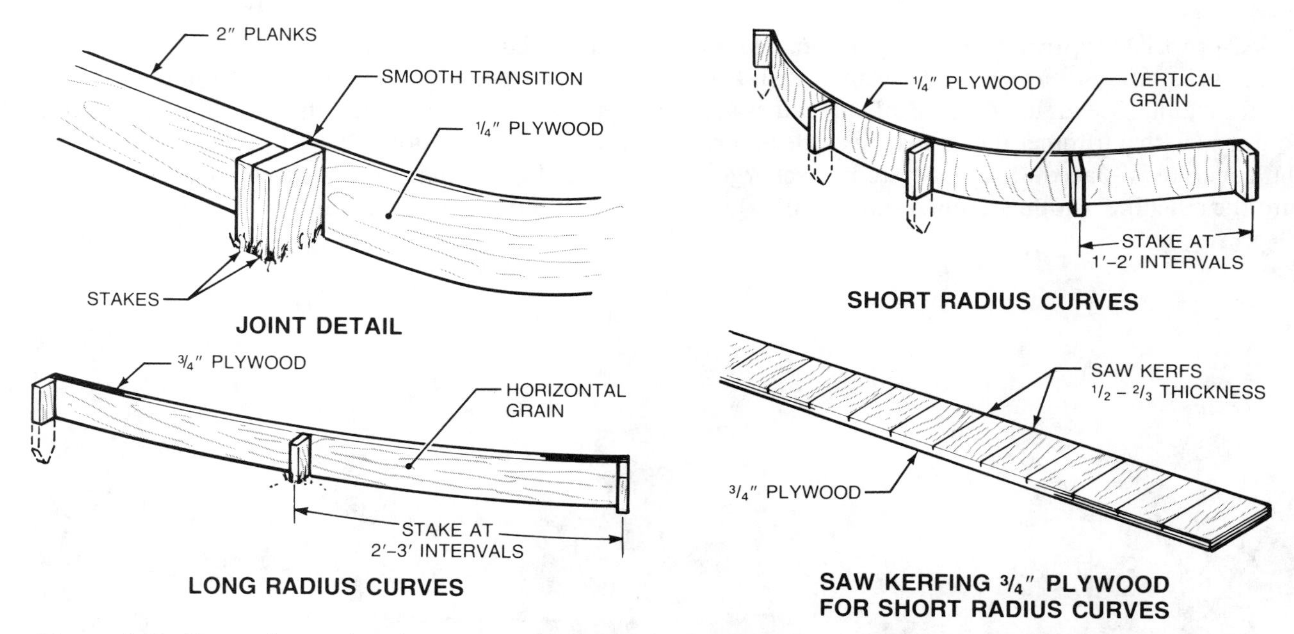

Figure 4-20. Plywood is used to form curves. Saw kerfs are made in 3/4" plywood to give it greater flexibility.

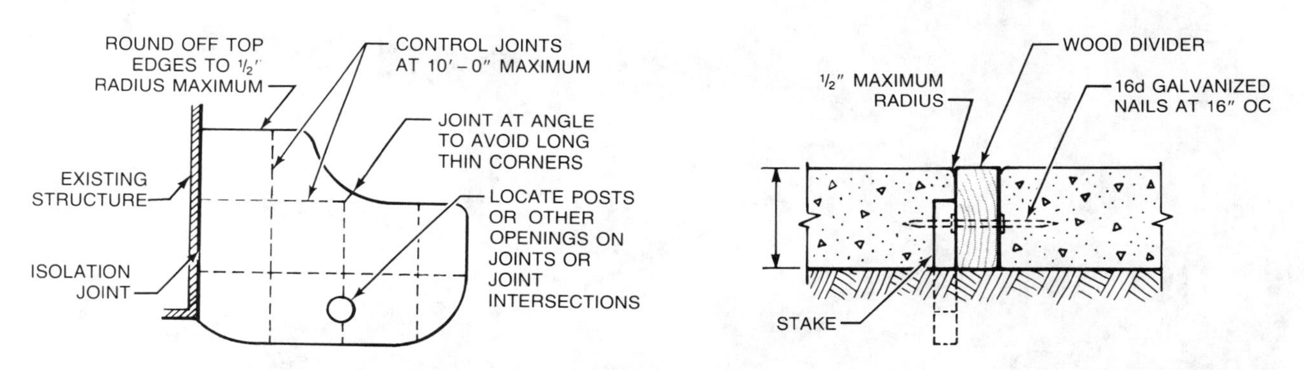

Figure 4-21. Patios serve as outdoor recreational areas in residential construction. Wood dividers add to the attractiveness of the patio and serve as screeds to strikeoff the patio.

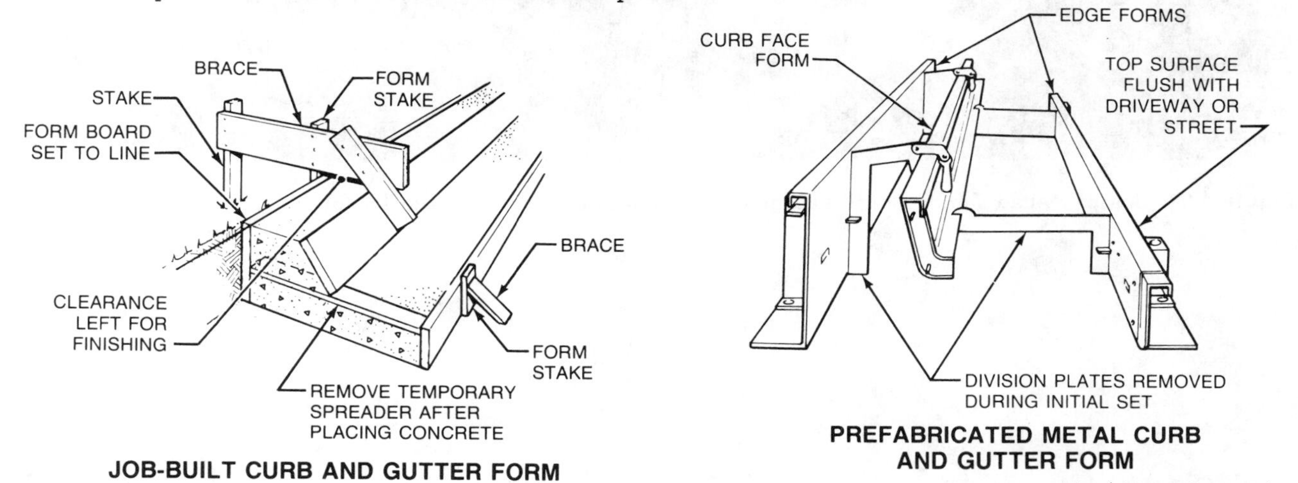

Metal Forms Corporation

Figure 4-22. Job-built or prefabricated curb and gutter forms are used to form curbs and gutters. The top of the curb is usually flush with the sidewalk surface and the top of the gutter is flush with the surface of the roadway.

Self-propelled slipform curb and gutter machines form curbs and gutters without formwork staked to the ground. See Figure 4-23. Many machines are equipped with a trimmer that trims the grade before the concrete is placed. The concrete is discharged into the machine's hopper from a transit-mix truck while the curb and gutter is formed. As the machine travels at speeds up to 75 feet per minute, the concrete is placed, vibrated, and consolidated in one continuous operation. The self-propelled slipform curb and gutter machines form curbs up to 42″ high and gutters with a maximum width of 60″.

Power Curbers, Inc.

Figure 4-23. Self-propelled slipform curb and gutter machines eliminate the need for forms.

Chapter 4—Review Questions

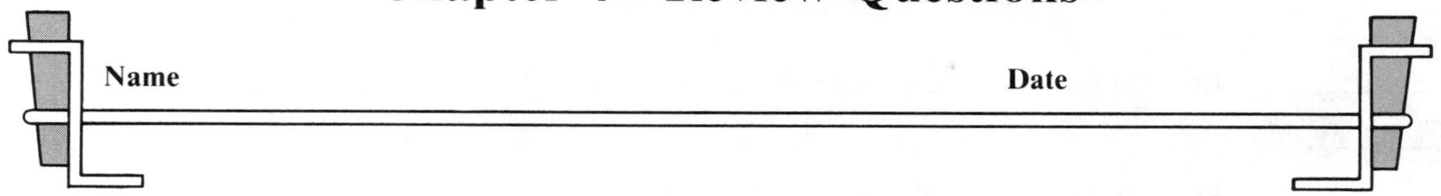

Completion

__________ 1. The floor of a slab-on-grade system is at the same elevation as the top of the __________.

__________ 2. The recommended minimum thickness of residential slab-on-grade floors is ________″.

__________ 3. The types of reinforcement used in slab-on-grade floors are ________ or ________.

____

__________ 4. Insulation along the perimeter of floor slabs should be at least __________″ thick and extend __________″ vertically or horizontally.

____

__________ 5. A(n) __________ joint is used when concrete is placed in sections for large floor slabs.

__________ 6. A(n) __________ system must be set up to maintain proper floor levels when placing concrete for a concrete slab.

__________ 7. A(n) __________ board is a straightedge placed between screeds to strikeoff the concrete.

__________ 8. __________ joints help confine and control the cracking of concrete slabs.

__________ 9. The spacing recommended for control joints in concrete slabs is __________′ to __________′.

____

__________ 10. __________ drawings of a foundation plan provide information regarding basement floors.

__________ 11. A(n) __________ joint is recommended between the slab perimeter and the foundation wall.

__________ 12. Asphalt-impregnated isolation strips are commonly __________″ thick.

__________ 13. Concrete floors should slope __________″ to __________″ per foot toward the front of the garage.

____

__________ 14. The total slope of a garage floor that is 20′ long with a $^1/_8$″ per foot slope is __________″.

__________ 15. The recommended thickness of driveways supporting truck movement is __________″.

__________ 16. The surface of a driveway butting against a garage floor should be __________″ below the garage floor.

__________ **17.** Single-car driveways range in width from __________' to __________'.

____

__________ **18.** Double-car driveways range in width from __________' to __________'.

____

__________ **19.** The top surface of a slab-on-grade floor is at least __________" above the finish grade.

__________ **20.** A 2" thick __________ for forming construction joints is staked at the edges of the concrete placement area.

__________ **21.** The letter D in 6 × 6—D2.9 × D2.9 welded wire fabric indicates that the wire is __________.

Multiple Choice

__________ **1.** Driveways should be flared and widened at the curbs to __________.
A. give it better appearance
B. conform with code requirements
C. allow for the back wheels of a vehicle to track a smaller radius than the front wheels
D. result in better water drainage

__________ **2.** Double-car driveways should be sloped __________.
A. ¼" per foot from one side to the other
B. from the outside edges toward the center
C. from the front to the rear of the driveway
D. from the center to the outside edges

__________ **3.** The spacing of control joints across the width of a driveway should not exceed __________'.
A. 6
B. 8
C. 10
D. 12

__________ **4.** Driveways that are 10' or more in width should have __________.
A. control joints spaced close together
B. a longitudinal control joint running down the center
C. greater pitch
D. none of the above

__________ **5.** Sidewalks are commonly __________" thick.
A. 3
B. 4
C. 5
D. 8

__________ **6.** Walks should be sloped __________" to __________" per foot.
A. ⅛, ¼
B. ½, ¾
C. ⅝, ¾
D. ¾, 1

________ **7.** The minimum width of front walks is commonly ________′.
A. 2
B. 3
C. 4
D. 5

________ **8.** A front walk should be placed ________″ below an entrance.
A. 4
B. 5 to 6
C. 6 to 8
D. 9½

________ **9.** ________ is recommended to form short radius curves.
A. Three-quarter inch plywood
B. Hardboard
C. Kerfed ⅜″ plywood
D. A 1½″ plank

________ **10.** A ________ controls the capillary rise of water and provides uniform structural support for a slab-on-grade floor.
A. vapor barrier
B. layer of fill
C. gravel base course
D. none of the above

________ **11.** A keyway is formed with a ________ strip for a horizontal construction joint.
A. recess
B. key
C. chamfer
D. all of the above

________ **12.** The W in welded wire fabric designated as 6 × 6—W4.0 × W4.0 indicates that the wire is ________.
A. rigid
B. deformed
C. smooth
D. thick

________ **13.** When positioning a polyethylene film vapor barrier, the joints should overlap ________″.
A. 1
B. 2
C. 4
D. 6

________ **14.** The depth of a control joint is approximately ________ the slab thickness.
A. ⅛
B. ¼
C. ½
D. ¾

__________ **15.** The top surface of a slab-on-grade floor should be __________″ above the finish grade level.

A. 2
B. 4
C. 6
D. 8

Identification

__________ **1.** Longitudinal spacing in inches

__________ **2.** Transversal spacing in inches

__________ **3.** Smooth wire

__________ **4.** Cross-sectional area of longitudinal wire

__________ **5.** Cross-sectional area of transversal wire

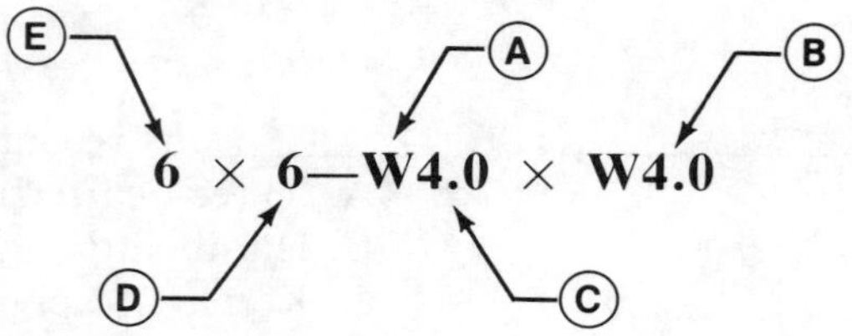

6 × 6—W4.0 × W4.0

__________ **6.** Reinforcement

__________ **7.** Vapor barrier

__________ **8.** Gravel base course

__________ **9.** Compacted fill

__________ **10.** Concrete slab

__________ **11.** Undisturbed soil

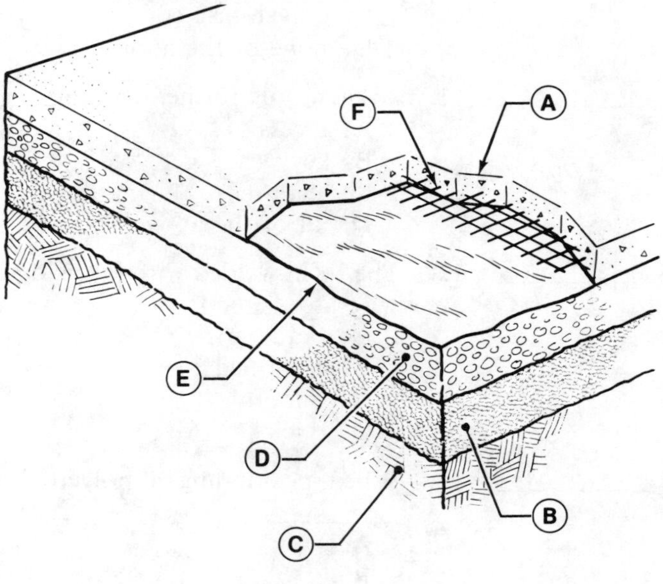

CHAPTER 5
Heavy Construction

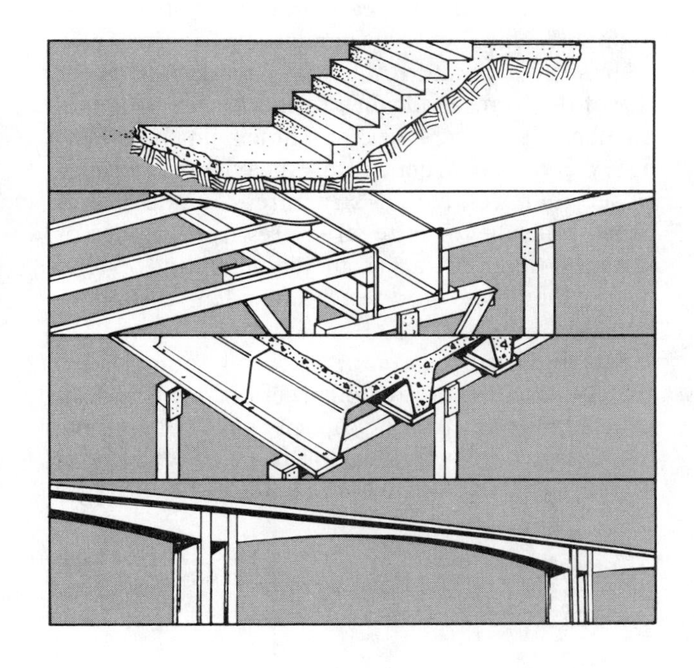

Heavy construction techniques are used to construct large concrete structures such as office and apartment buildings, hospitals, and highways. Heavy construction equipment is often required for deep excavations for large concrete structures.

Large concrete structures are erected using cast-in-place concrete and/or precast concrete members. Structural members of large concrete buildings include foundation footings and walls, floor slabs, columns, beams, and girders. Forms for cast-in-place concrete are designed to frame into each other to produce monolithic structural members. The forms also resist movement during the placement of concrete and are easily stripped without causing damage to the structural member.

Highway construction includes the construction and maintenance of highways, bridges, overpasses, and ramps. Although most of the paving and curbing of road surfaces is performed by mechanical slip-forming equipment, form construction is required for bridges, ramps, and overpasses. Formwork procedures and materials for highway construction are similar to the construction of other heavy concrete structures.

Safe and established construction procedures must be followed on heavy construction projects. Form construction may occur at great heights and near heavy construction equipment. Consult the American Concrete Institute (ACI) or Occupational Safety and Health Administration (OSHA) for information regarding safe construction procedures.

FOUNDATION FORMS

A foundation supports and transmits heavy loads to the soil below the structure. Building load and soil conditions are factors used to determine the type of foundation system. Foundation forms are laid out and built according to information supplied in the foundation plans and related section view drawings of a set of prints. See Figure 5-1.

Most heavy concrete structures rest either on T-foundations or grade beam and pile systems. The main difference between foundations for heavy concrete structures and residential foundations is the size of the forms used when forming large structural members. Stiffeners, ties, and bracing members for heavy concrete structures are heavier and spaced closer than residential structures. Grade beams beneath heavy concrete structures rest on piles or caissons that are larger in diameter and extend deeper than piers used beneath residential foundations.

Mat and *raft foundations* (floating foundations) are also used in heavy construction projects constructed over low bearing capacity soils or in seismic risk areas. A mat foundation is a thickened reinforced slab that transmits the load of the structure as one unit over the surface of the soil. A raft foundation is a thickened reinforced slab placed monolithically with the walls. See Figure 5-2.

Heavy concrete structure foundations often use *ground beams* that tie wall or column footings together. A ground beam is a reinforced beam running along the surface of the ground but does not rest on supporting piers or piles.

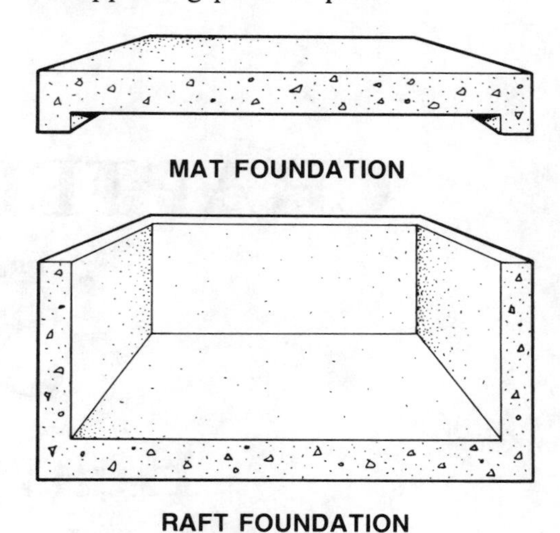

Figure 5-2. Mat and raft foundations support heavy concrete structures over low bearing capacity soils.

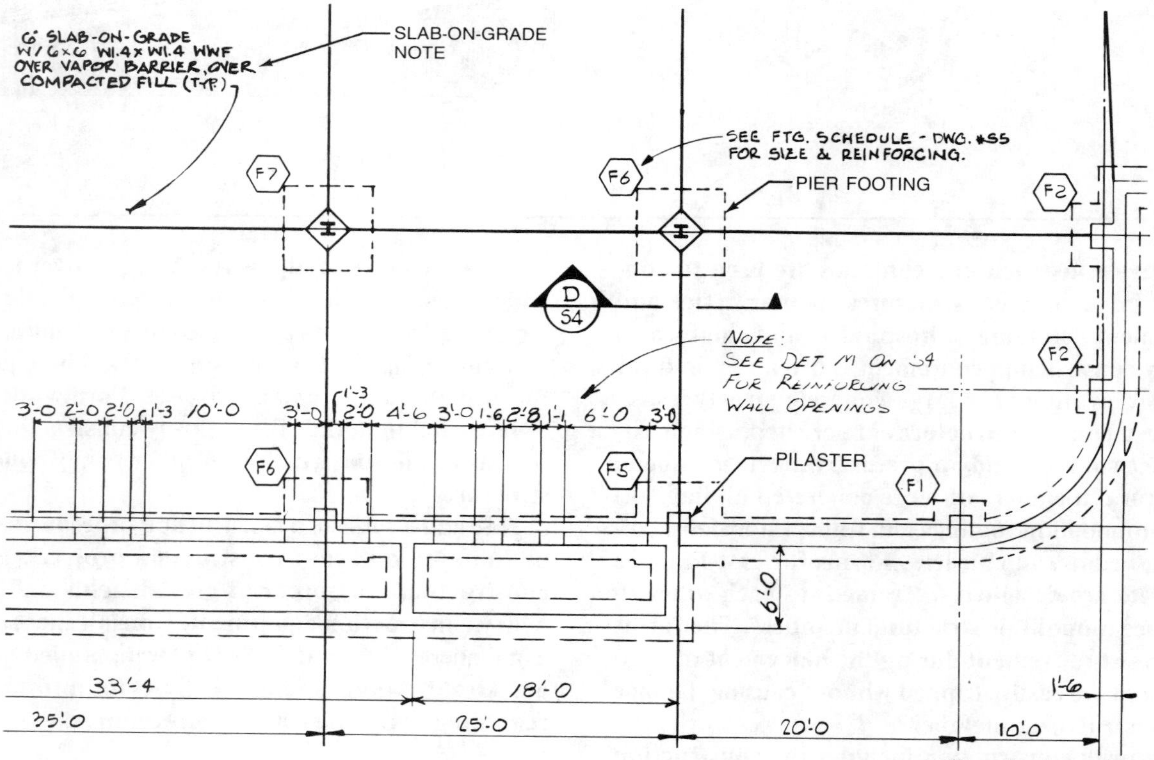

William Brazley and Associates

Figure 5-1. Foundation designs for heavy concrete structures may include pier footings, pilasters, and slab-on-grade floors.

Large concrete buildings may require deep excavation reaching load-bearing soil and providing space for a below-grade basement. Typical heavy equipment used for excavation includes power shovels, bulldozers, graders, and excavators. See Figure 5-3. Various mobile or stationary cranes are also used, which include *free standing* and *climbing tower cranes*.

Walsh Construction Company of Illinois

Figure 5-3. Heavy equipment is used for excavating heavy construction sites.

Free standing tower cranes are secured to a concrete pad next to the building. Climbing tower cranes are set in position during the foundation work and move up as the height of the building increases. Climbing tower cranes are raised to each new position with hydraulic jacks and are supported by steel collars resting on the floor slabs. They are often used in the construction of high-rises. See Figure 5-4.

Piles

Piles are long structural members that penetrate deep into the soil. Factors such as soil conditions and adjacent buildings are considered when determining whether a pile-supported foundation should be utilized. Soil conditions may be too unstable to allow a conventional deep excavation, or the proximity of adjacent buildings may limit the depth of an excavation.

Pile drivers equipped with either a drop, mechanical, or vibratory hammer drive solid piles or casings into the ground. Steam, diesel pistons, or pressurized hydraulic fluid is used as a power source for the pile drivers. Mobile cranes equipped with specialized machinery and attachments may also be used to drive piles. See Figure 5-5.

Bearing piles and *friction piles* are used to support foundations. Bearing piles are more frequently used and are driven completely through the unstable soil layers to rest on firm load-bearing soil. Friction piles do not have to penetrate to load-bearing soil. They must only be driven to a point where there is adequate soil resistance and pressure against the pile to support the imposed load.

Piles are placed beneath grade beams supporting bearing walls. *Grouped piles* may also be placed beneath concrete caps that act as a base for load-bearing columns. Grouped piles are piles driven in a close arrangement and are used when the main structural support of a building is provided by col-

Morrow Crane Co., Inc.

FREE STANDING TOWER CRANE

Portland Cement Association

CLIMBING TOWER CRANE

Figure 5-4. Tower cranes are used in the erection of heavy concrete structures.

Associated Pile and Fitting Corporation, Clifton, NJ

Figure 5-5. Pile drivers are used to drive wood, steel, or precast concrete piles.

Associated Pile and Fitting Corporation, Clifton, NJ

Figure 5-6. Pipe piles are driven into the ground using a pile driver or mobile crane. The pipe pile is filled with concrete after it has been driven.

umns, and the column load exceeds the bearing capacity of an individual pile.

Piles are constructed of wood, steel, or concrete. Wood piles are the oldest type and are used to support wharves, docks, and other structures built over water. Steel piles placed beneath buildings are most often *H-shaped* or *tubular*. Both types are driven into the ground with a pile driver. H-shaped piles are used as a foundation support and also in the construction of shoring around deep excavations. Tubular piles (pipe piles) are filled with concrete after they have been driven. See Figure 5-6.

Concrete piles are widely used beneath foundations of heavy concrete structures. They may be precast or cast-in-place. Precast piles are commonly fabricated in a plant and heavily reinforced with rebars or prestressed cables. Precast piles are delivered to the job site by truck and are driven into place with a pile driver.

The *pile head* is the upper surface of a precast pile in its final position. The large upper portion of the pile is the *butt*. The lower section of the pile is the *foot*, and the small lower end is the *tip*. The *driving head* is a metal device placed on top of the pile head to receive the pile driver's blows and protect it from damage. The *pile cutoff* is the portion of the pile head that is removed after the pile is in its desired position. A *pile shoe* is a metal cone placed over the tip of the pile to protect it from damage while the pile is being driven. The pile shoe also allows the pile to penetrate very hard materials in the ground. See Figure 5-7.

Grade beam forms are constructed over piles after they have been driven. The grade beams are secured to the piles by tying rebars placed in the grade beams to steel dowels projecting vertically from the piles. Holes for the steel dowels are drilled into the butt after the pile has been driven and cut off. The steel dowels are then inserted into the holes and grouted. The dowels often penetrate 4′ or more into the pile and extend the height of the grade beam above the pile.

Cast-in-place piles are placed using a *shell* or *shell-less* method. In the shell method, a metal casing is

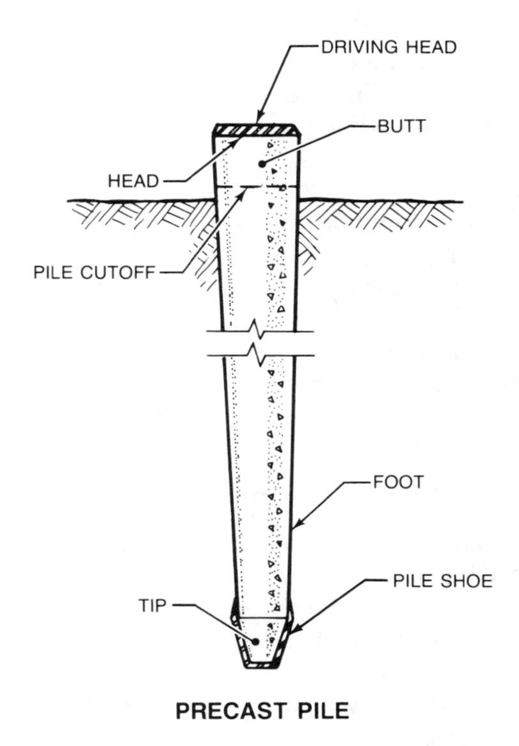

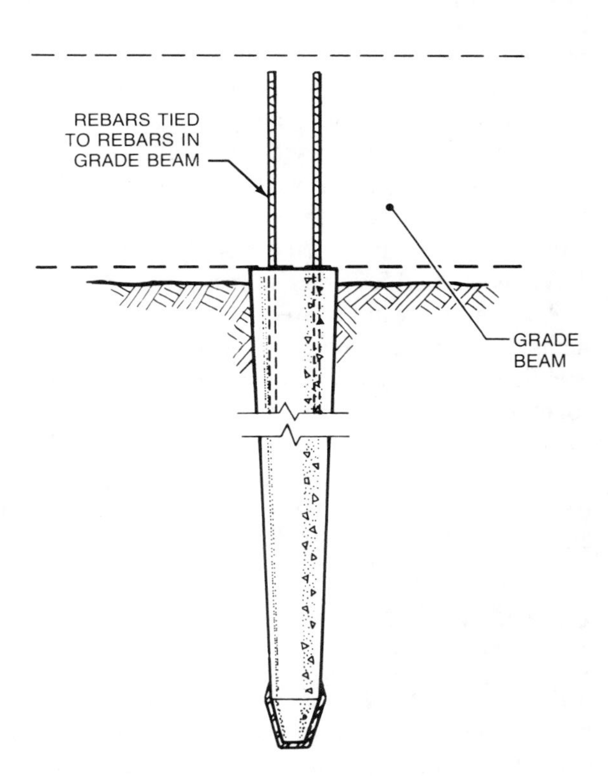

Figure 5-7. Precast piles are tied to grade beams with rebars after the piles are cut off.

driven into the ground and remains in position as the concrete is placed into the casing. In the shell-less method the metal casing is removed after being driven and concrete is placed into the bored hole.

Caissons

Foundation designs may require caissons instead of piles. Caissons are larger in diameter and extend to greater depths than piles. Caissons are used where the building design and/or soil conditions make pile driving difficult or inadequate. For example, buildings constructed with columns spaced a great distance apart provide greater floor space within the building. Columns, rather than walls, carry the main vertical loads transmitted to the foundation and soil below. The depth of soil penetration needed to support the column loads may not be possible using a grouped pile system. Caissons may be used because of their large surface area supporting the loads.

A caisson is constructed by placing a cylindrical or box-like metal casing into the ground and filling it with concrete. A large piece of equipment capable of drilling large diameter holes in the ground is used to excavate the hole. Equipment with a large auger or a reamer and bucket device bores holes up to 12′ in diameter and to depths of 150′. See Figure 5-8. After the holes are bored, a cylindrical metal casing is positioned in the hole. The casing is then filled

Calweld, Inc.

Figure 5-8. Caissons are larger in diameter and extend to greater depths than piles. Holes for the caissons are bored with special drilling rigs.

with concrete. Casings for deeper holes are constructed of sections that are added as the drilling proceeds.

A *belled caisson* features a greater bearing area at the base of the caisson. After the caisson hole has been bored to the desired depth, a belling tool is attached to the drilling head. The bottom of the hole is then dug out to the bell shape. See Figure 5-9.

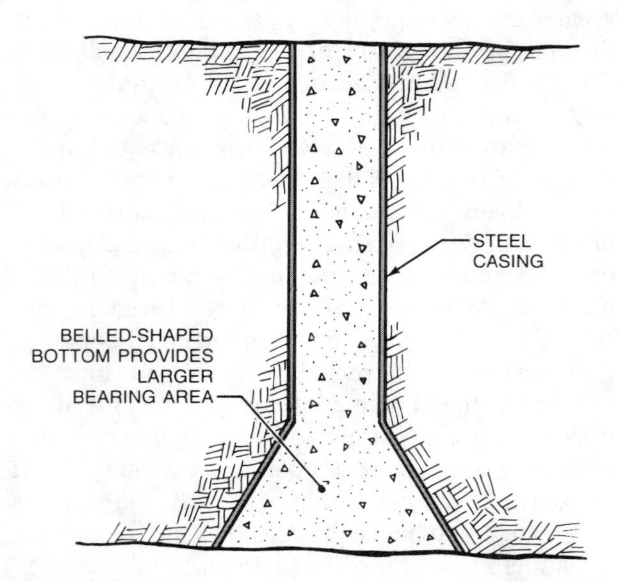

Figure 5-9. A belled caisson provides a larger bearing area at the base.

WALL FORMS

Foundation wall forms are built over footings or grade beams. Basic form designs and methods used in large concrete buildings are similar to foundation wall designs for residential and light commercial construction. However, heavy construction requires much larger forming units, such as *large panel forms* or *ganged panel forms,* and mechanical equipment to position the forms. Large panel forms are wall forms constructed in large prefabricated units. Ganged panel forms are wall forms constructed of many small panels bolted together. Materials commonly used for the wall forms are plywood panels with wood stiffeners, metal-framed plywood panels, and all-metal panels and frames.

Foundation wall forms and wall forms for floor levels above the foundation differ in the way the form bottoms are secured. The bottom of an outside foundation wall form is commonly fastened in position by securing a plate to the top of the footing. The bottom of an outside wall form placed at floor levels above the foundation is fastened with bolts or ties provided in the previous *lift*. A lift is the concrete placed between two horizontal construction joints.

Information related to the layout and design of the walls is contained in the floor plans and working drawings of the prints for a building. On large construction jobs, formwork detail drawings and schedules are developed from the working drawings to aid in form construction. Detail drawings of prefabricated forms are commonly furnished by the form manufacturer. See Figure 5-10.

Large Panel and Ganged Panel Forms

Large panel forms and ganged panel forms, also referred to as *climbing forms,* are most efficient in constructing high walls covering large areas. Cranes and other lifting equipment raise and position the forms. After the concrete has been placed and has gained sufficient strength, the forms are released and raised for the next lift.

Ganged panel forms are large panels constructed by fastening a series of smaller panels together. Steel-framed plywood panels or all-steel panels are used to construct the ganged panel forms that may range in sizes up to 30′ × 50′. The inner and outer form walls are held together with internal disconnecting ties that are easily unscrewed when the ganged panel form is ready to be released from the wall. Ganged panel forms can be taken apart at the end of a job and reassembled for other shapes and sizes of walls.

A layout drawing is used to identify the components of a ganged panel form. The individual panels are placed on sleepers laid in a flat area and bolted together according to the layout drawing. Walers are then attached to the panels, and strongbacks are fastened to the walers. Lifting brackets are bolted to the ganged panel form for the crane attachment. See Figure 5-11.

Curved Wall Forms

Curved wall forms may be required for corners and wall sections of buildings that have an arc or other circular elements. Curved forms for completely circular walls may be required for projects such as storage tanks or silos. Curved wall forms are built-in-place or constructed with sections that are fabricated on the job. The forms for complex circular structures are often custom built in fabricating

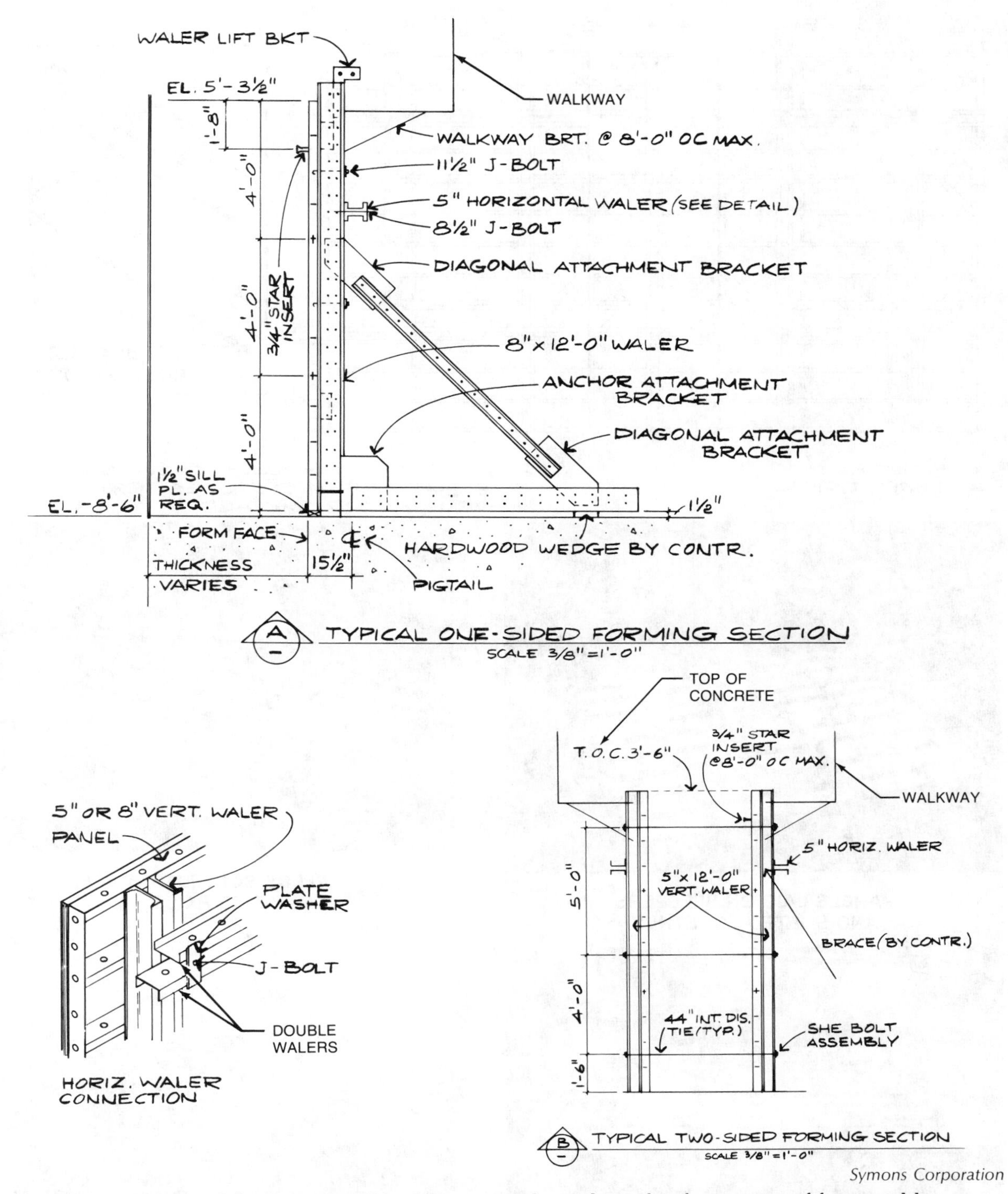

Figure 5-10. Formwork detail drawings provide information to the tradesworker for erection of forms and braces.

plants using all-wood, all-metal, or metal-framed plywood sections. Large prefabricated curved wall forms may be constructed of ganged panels. Curved walls may also be formed with all-steel panels consisting of a ⅛″ flexible skin and supported by 4″ wide vertical stiffeners. The panels are adjustable to form curved walls with a minimum radius of 5′-0″.

Plywood sheathing is commonly used to construct built-in-place curved wall forms. See Figure 5-12.

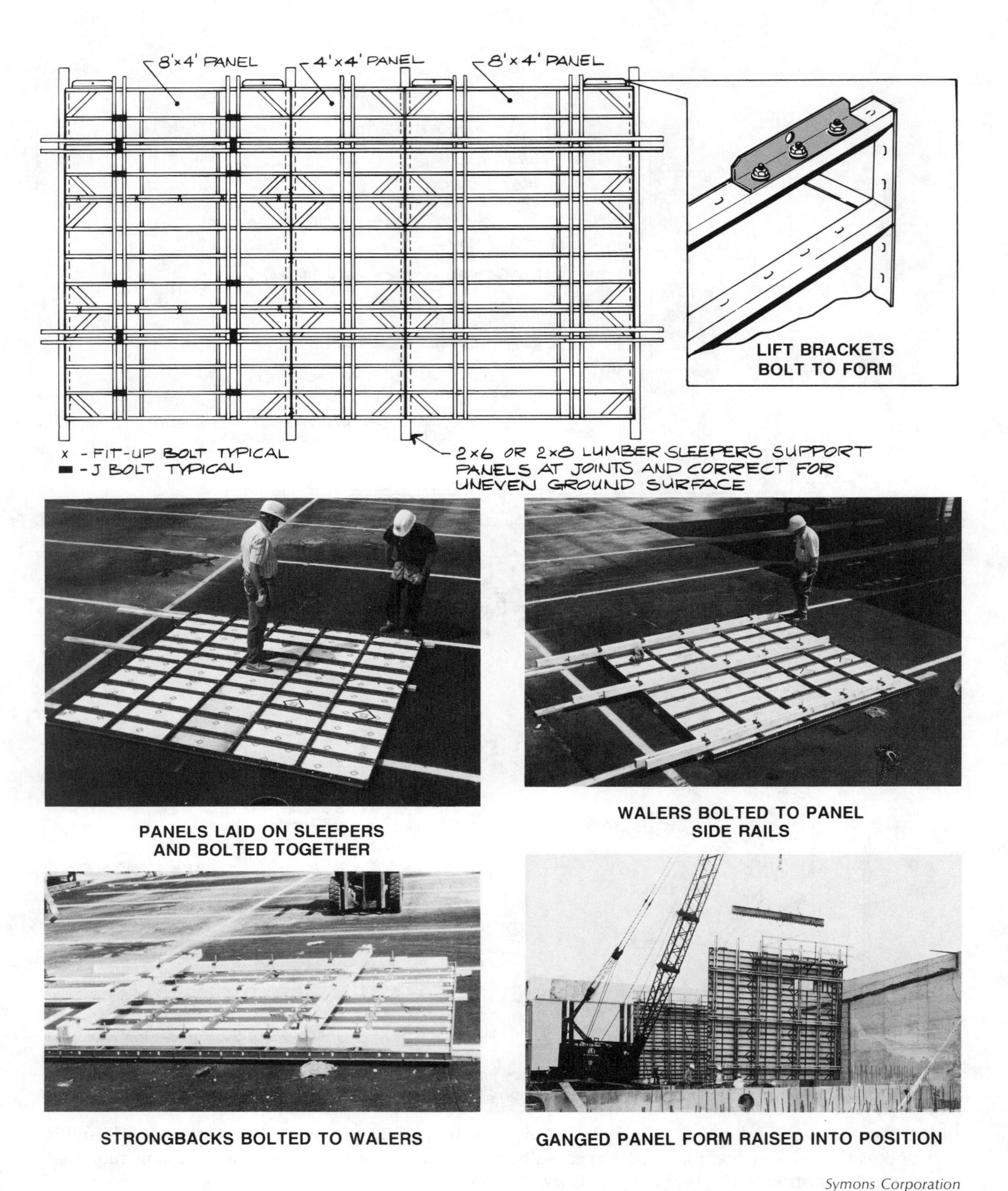

Symons Corporation

Figure 5-11. A ganged panel form is assembled from smaller panels and lifted in position by crane.

The thickness of plywood used for curved wall form sheathing depends on its bending capacity in relation to the radius of the curve. Short radius curve forms can be sheathed by using two or more layers of ¼″ plywood panels. Thicker panels may be kerfed partially through to increase its bending capacity. The kerfs should be close, evenly spaced cuts. Although a kerfed thicker panel may be used, thinner panels should be used because the kerfs weaken the thicker panel. Built-in-place curved wall forms can also be formed by using horizontal and vertical boards as form walls.

Laying out and establishing a line for a curved wall is similar to laying out a curved sidewalk. See Figure 5-13. Marks are made on top of the floor slab or footing and the wall lines are snapped. A common method for constructing curved wall forms uses 6″ wide top and bottom plates cut from ¾″ plywood. Studs spaced 8″ OC are nailed to the top

American Plywood Association

Figure 5-12. Plywood sheathing is used to construct curved wall forms.

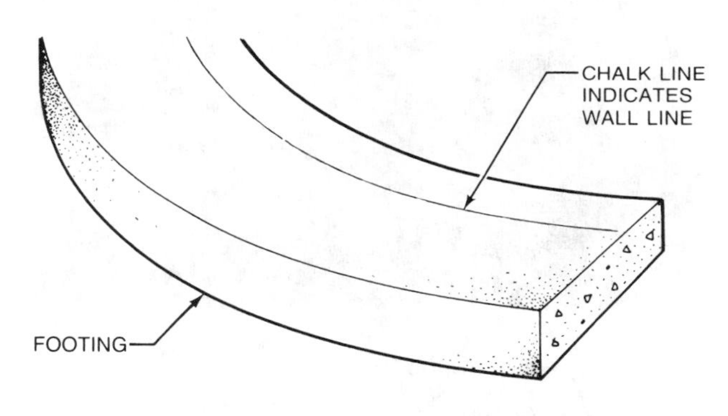

1. Establish the position of the wall on top of the footing and snap a chalk line.

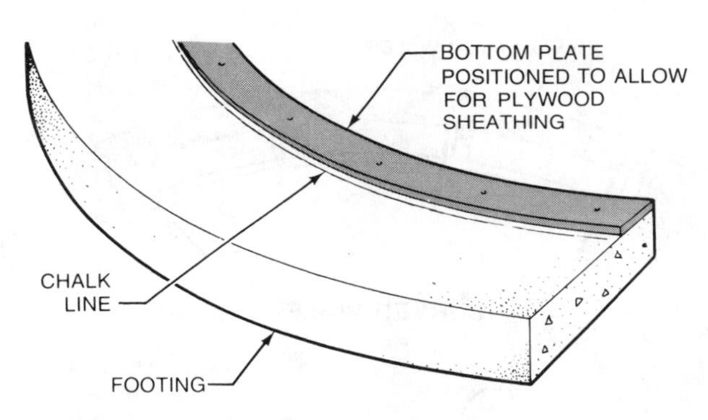

2. Position the bottom plate back from the line the thickness of the plywood sheathing. Nail the plate into position.

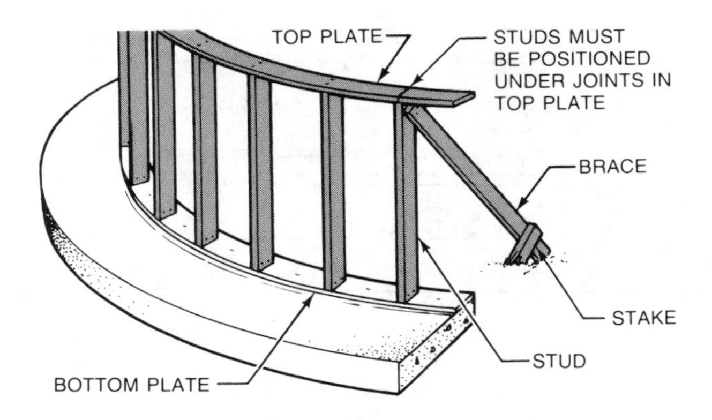

3. Toenail the studs to the bottom plate. Endnail the studs to the top plate. Plumb and brace the studs.

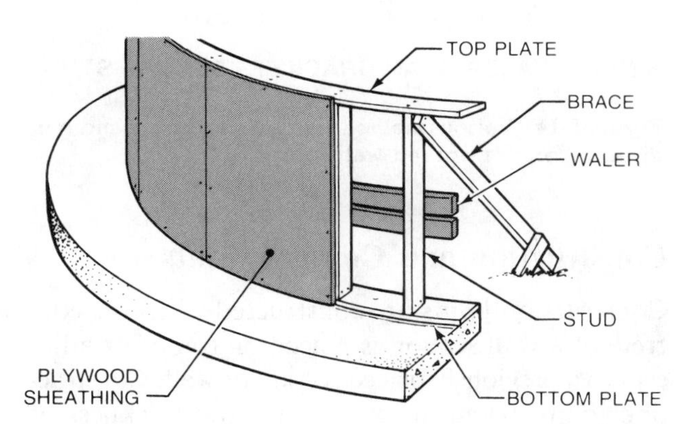

4. Nail walers and plywood sheathing to the studs.

Figure 5-13. Curved wall forms are constructed using plywood sheathing reinforced with studs and walers.

and bottom plates. Walers are fabricated with 1″ pieces of laminated plywood and placed flat against the studs. Ties are placed midway between the studs and should not be overtightened. Overtightening may cause distortion. Curved pieces ripped out of 2″ thick stock can also be used as walers. In this case, the ties should be placed next to the studs to produce a smooth curve. See Figure 5-14. When constructing curved wall forms, the inner wall form is constructed first to facilitate rebar placement.

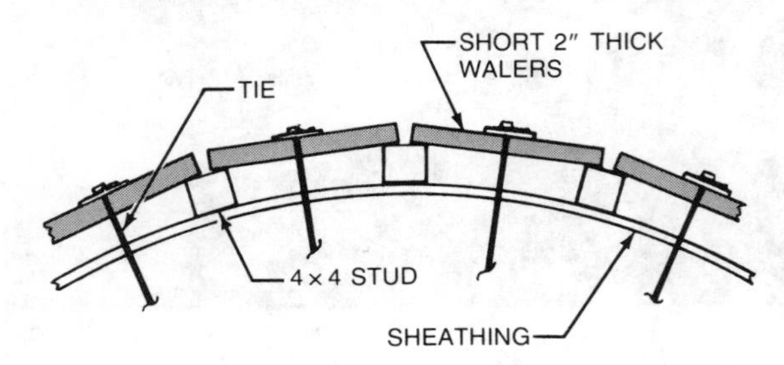

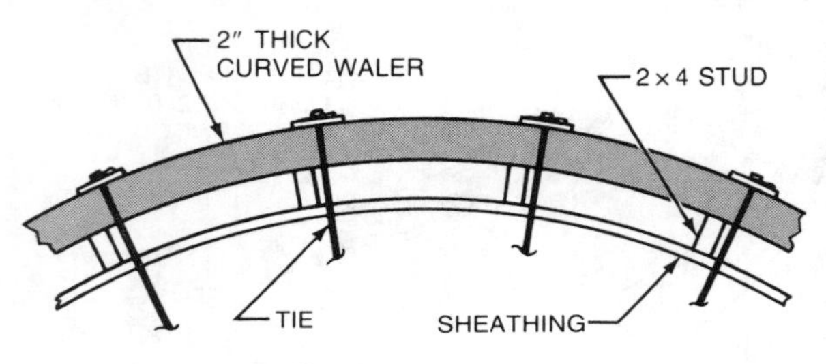

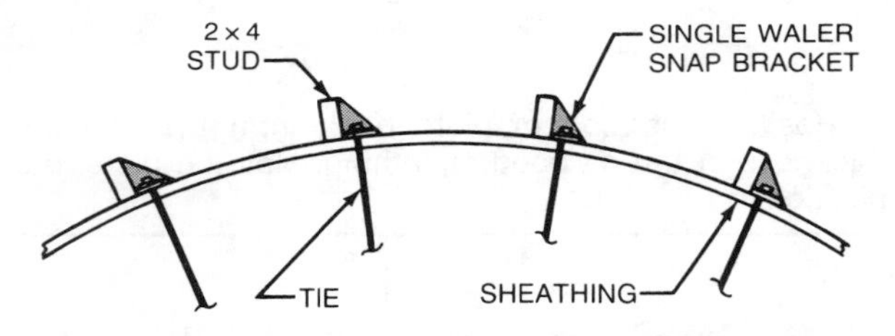

Figure 5-14. Various methods are used to secure and reinforce walers for curved wall forms.

Construction and Control Joints

Construction joints are constructed when the concrete of a wall section is placed on top of or adjacent to a previously placed section of wall. Construction joints extend through the entire thickness of a wall. Horizontal construction joints are constructed for walls placed in two or more lifts. Control joints are shallow grooves placed in the wall to control cracking that results from expansion and contraction in the set concrete.

Vertical Construction Joints. A vertical construction joint requires that a bulkhead be constructed inside the form at the end of the section of concrete being placed. Bulkheads are constructed between form walls with short boards nailed horizontally against vertical cleats. The horizontal boards are notched around rebars that extend past the bulkhead. See Figure 5-15.

Keyways may be required to prevent lateral movement between walls. Keyways are formed by attaching a tapered key strip to the bulkhead.

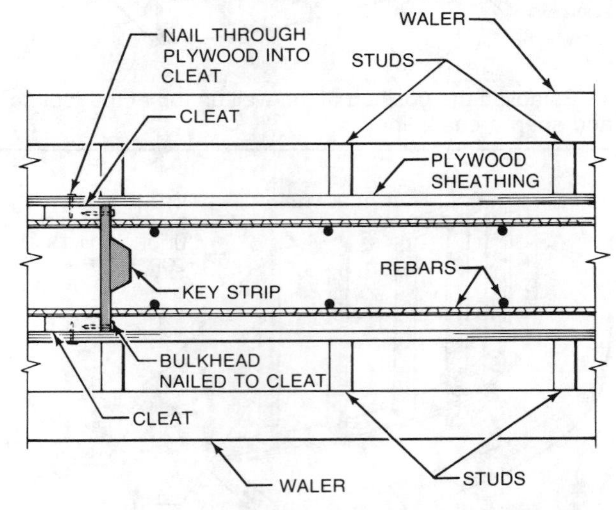

Figure 5-15. A vertical construction joint requires a bulkhead constructed of short boards.

A waterstop made of rubber, neoprene, polyvinyl chloride (PVC), or other plastic is installed to prevent water leakage at a vertical construction joint. Center-placed waterstops (waterstops placed at the center of the wall) are available in various designs, including single piece, split fin, labyrinth, and cellular. They are placed and attached to the bulkhead before the first placement of concrete. See Figure 5-16.

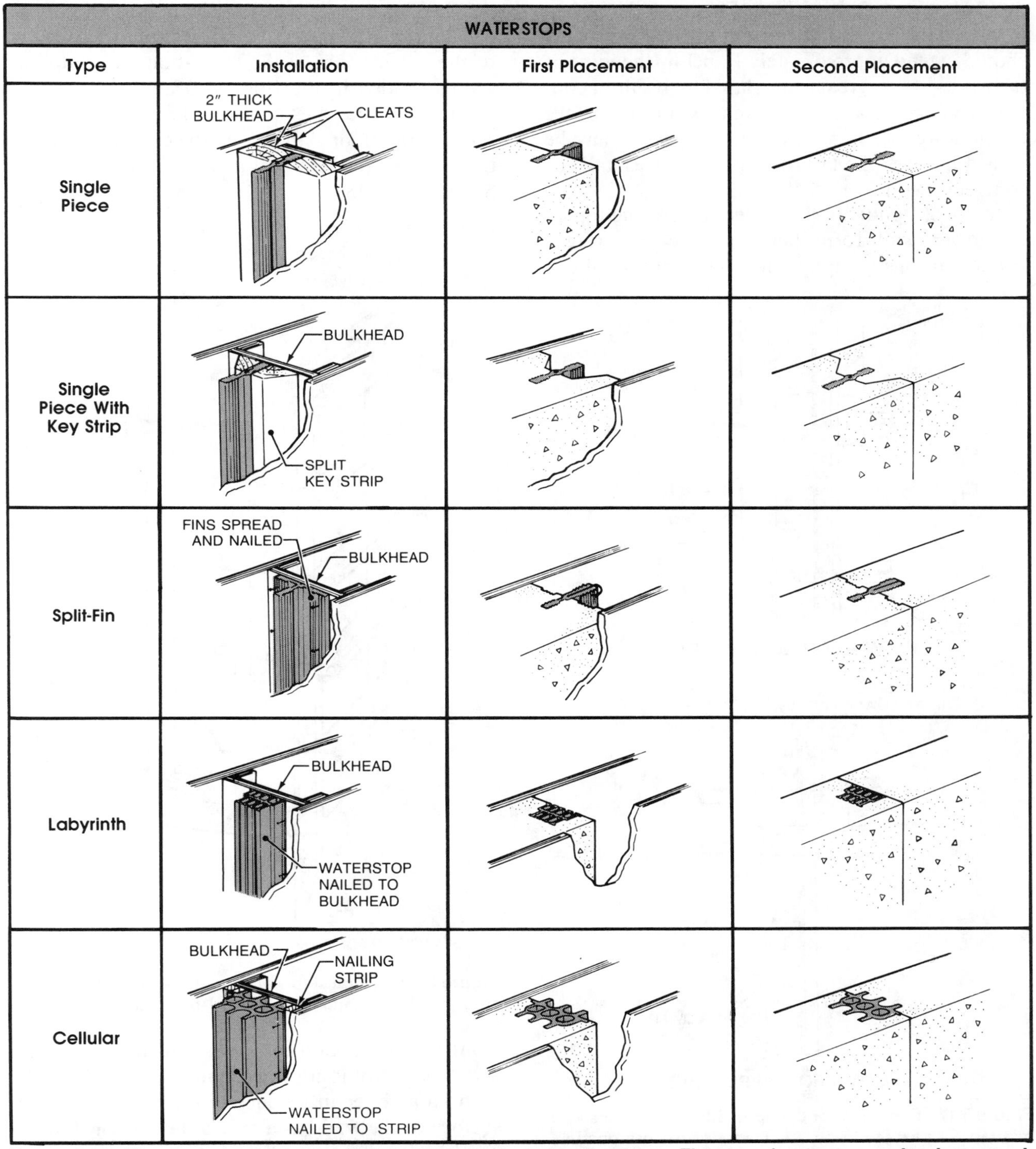

WATERSTOPS			
Type	**Installation**	**First Placement**	**Second Placement**
Single Piece	2" THICK BULKHEAD; CLEATS		
Single Piece With Key Strip	BULKHEAD; SPLIT KEY STRIP		
Split-Fin	FINS SPREAD AND NAILED; BULKHEAD		
Labyrinth	BULKHEAD; WATERSTOP NAILED TO BULKHEAD		
Cellular	BULKHEAD; NAILING STRIP; WATERSTOP NAILED TO STRIP		

Figure 5-16. Waterstops prevent water leakage in vertical construction joints. The type of waterstop used is determined by water pressure, wall thickness, and anticipated wall movement.

Horizontal Construction Joints. When constructing walls with climbing forms, the bottoms of the upper lift forms are secured toward the top of the lower lift of concrete. A row of tie rods or bolts are embedded 4″ below the top of the lower lift. The bottom row of ties for the upper lift should be placed 6″ above the joint. The bottom of the form panel should extend approximately 1″ below the joint of the lower lift. A greater overlap may result in concrete leakage because of unevenness in the surface of the lower wall. A compressible gasket may be placed beneath the lap to help prevent leakage of concrete. See Figure 5-17.

A 1 × 2 "pour" strip tacked toward the top of the lower wall form helps produce a straight horizontal construction joint. Wall forms should be filled until the concrete is slightly above the bottom of the strip. The strip is removed when the concrete sets. A horizontal construction joint can be made less apparent by creating horizontal grooves at the joint by nailing tapered pieces inside the wall form.

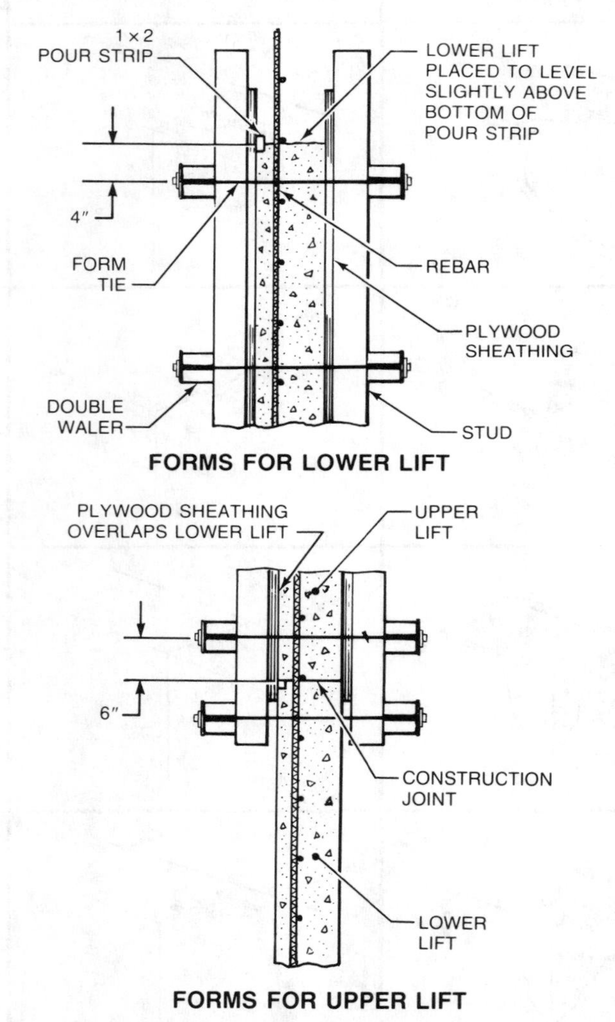

Figure 5-17. Formwork for the upper lift of a concrete wall overlaps 1″ past the lower lift. Tie rods are positioned 4″ from the top of the lower lift and 6″ from the bottom of the upper lift.

Where form panels are raised from floor to floor, a waler rod attached to a J-bolt secures the bottom of the outside form wall. The J-bolt and coupling are embedded in the floor slab and a waler rod secures the form panel in place. The inside form wall rests on the floor slab. When the concrete has set, the waler rod is removed and the hole is grouted. See Figure 5-18.

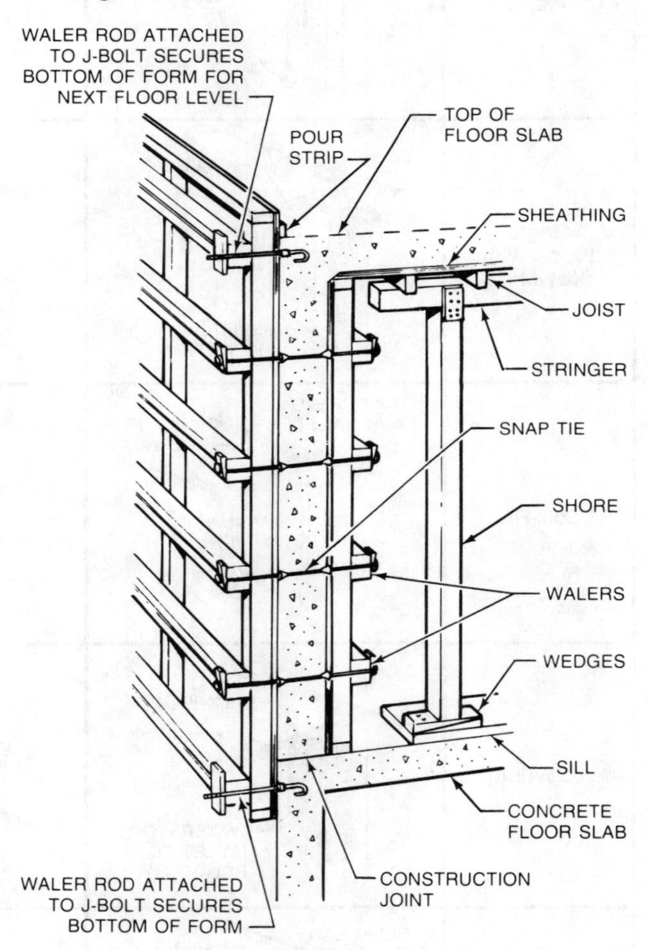

Figure 5-18. An outside wall form for a concrete wall at an upper level is attached using J-bolts and waler rods.

Control Joints. Control joints control cracking in a wall. Control joints are usually required in walls with an architectural concrete finish in which the surface of the wall has a special texture or design. The location and spacing of control joints are shown in working drawings and are commonly part of the decorative texture or pattern. Control joints are formed by attaching a beveled strip of wood, metal, rubber, or plastic to the sheathing of the wall form. See Figure 5-19. The strips are removed after the concrete has set and the forms have been stripped. Control joints are usually caulked after the strips have been removed.

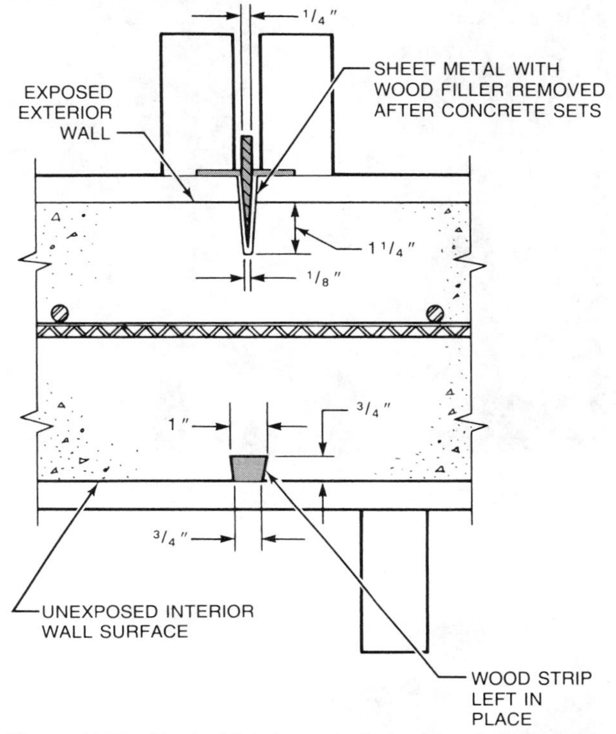

Figure 5-19. Control joints control cracking in a concrete wall. Control joints are formed with sheet metal, wood, rubber, or plastic strips.

COLUMNS, GIRDERS, BEAMS, AND FLOOR SLABS

Columns, girders, beams, and floor slabs are combined to form integral structural units of a concrete building. Columns support girders and beams that hold up the floor slabs. Girders are heavy horizontal members that support beams and other bending loads. Beams are horizontal members that support a bending load over a span, such as from column to column. See Figure 5-20. Girders and beams are also lateral ties between the outside walls of the building. The columns, girders, beams, and floor slabs must be tied together and reinforced with rebars.

Columns, girders, and beams provide intermediate support for the floor slab when the perimeter of the floor is tied into and supported by the outside walls. Columns and girders may be formed in the perimeter of the building with additional beams and girders providing interior floor support. The space between the girders and columns at the perimeter of the building are filled with panes of glass or *curtain walls*. A curtain wall is a light non-load-bearing section of wall made of metal or precast lightweight concrete that is attached to the exterior framework of a building.

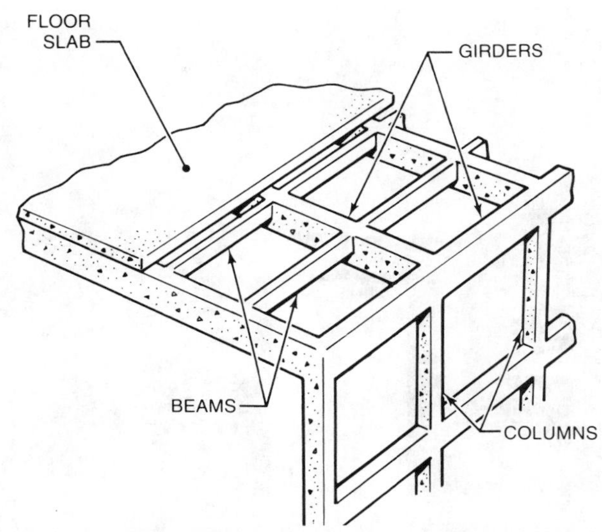

Figure 5-20. Columns, girders, beams, and floor slabs are combined to form integral structural units of a concrete building.

Flat slab and *flat plate* systems eliminate the use of beams and girders. In a flat slab system, the floor slab is directly supported by columns and *drop panels* (thickened area over a column). A *capital* (flared section at the top of a column) may also be used beneath the drop panel. In the flat plate system, the columns tie directly to the floor above without using drop panels or capitals.

One- and *two-way joist and floor slab systems* also eliminate beams and girders. The systems have thin slabs integrated with supporting girders, beams, and columns. One-way joist and floor slab systems (ribbed slabs) have cast-in-place joists running in one direction. Two-way joist and floor slab systems (waffle slabs) have joists running at right angles to each other. See Figure 5-21.

Column Forms

Column forms are subject to greater lateral pressure than wall forms because of the small cross-sectional area in relation to the height. Column forms require tight joints, adequate bracing, and strong anchorage at the base of the form. A cleanout opening at the

FLAT SLAB FLOOR WITH DROP PANEL AND CAPITAL OVER COLUMN

FLAT PLATE FLOOR WITH COLUMN

ONE-WAY JOIST SYSTEM

TWO-WAY JOIST SYSTEM

Portland Cement Association

Figure 5-21. Flat slab and flat plate construction methods eliminate the use of beams and girders. One- and two-way joist systems integrate floor slabs with supporting girders, beams, and columns.

base of the column form is used to remove debris before the concrete is placed. A compressed air hose is lowered into the form and the debris is blown out. In high column forms, a pocket or window is placed midway in the height of the form to place and consolidate the concrete in the bottom section of the form. The pocket or window is nailed shut when the concrete reaches the bottom of the pocket or window.

Most columns are square, rectangular, or round. L-shaped or oval columns are less frequently used. See Figure 5-22. Square and rectangular columns

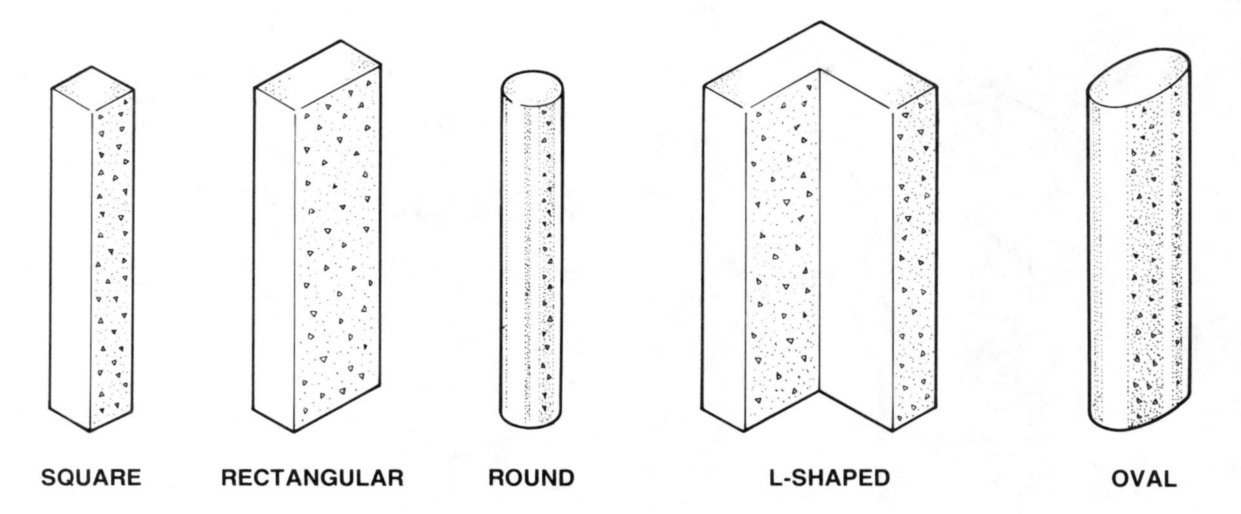

Figure 5-22. Typical column designs used in heavy construction include square, rectangular, round, L-shaped, and oval columns.

are usually constructed with plywood, prefabricated metal-framed plywood forms, or all-steel custom-made forms. Most round columns are constructed with tubular fiber forms or all-steel custom-made forms.

Square and Rectangular Column Forms. When plywood is used to construct square or rectangular column forms, two sides are cut to the width of the column, and the other two sides are cut to the width plus twice the thickness of the plywood. Light column forms up to 12″ square may be stiffened using battens and ties, or adjustable metal clamps placed directly against the plywood. Heavy column forms are stiffened with vertical 2 × 4s nailed to the plywood. Hinged adjustable metal scissor clamps are placed around the stiffeners and tightened with wedges. See Figure 5-23.

A cleanout opening is cut out of one side of the column form before the form is assembled. The plywood and stiffener attached to the cleanout door should be cut at a 45° angle. The cleanout door is replaced and clamped in position after the debris inside the column form has been removed. *Chamfer strips* (narrow strips of wood ripped at 45° angle) should be placed at all four corners of the form. The chamfer strips produce beveled corners for the finished concrete columns, making them less susceptible to chipping and other damage. When erecting a square or rectangular column form, the bottom is secured in position by a template aligned with center lines marked on the slab or footing. See Figure 5-24.

If the rebars for a column are in place, three sides of the form are nailed together and set in position. The fourth side is then nailed into place and the column form is braced. If a rebar cage is used the four sides of the column form are nailed together and the form is positioned. The rebar cage is then lowered into position with a crane. When columns are required in the floor above, the rebars should project above the form to later tie into the rebars placed in the column above.

Steel-Framed Plywood Column Forms. Steel-framed plywood column forms (steel-ply or hand-set column forms) are commonly used to form standard size columns ranging from 8″ × 8″ to 24″ × 24″ in 2″ increments. Columns with odd dimensions are formed by adding filler pieces to the column form. Steel-framed plywood column forms do not require clamps or additional stiffeners and are often designed with hinged corners that facilitate erecting and stripping. See Figure 5-25. Steel-framed plywood column forms are also used for large columns. They are set in place by crane and may require additional ties and stiffeners.

Round Column Forms. Tubular fiber forms are used to form standard size round columns ranging from 6″ to 48″ in diameter and up to 18′ in length. Longer lengths are also available. Tubular fiber

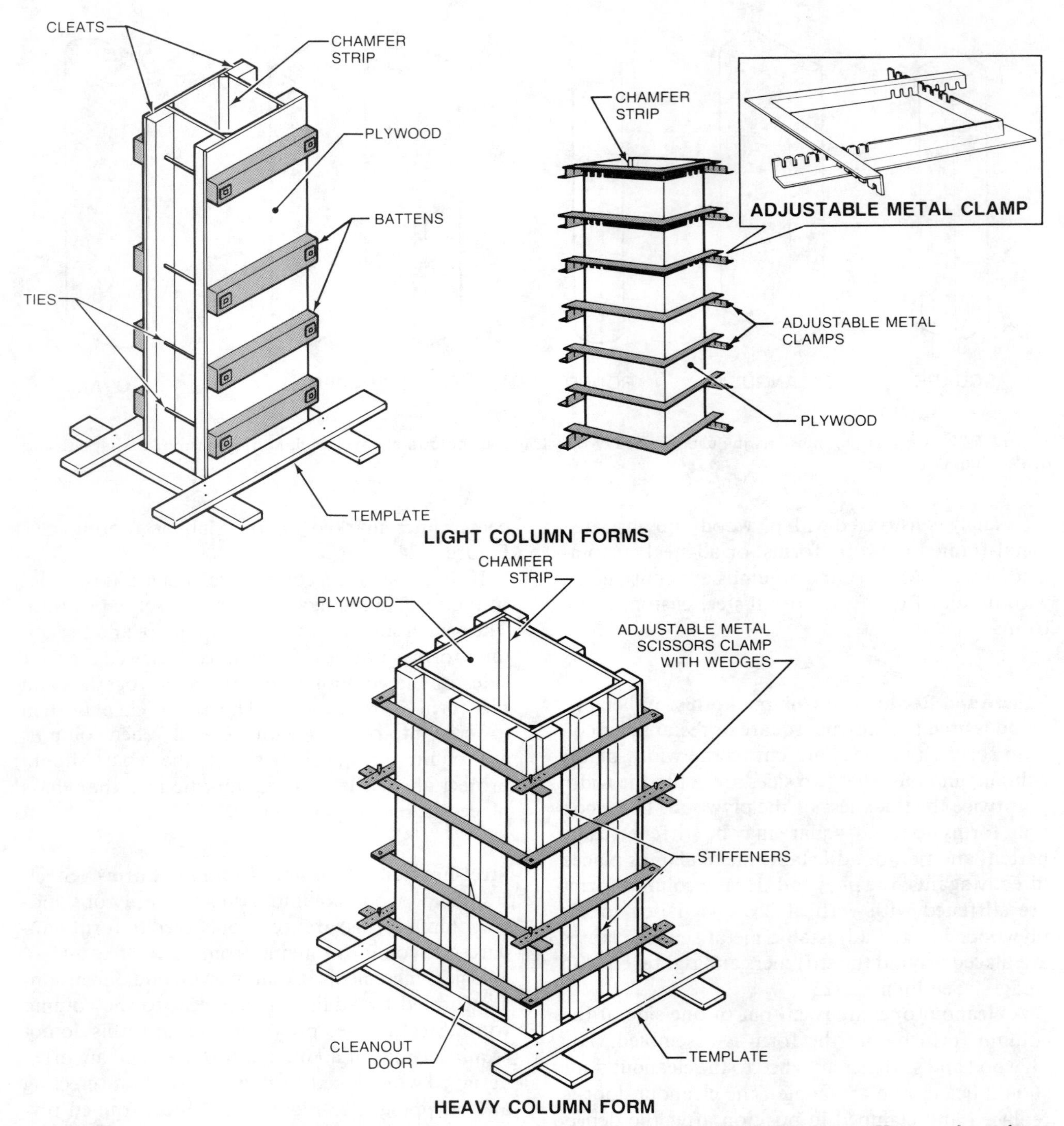

Figure 5-23. Light column forms are stiffened using battens and ties, or adjustable metal clamps. Heavy column forms are stiffened with vertical 2 x 4s and adjustable metal scissors clamps.

forms are made of spirally constructed fiber plies and are available with wax-impregnated inner and outer surfaces for weather and moisture protection.

Tubular fiber forms are positioned after the rebar cage is in place. See Figure 5-26. Small column forms can be lowered by hand or with a block and tackle. Large column forms may require a crane to position the form. The interior surface of the form should not be damaged when lowering the form. The base of the form is secured with a wood template and the top is plumbed and secured with braces nailed to a wood collar.

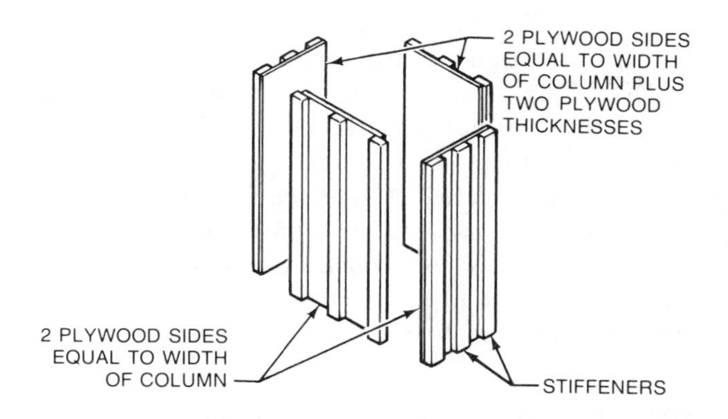

1. Cut four pieces of plywood to dimension. Nail 2 x 4 stiffeners to the plywood.

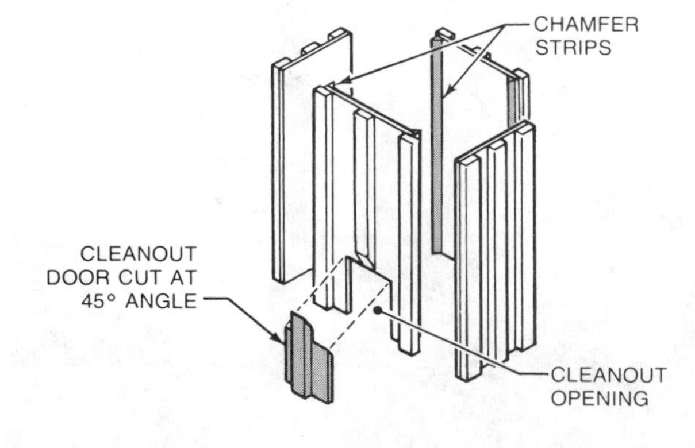

2. Nail chamfer strips to two sides. Cut a cleanout door in one of the sides.

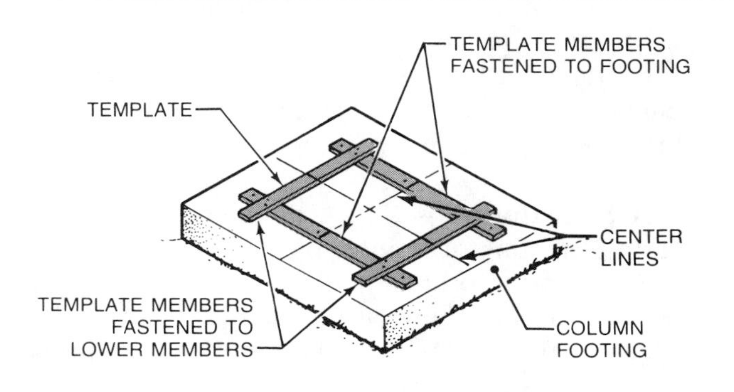

3. Establish center lines on the column footing. Fasten a template down to hold the column form bottom.

COLUMN FORM
CHAMFER STRIP
BRACE
COLUMN CLAMPS
TEMPLATE
CLEANOUT DOOR REPLACED IN OPENING

4. Nail column form together and position in template. Clean debris from form and replace door. Tie column forms together with column clamps. Plumb and brace form.

Figure 5-24. Square and rectangular column forms are constructed with plywood and 2 x4 stiffeners. A cleanout door is provided for access to debris on the inside of the form.

Symons Corporation

Figure 5-25. Standard size columns are often constructed with prefabricated steel-framed plywood panels.

The Burke Company

Figure 5-26. Tubular fiber forms are used to form round columns. The tubular fiber forms are positioned after the rebar cage has been set in place.

Tubular fiber forms can also be used to construct oval-shaped columns. A rectangular column form is constructed and a tubular fiber form that is cut in half lengthwise is inserted at both ends. The edges of the tubular fiber form and rectangular form should be flush to ensure a smooth transition. See Figure 5-27.

Tubular fiber forms are notched with a saw for beam openings or utilities, such as light switches and electrical outlets. Wooden blocks are nailed to the inside of a tubular fiber form to form slots for beams or joists if required.

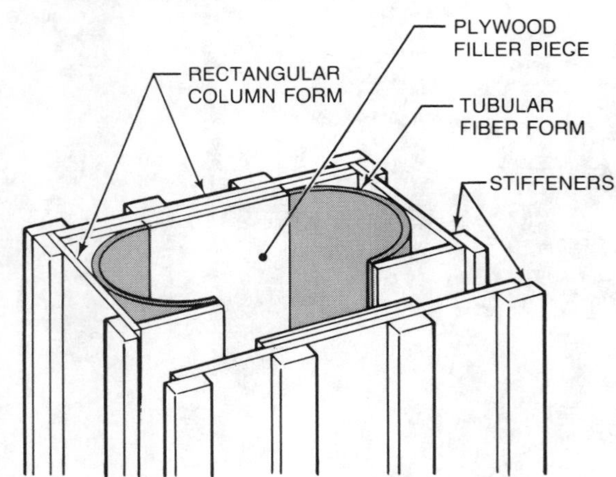

Figure 5-27. Oval column forms are constructed with tubular fiber forms combined with a rectangular column form.

A portable electric handsaw or knife may be used to strip the tubular fiber forms after the concrete has set. Care must be taken not to mar the concrete column. A portable electric handsaw is adjusted to the thickness of the form and two vertical cuts are made the complete length of the form on opposite sides. The two sections of tubular fiber form are then removed. When using a sharp knife, a 12″ slit is made and broad-bladed tool is then used to pry the tubular fiber form.

Round Steel and Fiberglass Column Forms. Round steel and fiberglass forms are used to construct large columns for heavy construction projects. Round steel column forms are available in diameters ranging from 14″ to 10′. The sections that make up the form range in lengths from 1′ to 10′. Bracing is built into the round steel column forms so additional bracing is not required except for plumbing the form.

A Springform® is a type of fiberglass column form consisting of a single piece of molded fiberglass. A Springform® is pulled apart and placed in position around previously installed rebars. The edges are then secured with bolts at closure flanges reinforced with a predrilled steel bar. The form is plumbed and secured with braces tied to a steel bracing collar. Springforms® provide a smooth architectural finish and can be easily stripped. Springforms® can be conveniently combined with a two-piece capital form to construct a column with a capital. See Figure 5-28.

Beam and Girder Forms

Beam and girder forms are constructed after the column forms are positioned and braced. In general, the concrete for beams, girders, and columns is placed monolithically at each floor level. Therefore, beam, girder, and column forms must be framed and tied to each other.

Two methods are commonly used to frame beam and girder forms to column forms. In one method, the beam or girder forms rest on or butt against the top of the column form. In the second method, the beam or girder forms frame into a pocket in the side of the column form. See Figure 5-29.

A bottom and two sides are the main components of beam and girder forms. Chamfer strips are used to produce beveled edges where the sides and bot-

OPENING SPRINGFORM® PRIOR TO PLACEMENT

RAISING SPRINGFORM® INTO POSITION

STRIPPING SPRINGFORM®

SPRINGFORM® WITH CAPITAL FORM

Symons Corporation

Figure 5-28. A Springform® is a molded one-piece fiberglass column form. A Springform® produces a smooth finish with one vertical seam.

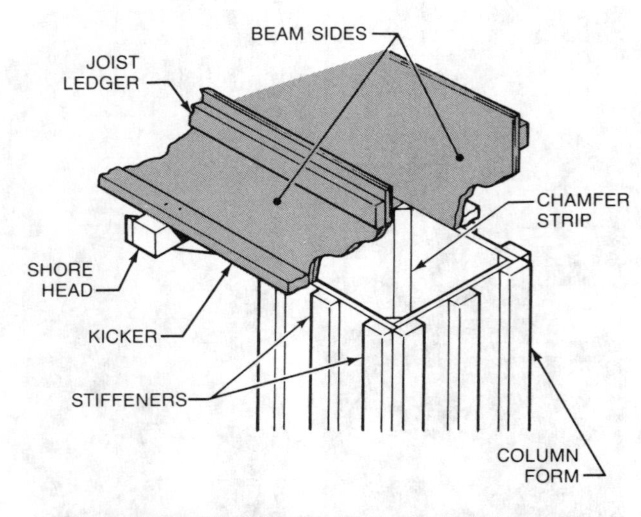

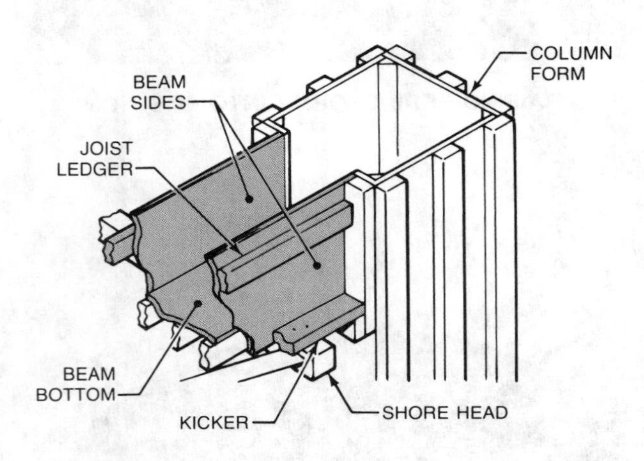

Figure 5-29. Beam and girder forms rest on or butt against a column form, or frame into a pocket cut into the side of the column form.

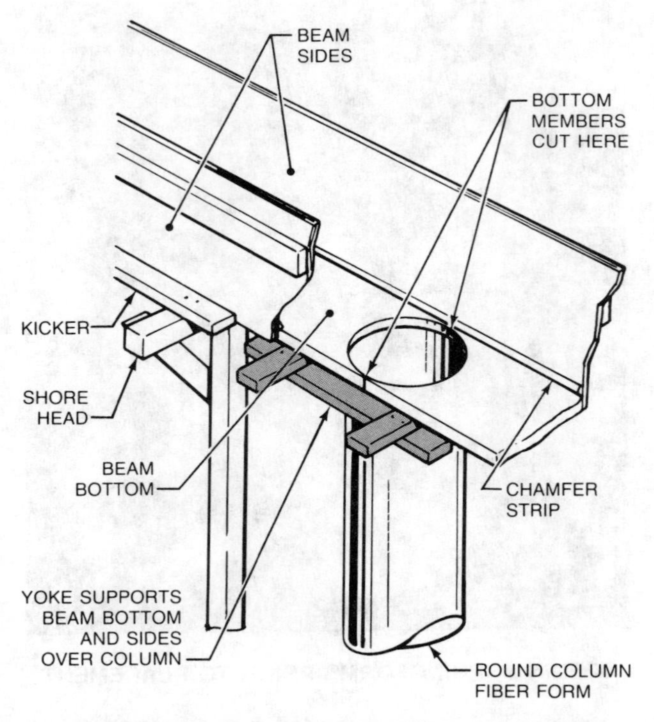

Figure 5-30. A yoke supports the beam sides and bottom over a round column fiber form.

tom join. Beam and girder forms are supported by T-head shores, double-post shores, or scaffold shoring placed beneath the bottom of the form.

A beam or girder form framing into a tubular fiber form is supported by a yoke or collar placed at the top of the column form. A half circle with a diameter equal to the column must be cut at the ends of the bottom piece to facilitate construction. See Figure 5-30.

Beam and Girder Bottoms. Beam and girder bottoms are constructed of 2″ planks or plywood stiffened with 2 × 4s. The width of the bottom piece should be the same as the width of the finished soffit. The bottom piece should be long enough to butt against the column form or rest on top of the column form. If the beam or girder forms are designed to be stripped before the column form is stripped, the bottom piece should be butted against the column form. If the ends rest on the top of the column form, a 45° bevel is cut to provide a chamfer where the finished beam or girder meets the column. Joist ledgers are nailed against the columns and beneath the ends of the beam bottoms for additional support. See Figure 5-31.

Beam and Girder Sides. Plywood is commonly used to form beam and girder sides. The bottoms of the beam sides may be nailed against or set on top of the bottom piece. In either case, a kicker is nailed against the bottom of the sides to secure it in position. A joist ledger is nailed against the column form and below the ends of the beam bottoms for additional support. Blocking between the joist ledger and kicker may also be added to stiffen the sides and support the ledger.

Studs, bracing, walers, and ties are used to construct and reinforce the sides for large beams and

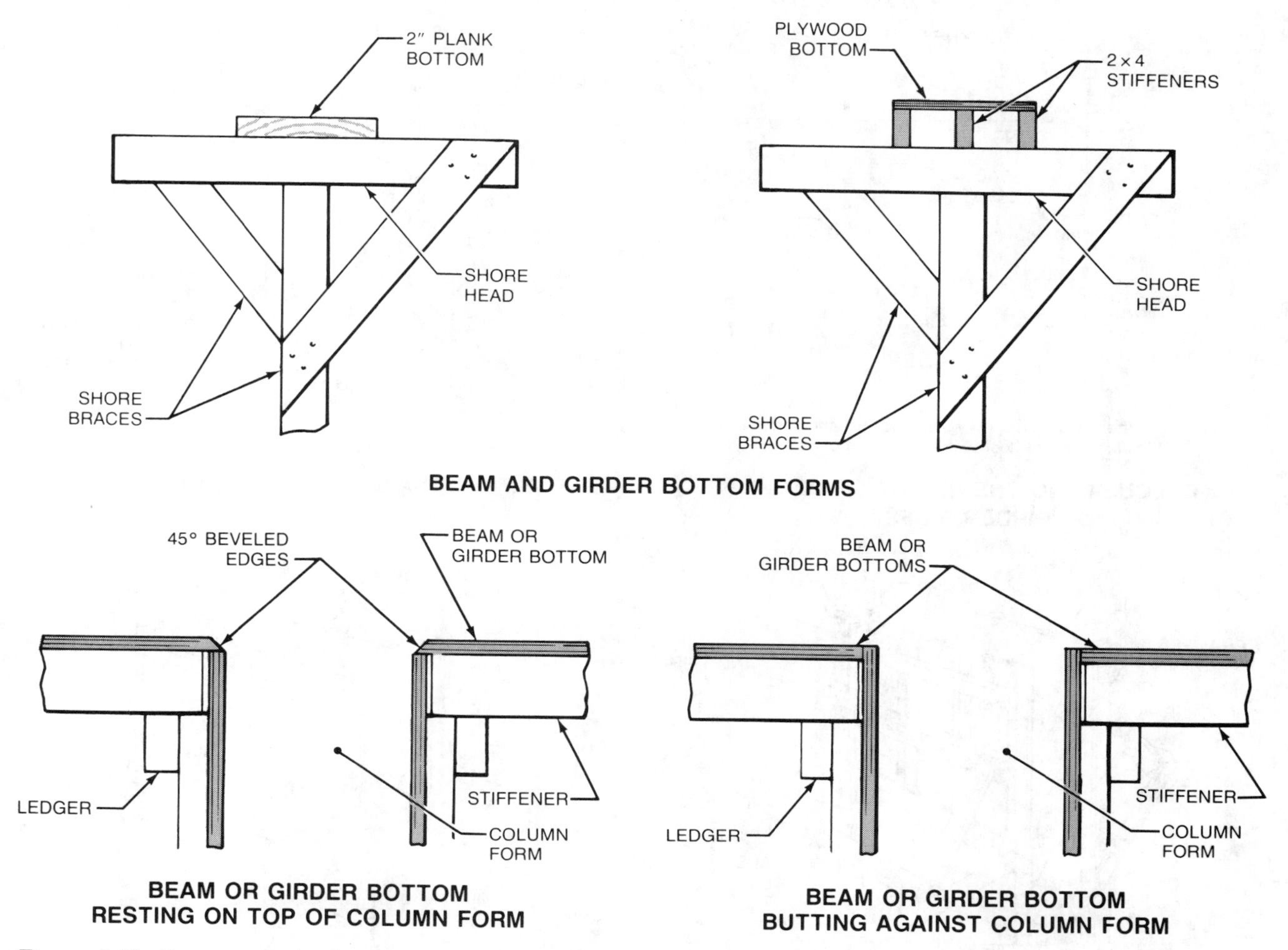

Figure 5-31. Beam or girder bottoms are formed with 2″ thick planks or plywood reinforced with 2 ×4s. The bottom rests on a column form or butts against it.

girders. The height of the beam or girder sides is determined by the form framing method used. When the sides are nailed against the beam bottom and the slab sheathing rests on top of the beam side, the total height of the sides is equal to the height of the beam plus the thickness of the beam bottom and the width of the stiffeners, minus the concrete slab and slab sheathing thickness. The length of the beam or girder sides depends on whether the beam sides butt against or pass beyond the sheathing of the column. See Figure 5-32.

Constructing Beam and Girder Forms. Shores are braced horizontally between two columns or between a wall and column to support beam and girder forms. The beam and girder forms may be prefabricated on the ground and lifted in place or constructed on top of shores. When constructing the forms on shores, the bottom is positioned on the shores, and the sides are then attached. Studs are nailed to the form sides and a joist ledger is nailed to the studs. The sides are braced between the studs and the shore heads. See Figure 5-33. After the slab forms have been constructed, the beam or girder forms are adjusted to their correct height by raising or lowering the shores with wedges or other adjusting devices.

Framing Beam Forms to Girder Forms. Beam forms are framed to girder forms by cutting a beam pocket in the side of the girder form. If the sides and bottom of the beam form butt against the girder form, the pocket is cut to the size of the finished beam. If the sides and bottom of the beam form extend past the sheathing of the girder form, the pocket is cut to accommodate the thickness of the beam bottom and sides, plus a small allowance for

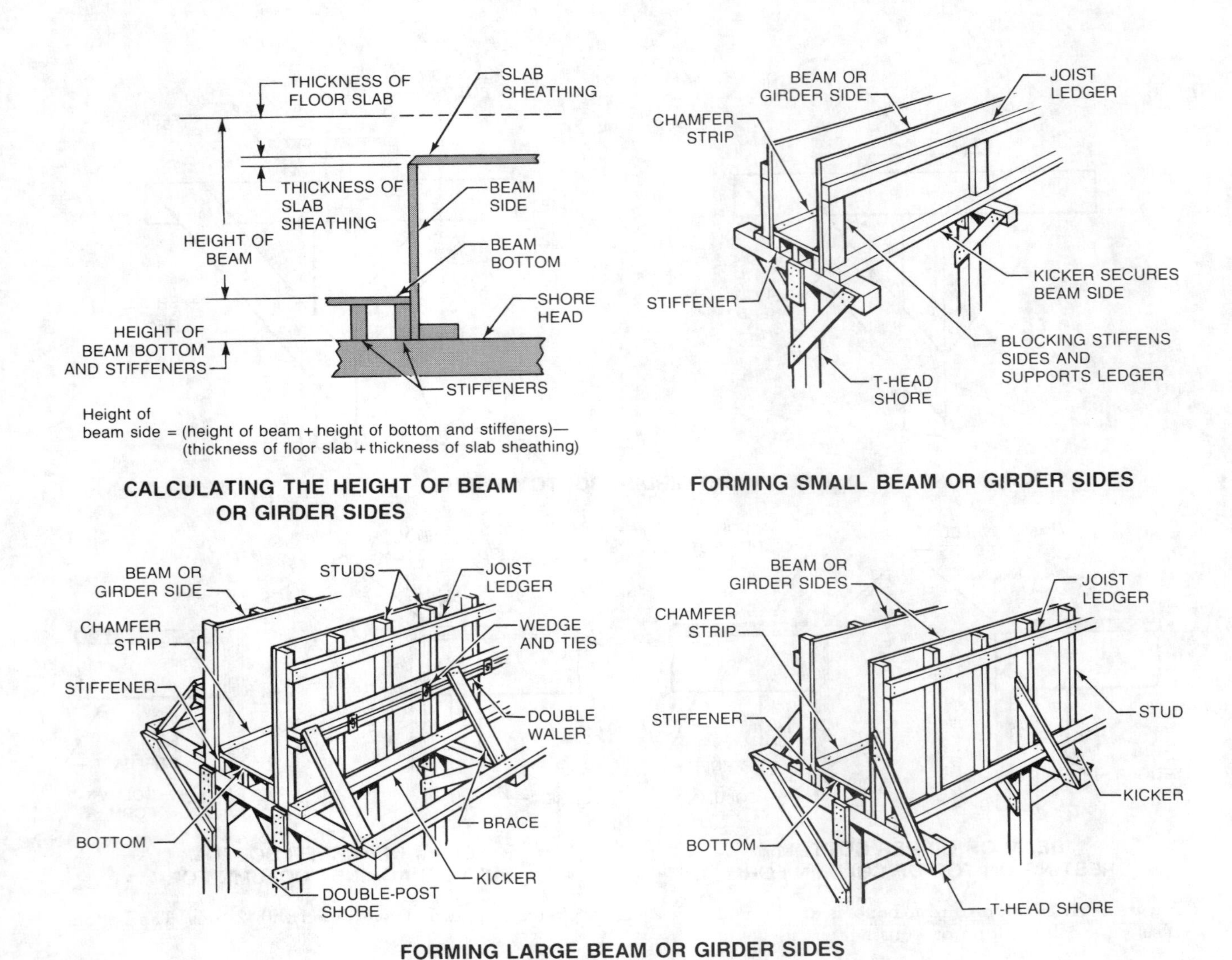

Figure 5-32. Beam and girder sides are reinforced with blocking, kickers, studs, walers, ties, and braces.

easy fitting. The opening should be reinforced with cleats and a beam ledger. See Figure 5-34.

Spandrel Beams. Spandrel beams are located in the outer walls of a building and tie into the floor slab above. Spandrel beam forms consist of a bottom piece and two sides supported on extended shore heads that are supported with double posts. If a walkway is required, the shore head should extend beyond the outside wall. Knee braces extend from the shore head to the outside wall. Kickers are nailed against the bottoms of the beam sides. A joist ledger is nailed toward the top of the inside wall to support the joists of the slab form.

Spandrel beams often require intermediate ties and walers because of their depth. A *spandrel tie* is driven into the deck sheathing at the top of the form. A spandrel tie is a type of snap tie with a hooked end. See Figure 5-35.

Suspended Forms. Suspended forms are used to form concrete slabs that are supported by steel beams or girders. Suspended forms are secured with U-shaped snap ties or coil hangers that slip over a steel beam or girder, eliminating the need for shores to support the forms. If the steel beam or girder is to be encased in concrete for fireproofing, the hangers support formwork consisting of a bottom member and two sides. The ends of the hangers extend through short walers placed beneath the form-

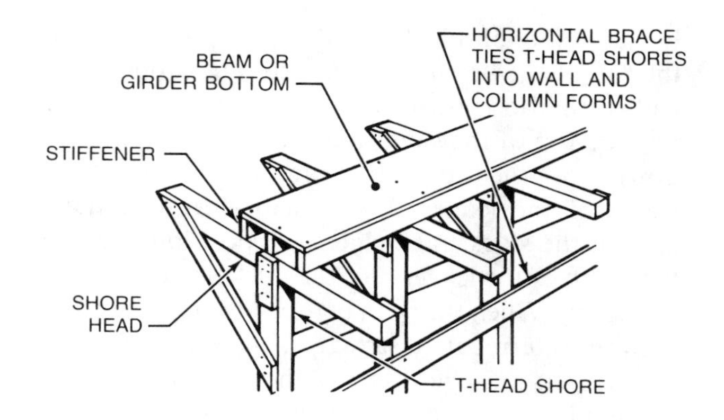

1. Position T-head shore and nail a horizontal brace across them. Nail the beam or girder bottom to the top of the shore head.

2. Nail the form sides to the bottom. Nail ties across the top of the form sides. Nail studs to the form sides. Nail chamfer strips to the beam or girder bottom.

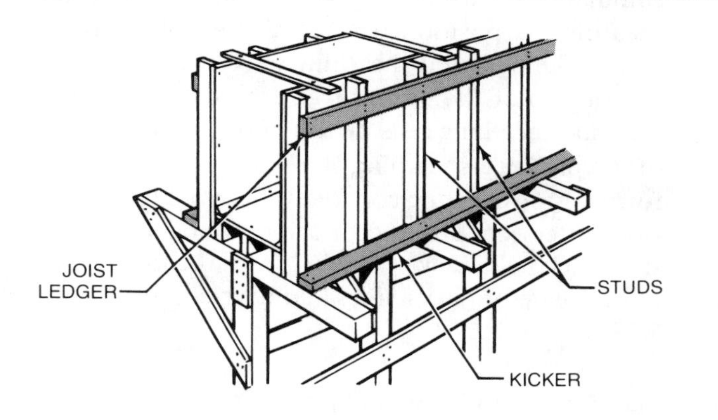

3. Nail a kicker and joist ledger against the studs.

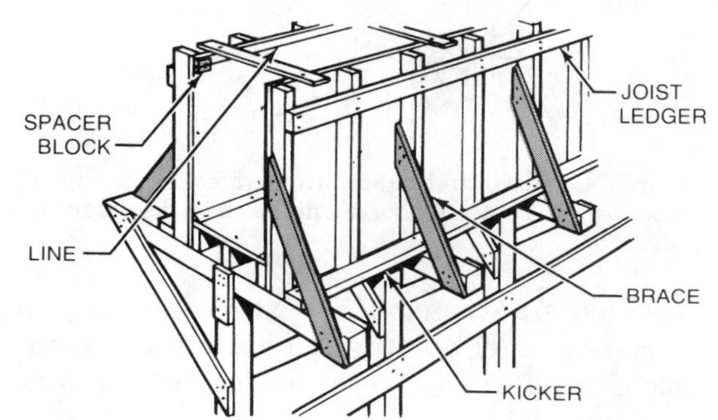

4. Set up a line held away from the form sides with a spacer block. Align and brace the sides.

Figure 5-33. Beam and girder forms are constructed on well-braced T-head shores. The sides of the beam or girder forms are braced to resist lateral pressure.

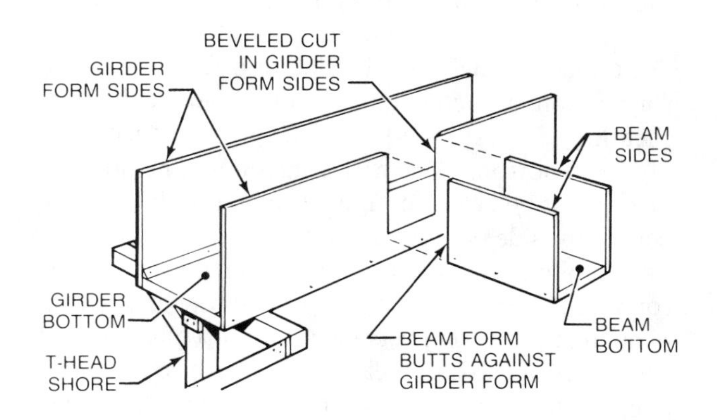

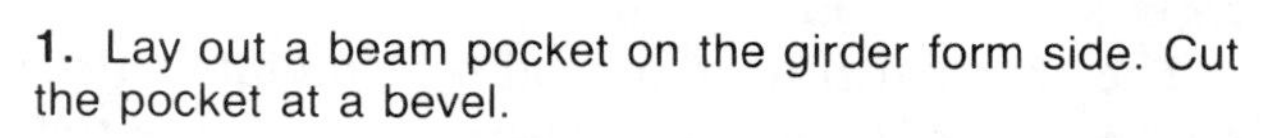

1. Lay out a beam pocket on the girder form side. Cut the pocket at a bevel.

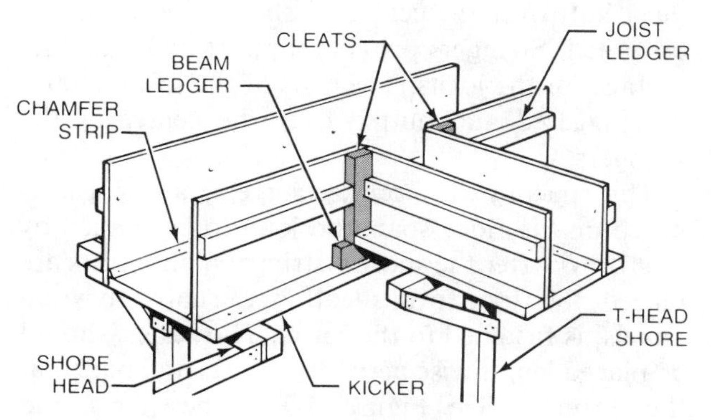

2. Nail a kicker across the shore heads. Frame the beam pocket with cleats and a beam ledger. Shore the beam form and nail into position.

Figure 5-34. Beam forms are framed to girder forms with cleats and a beam ledger.

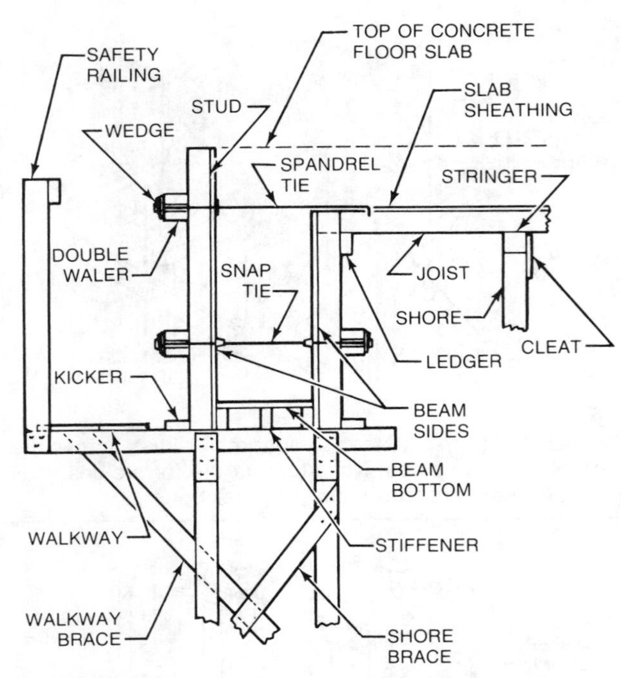

Figure 5-35. Spandrel beams are located in the outside walls of a concrete structure and tie into the floor slab above.

work and are secured with wedges or bolts. The hangers for steel beams or girders not encased in concrete secure the joists supporting the slab form. See Figure 5-36.

Floor Form Construction

Floor forms are constructed after the column, beam, and girder formwork has been completed. Shores are used to support stringers. The ends of stringers must butt over the center of shores and each end toenailed. Stringers provide support and a nailing surface for the joists. The ends of the joists should be staggered and butted over the center of the stringers.

The spacing of shores, stringers, and joists is based on the floor span and load to be carried by the form. After the shores, stringers, and joists are placed, the floor form sheathing, usually plywood panels, is fastened to the joists. The panels should be placed lengthwise across the joists and nailed at the corners. See Figure 5-37. The joists and sheathing can also be prefabricated in large panels and hoisted and set in place over the stringers.

Beam, Girder, and Slab Floor Systems. Concrete for beam, girder, and slab floor systems is commonly placed monolithically. Ledgers are nailed toward the top of the beam or girder sides to support one end of the floor form joists. If the slab sheathing rests on top of the sides, the ledgers are positioned down from the top of the beam or girder sides the width of the ledger. If the slab sheathing butts against the sides, the ledgers are positioned down the width of the ledger plus the thickness of the slab sheathing. When the span between beams or girders requires intermediate stringers, the top surface of the stringers must be level with the top of the ledgers. To simplify the stripping operation only a few nails are used to secure the joists.

Block bridging is placed between the joists to prevent the joists from tipping. Block bridging may be eliminated by using 4 × 4s for joists. Plywood sheathing is positioned after all the joists are in place. The plywood sheathing may butt against or rest on top of the beam or girder sides. When the plywood sheathing rests on top of the sides, the edge of the plywood sheathing should not be allowed to form a groove in the concrete where the beam and floor intersect. The end of the plywood sheathing should be held back ¼″ to ½″ from the inside face of the beam or girder sides and beveled. See Figure 5-38.

Flat Plate and Flat Slab Floors. Floor joists for flat plate and flat slab floor systems are supported entirely by stringers and shores. The formwork for flat plate and flat slab floors is similar to beam and slab formwork. Slab sheathing is cut to the shape of the column form, beveled, and held back ¼″ to ½″ from the inside face of the column form.

The column forms and flat plate floor forms are tied to one another. For flat slab floors, a drop panel form is constructed over the column before the slab sheathing is positioned. The drop panel form is a plywood bottom piece with four sides. The bottom piece must be large enough to accommodate cleats behind the side pieces. An opening the size and shape of the column is cut out of the bottom piece to allow for rebars extending from the column. The drop panel form rests on joists placed on top of stringers. The edges of the slab sheathing are beveled and held back slightly when placed on top of the drop panel form sides. See Figure 5-39.

The drop panel form may also be prefabricated on the floor or ground and positioned on top of the joists. The slab sheathing may butt against or rest on top of the drop panel form sides.

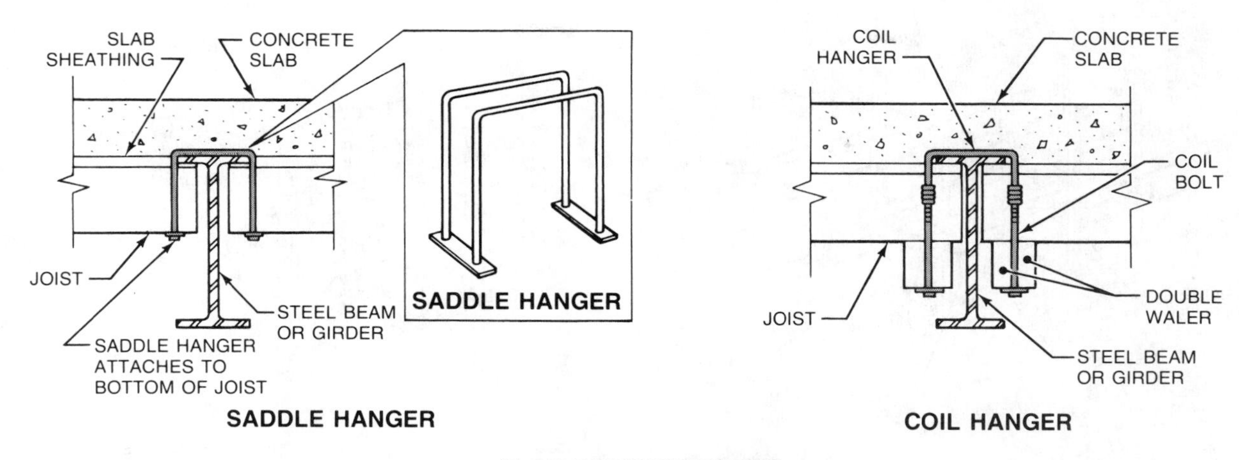

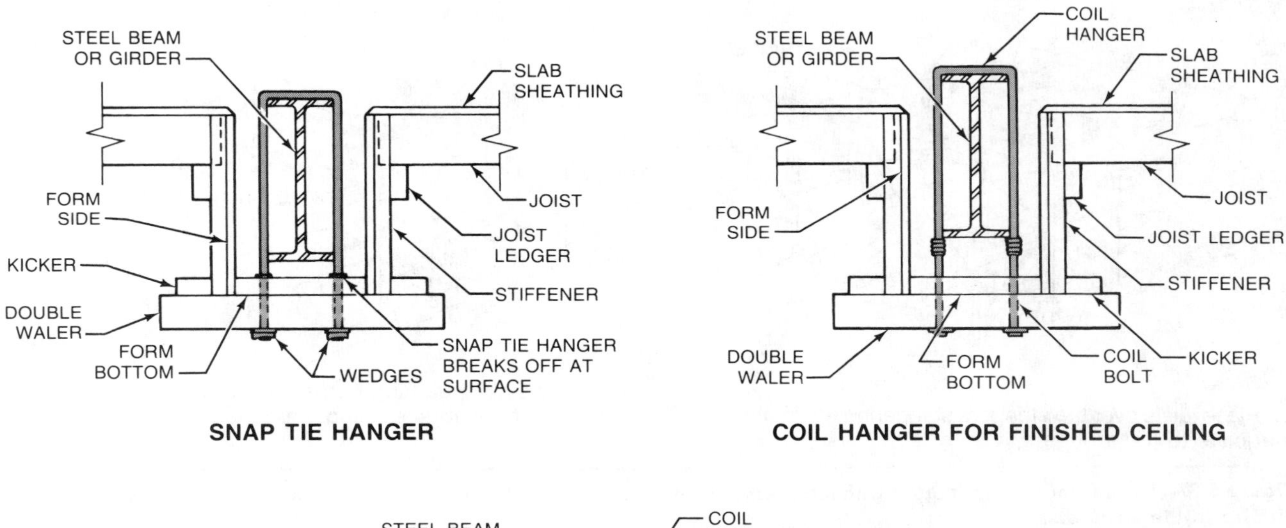

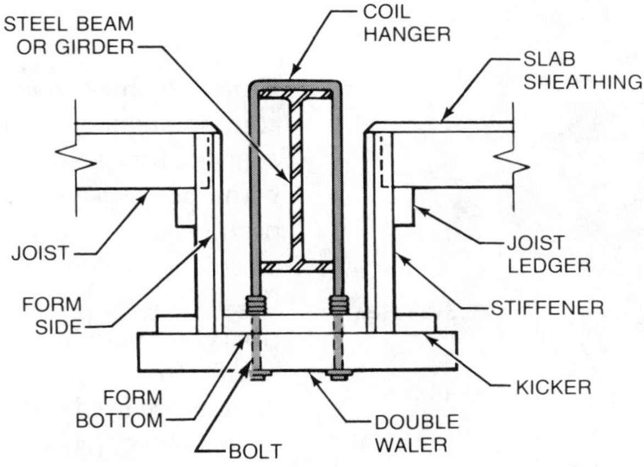

BEAM ENCASEMENT FORMS

Figure 5-36. Suspended forms are used to form a concrete slab supported by steel beams or girders. The steel beams or girders may be encased in concrete or exposed.

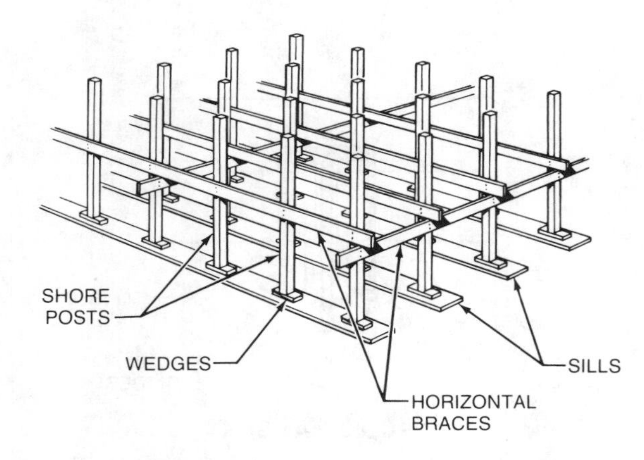

1. Place sills firmly over soil. Erect shore posts and tie together with horizontal braces.

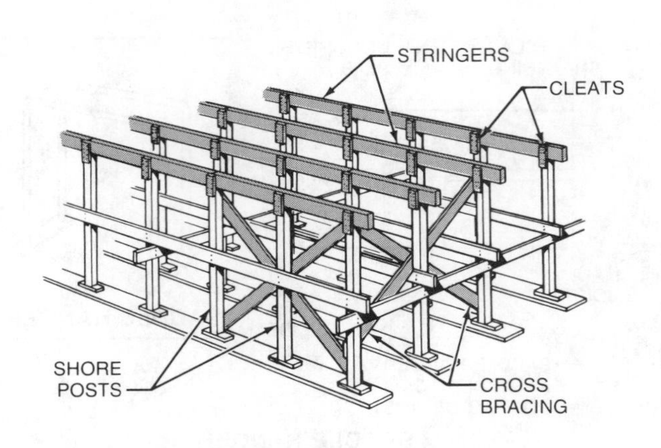

2. Fasten cross bracing to the shore posts. Secure stringers to the top of the shore posts with plywood cleats or steel angles.

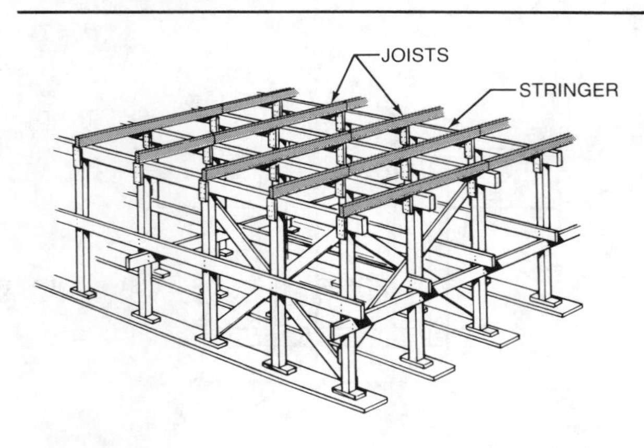

3. Place joists over the stringers staggering the joints. Nail the joists to the stringers.

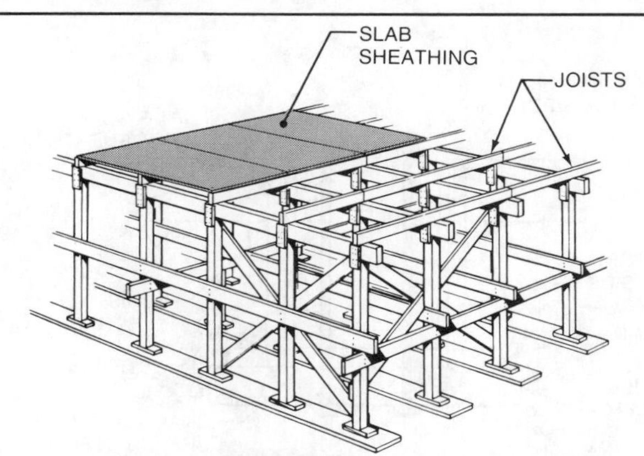

4. Place slab sheathing perpendicular to the joist direction. Nail the slab sheathing into place by nailing at the corners.

Figure 5-37. Floor forms are supported with stringers and shores. Plywood sheathing is fastened to joists positioned perpendicular to the stringers.

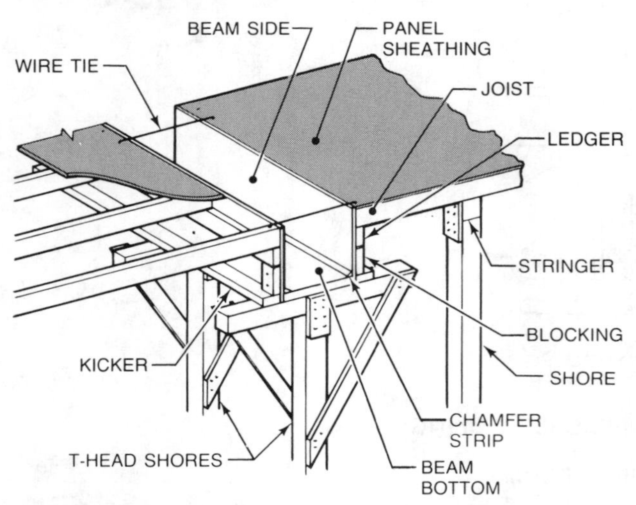

Figure 5-38. The plywood sheathing for a floor form is secured to the beam or girder side when forming a monolithic beam or girder and floor slab system.

Concrete Joist Systems. Concrete joist systems combine concrete joists with a concrete slab. The concrete for the joist systems is placed monolithically with beams, girders, and columns. Reusable prefabricated pans are placed at regular intervals on soffits to form the concrete joist system. The soffits are supported by stringers and shores. Steel or fiberglass pans are commonly predrilled for nailing to the soffit. *Nail-down* and *adjustable* pan designs are used to form the concrete joist systems.

Nail-down pans are the easiest to install because the flanges allow them to be nailed from the top side into the soffit. Standard pans for one-way joist systems are positioned with the flanges parallel to the soffit. *Long pans* are secured in place with the flanges perpendicular to the soffits. Long pans reduce the number of seams and produce a smooth

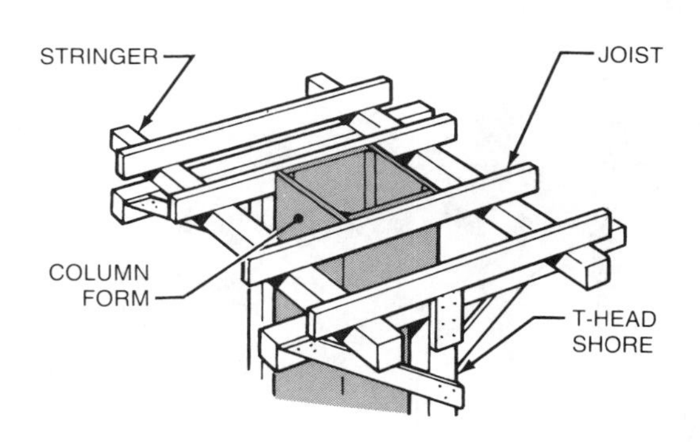

1. Place shores and stringers around column form. Place joists across the stringers.

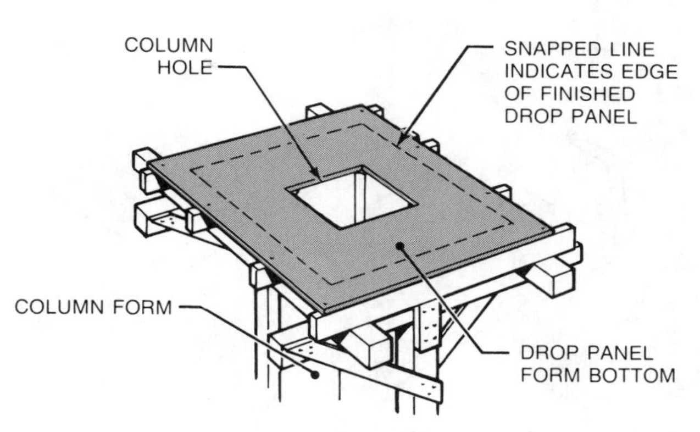

2. Cut a hole in the drop panel bottom and nail to the joists. Lay out and snap a line to indicate the edges of the completed drop panel.

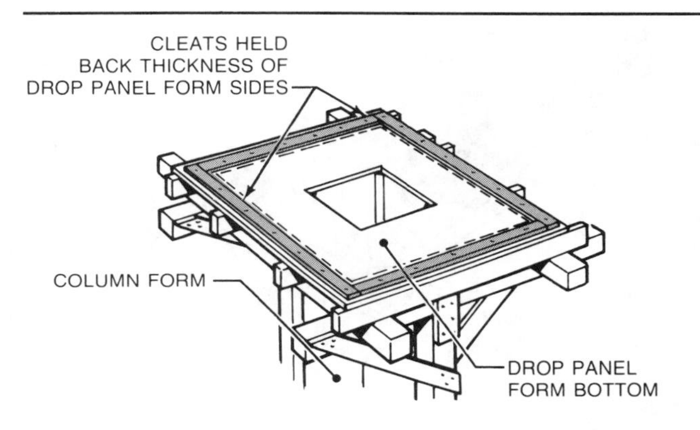

3. Nail cleats to the bottom holding them back the thickness of the drop panel form sides.

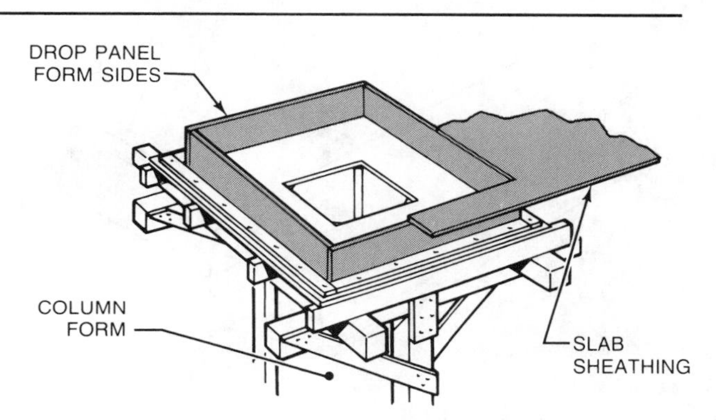

4. Nail the drop panel form sides to the cleats. Fit the slab sheathing to the inside edge of the drop panel form.

Figure 5-39. A drop panel form is positioned over a column form. Floor slab sheathing is fitted around the drop panel form.

exposed surface. The adjoining flanges clamp together to form the concrete joist. See Figure 5-40.

One-way joist pan forms are available in standard size widths of 20″, 30″, and 40″, and depths ranging from 8″ to 20″ in 2″ increments. The concrete joists formed range in size from 4″ to 8″. Concrete slabs incorporated with the joists systems are from $2\frac{1}{2}$″ to $4\frac{1}{2}$″ thick.

Pan forms commonly frame into girder forms. Shores, stringers, and soffits are positioned and tapered end pans are placed against end caps. The pans are then placed from the ends and progress toward the center with the pans overlapping 1″ to 5″. A filler piece is placed at the center to fill the open area. See Figure 5-41. Plywood sheathing may also be used as a base for nail-down pans. Plywood sheathing provides a convenient working surface and simplifies one-way joist system construction.

Adjustable pans for one-way joist systems are nailed to the sides of the soffit. Adjustable pans produce a smoother finish than nail-down pans because flange nail head impressions are not left in the exposed concrete. Adjustable pans are often used to form exposed ceilings.

Flanged and unflanged adjustable pans are also used to form one-way joist systems. When using unflanged adjustable pans, a soffit the width of the bottom of the joist is supported by a stringer. A template is used to hold the pans to their correct height while the pans are fastened to the soffit with double-headed nails.

When using flanged adjustable pans, stringers support a joist that is placed on edge. Soffits the width of the bottom of the concrete joist are positioned on top of the joists. Spreaders are placed between the joists with ledgers nailed along the sides

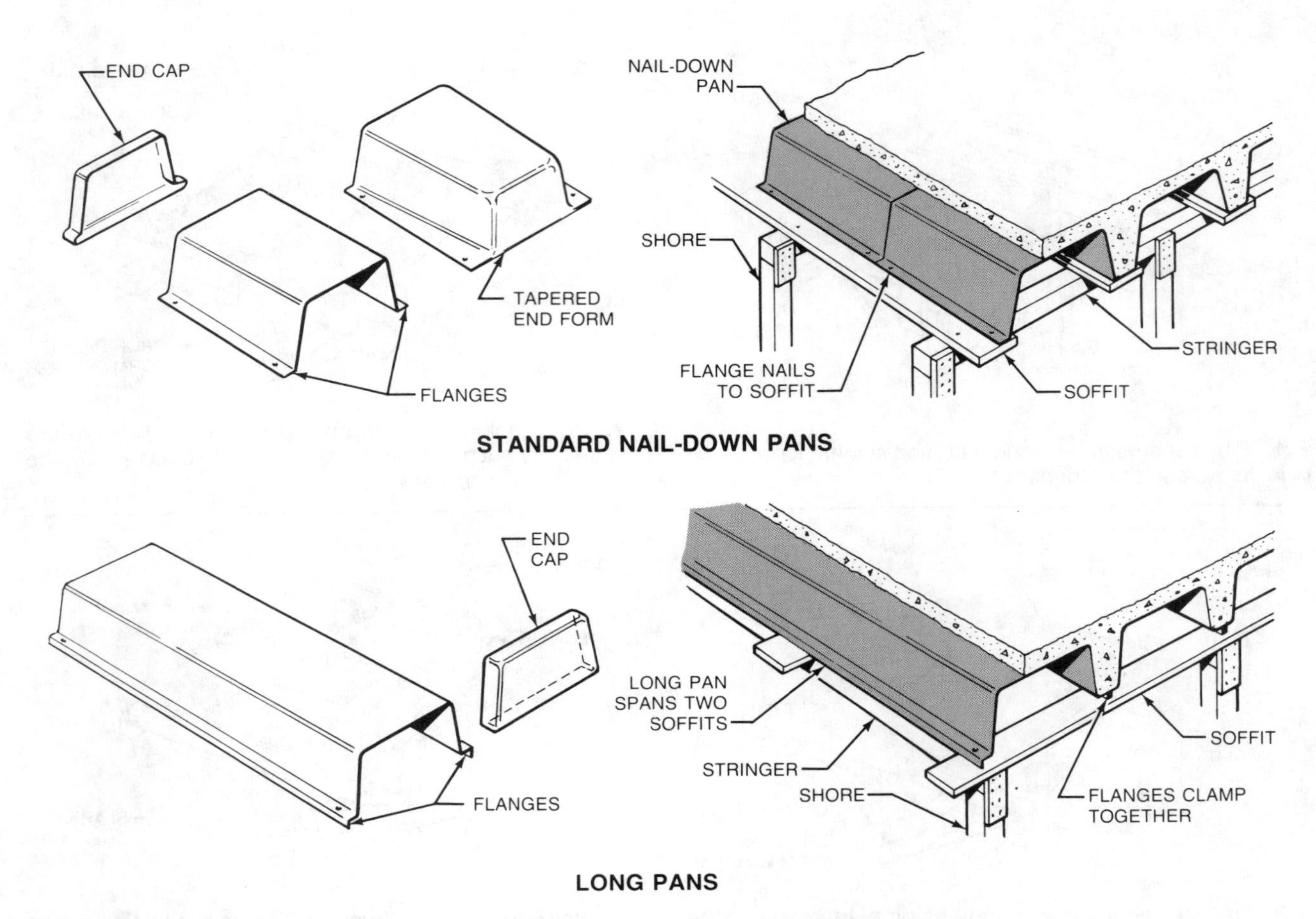

Figure 5-40. Standard nail-down pans and long pans are used to construct one-way joist systems.

of the joists. A shorter spreader, equal to the distance between joists less twice the flange width and pan thickness, is positioned on top of the bottom spreader to hold the pan to its correct width. See Figure 5-42.

A two-way joist system is constructed in a similar manner to the nail-down pan method. *Dome pans* are placed on the soffits or plywood sheathing after the shores and stringers have been set up. Dome pans are square prefabricated pan forms nailed in position through holes in the flanges. Most dome pans are designed so the flanges butt together to produce the required joist size. If a wider joist is required, the pans are set to chalk lines snapped on the soffit or plywood sheathing. Dome pans are available in 2′, 3′, 4′, and 5′ widths and 8″ to 24″ depths. See Figure 5-43.

Two-way joist systems usually require a solid area around a column equal to the thickness of the slab and joist system. Solid deck sheathing is placed in this area and the dome pans are omitted.

Vertical Shoring

Vertical shoring provides the main support for beam, girder, and slab formwork. Wood posts, adjustable metal shores, or metal scaffolding may be used as vertical shoring. The vertical and lateral pressures during and after concrete placement and the weight of the form materials, machinery, and construction workers are considered in designing shoring systems. Shoring system design should adhere to guidelines established by the American Concrete Institute (ACI) and U.S. Occupational Safety and Health Administration (OSHA). (See ACI and OSHA Shoring Standards in Appendices C and D.)

All shoring members must be straight and true. Cuts made at the bearing ends and splices of the shoring members should be square. Vertical shores must be placed in a plumb position and secured with braces so they cannot tilt. See Figure 5-44. Inclined shores must be securely braced to prevent slippage. Shoring systems must be properly braced to ensure

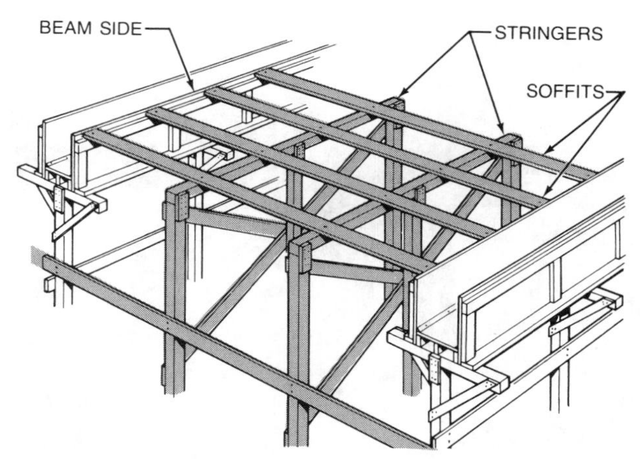

1. Set up and adjust the shores and stringers to the required height. Nail soffits to the top of the stringers.

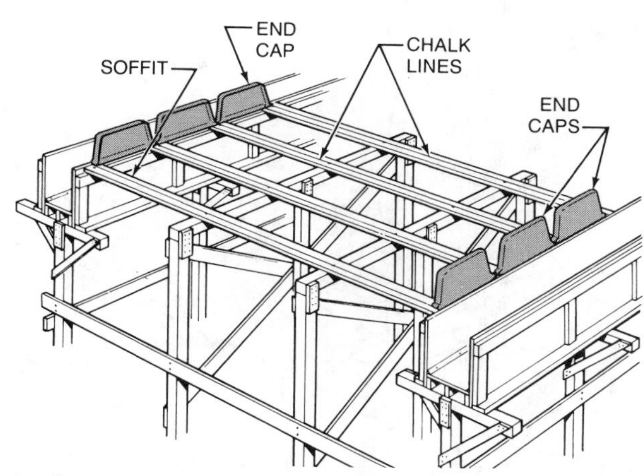

2. Snap chalk lines on the soffits. Nail end caps to the soffits at both ends.

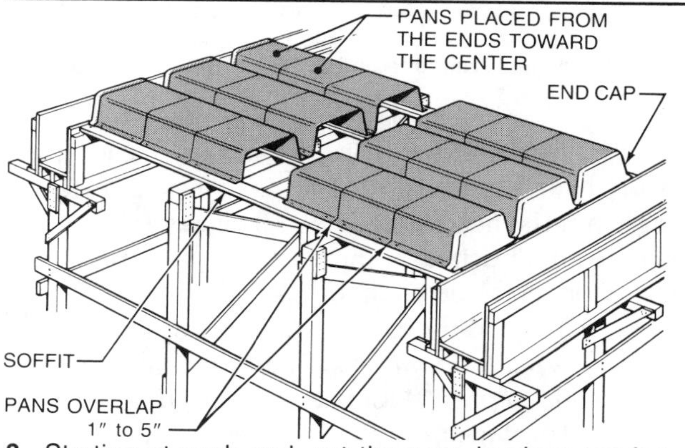

3. Starting at each end, set the pans in place moving toward the center of the span. Overlap the pans a few inches.

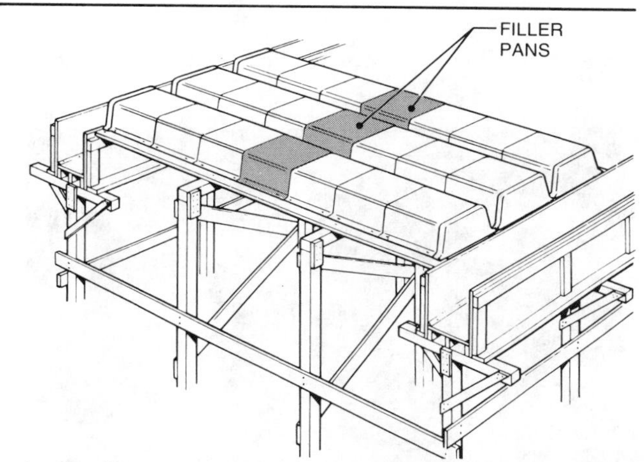

4. Set filler pans in position at the center of the span.

Figure 5-41. Pan forms are supported by soffits or solid deck sheathing. Pan placement starts at both ends with the pans overlapping 1″.

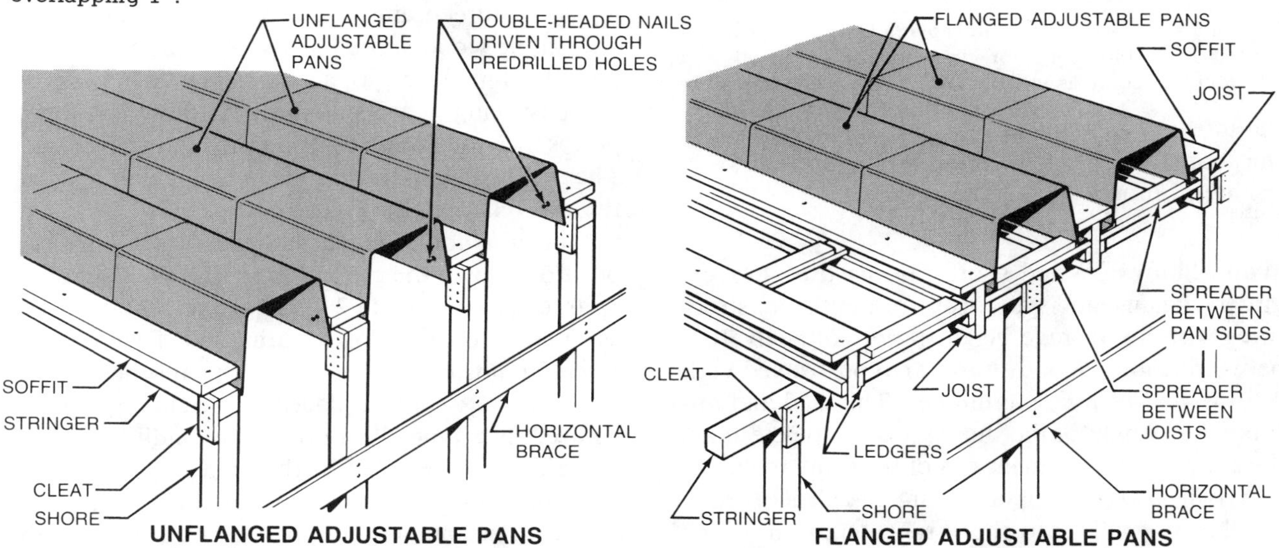

Figure 5-42. Adjustable pans are used to construct one-way joist systems. Unflanged adjustable pans are nailed to the sides of the soffits. Flanged adjustable pans are supported and secured by spreaders.

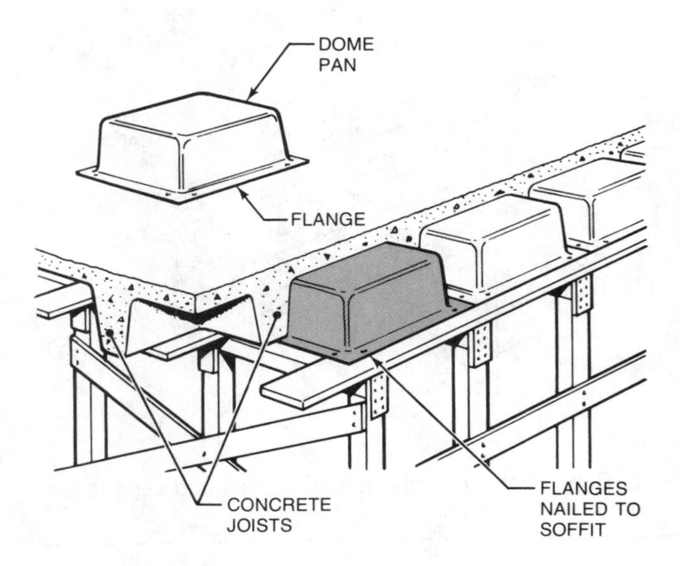

Figure 5-43. Dome pans are used to form a two-way joist system. The dome pans are placed on soffits or plywood sheathing.

Portland Cement Association

Figure 5-44. Vertical shoring is placed beneath stringers that support the joists for floor forms. Diagonal and horizontal bracing is used to secure the shores.

the integrity of the shoring. Horizontal braces tie the shores together. Diagonal braces prevent sway or lateral movement of the shores.

Wood Shores. Wood shores are used to support girders, beams, and slab forms for multistory structures with an average height (approximately 10′) between floors. Wood shores are constructed with 3 × 4, 4 × 4, or 6 × 6 lumber. The load and unsupported height of the shore are used to determine the cross-sectional dimensions of the shore required.

A *T-head shore* supports beam and girder forms. The head of a T-head shore is centered on top of a vertical post. Braces are nailed between the post and head. An *L-head shore* is commonly used under spandrel beam forms. The head of the L-head shore is offset and braced with 1 × 4s or 1 × 6s extending from the head to the post. The heads of wood shores are attached to the posts by using plywood cleats or metal or angle brackets.

Type of wood shores used for vertical shoring are the *single-post*, *double-post*, and *two-piece adjustable wood shores*. A single-post wood shore is a single vertical member placed beneath stringers supporting floor slab forms. Double-post wood shores consist of a head placed over two vertical posts. Cross bracing may be used to reinforce the shore. Double-post wood shores support heavy girder loads, spandrel beams, and drop panels. A two-piece adjustable wood shore has two overlapping wood posts held in place with a *post clamp (Ellis clamp)*. The post clamp is nailed to the lower post. The upper post is then raised into position with a portable jack. The two posts are held in place by friction against the post clamp. See Figure 5-45.

Mudsills should be placed beneath wood shores positioned over the ground to spread the load over a large area. A single 2 × 10 plank may be used as a mudsill on good load-bearing soil. Two or three 3 × 10 mudsills placed next to one another with a 4 × 4 or 4 × 6 plate placed at a right angle may be used as a mudsill in poor soil conditions. The shore post is toenailed to the plate.

Wood shores are generally cut shorter than required to accommodate a pair of wedges beneath the posts. The wedges are used to adjust the height of the shores when aligning the formwork. When

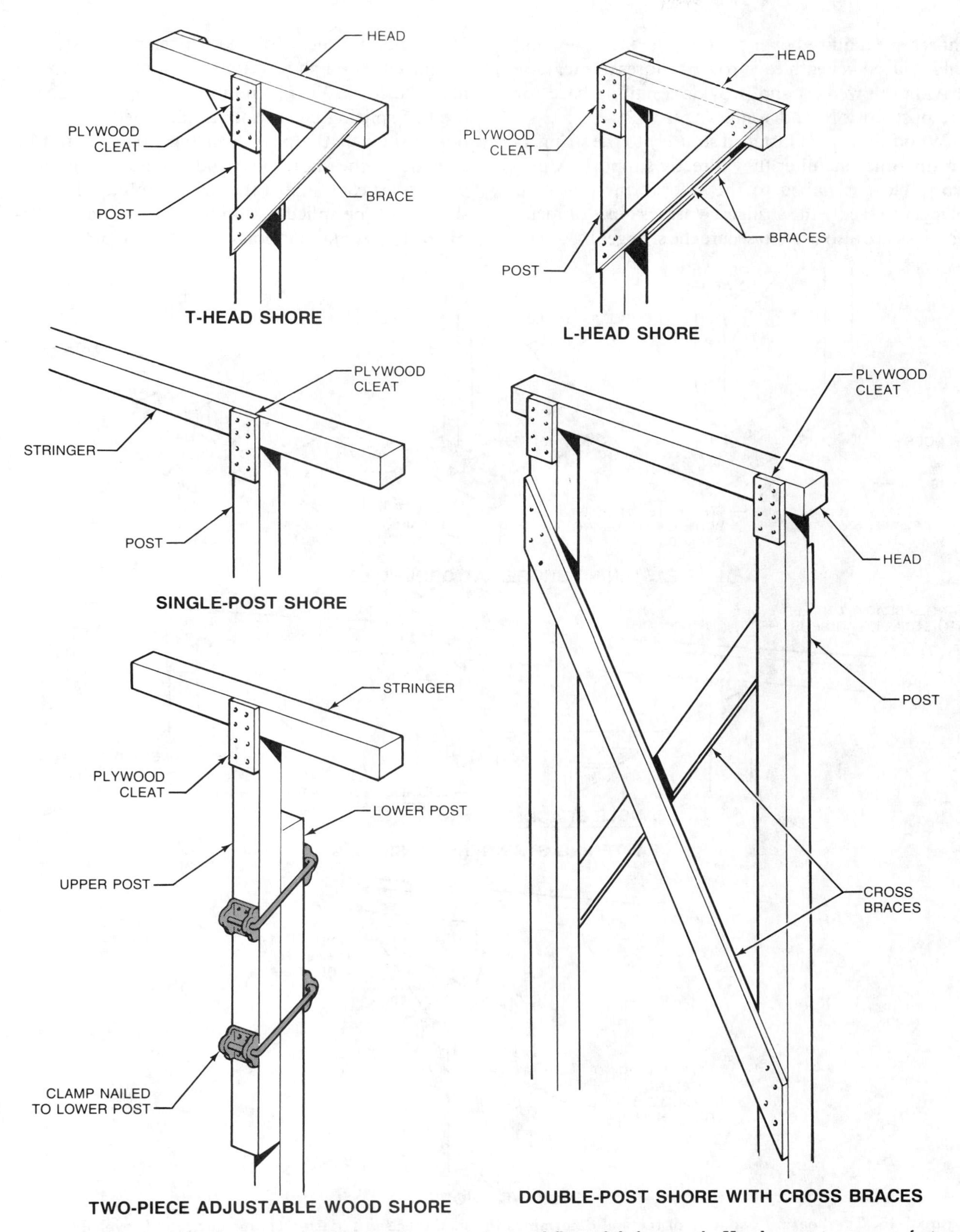

Figure 5-45. Wood shores are used to support beam, girder, and floor slab formwork. Heads or stringers are fastened to the tops of the posts to support formwork.

shores are required over concrete surfaces, a wood sill is placed beneath each row of shores to facilitate driving the wedges and provide a nailing base for the post bottom and wedges.

Wood shores are fastened securely to the stringers or any other member they directly support. A plywood cleat is nailed to the side of the post and stringer to secure the stringer. Various types of metal brackets are also used to secure the stringer by screwing or nailing them to the stringer and post.

Spliced wood shores are commonly used to cut material costs. Two-inch lumber or ⅝″ plywood is used as splicing cleats and are fastened to the opposite sides of the post. Splicing cleats should be as wide as the shore post and extend a minimum of 12″ past each side of the splice. Unbraced shores should not be spliced at midheight or midway between horizontal supports. See Figure 5-46.

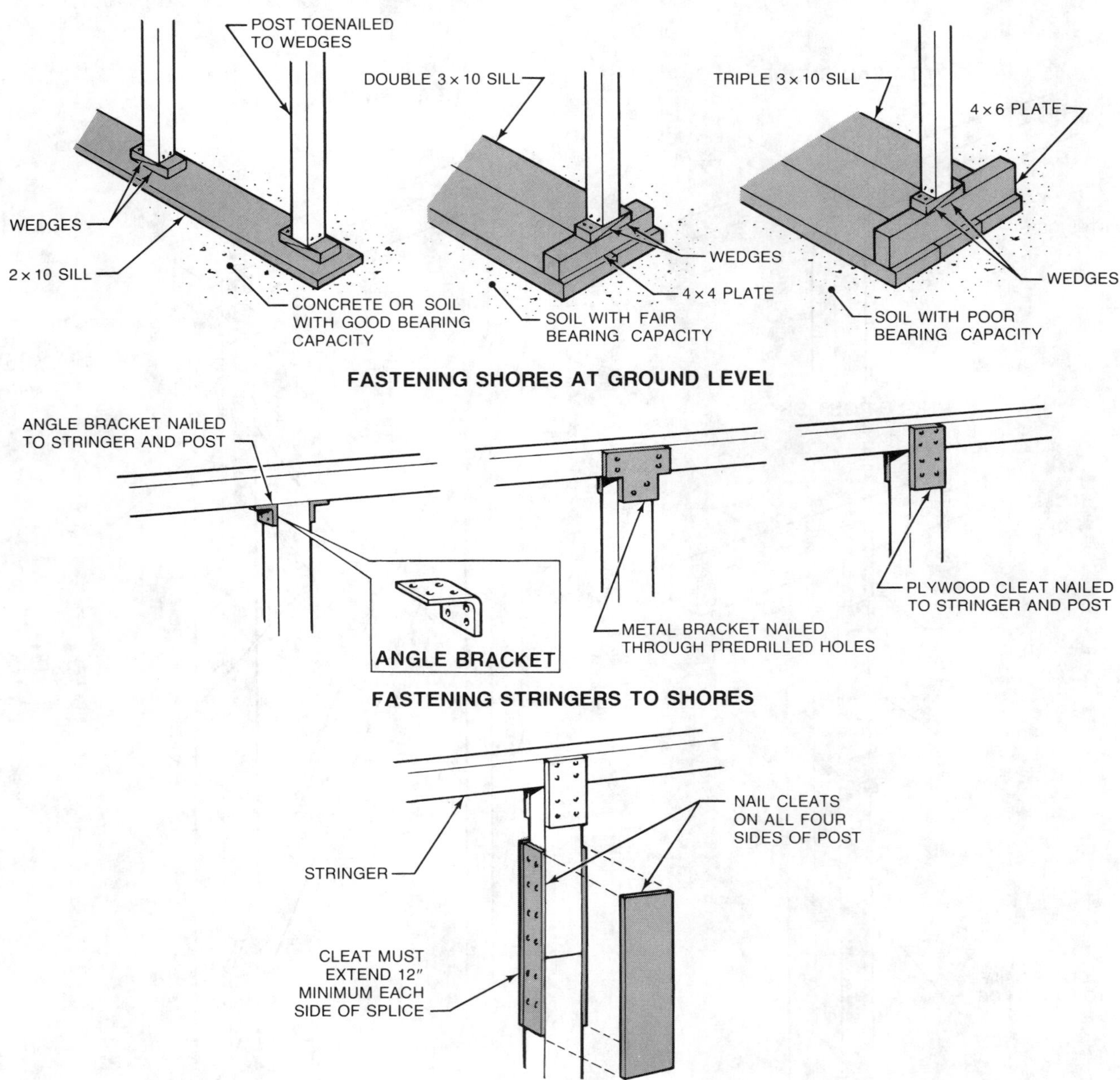

Figure 5-46. The bearing capacity of the soil determines the method used to fasten shores at ground level. Metal or angle brackets, or plywood cleats secure stringers to posts. When splicing wood shores the splicing cleats extend 12″ beyond each side of the splice.

Adjustable Metal Shores and Shore Jacks. Adjustable metal shores, used for vertical shoring, are constructed of tubular steel. The tubular steel is open at both ends to prevent accumulation of water and rust. The upper tube is adjusted to the approximate height required and a locking pin is inserted through a hole in a slot above the adjustment collar. The adjustment collar is turned to make the final adjustments. Flanges secure the top and bottom of the adjustable metal shore to a stringer and pad or sill. Braces may be attached to the shores by using nailing brackets or other devices.

Shore jacks are used to adjust the height of wood shores without using wedges. A metal fitting slips over a 4 × 4 or 6 × 6 post and is nailed into place. The jack is then adjusted to the final height. See Figure 5-47.

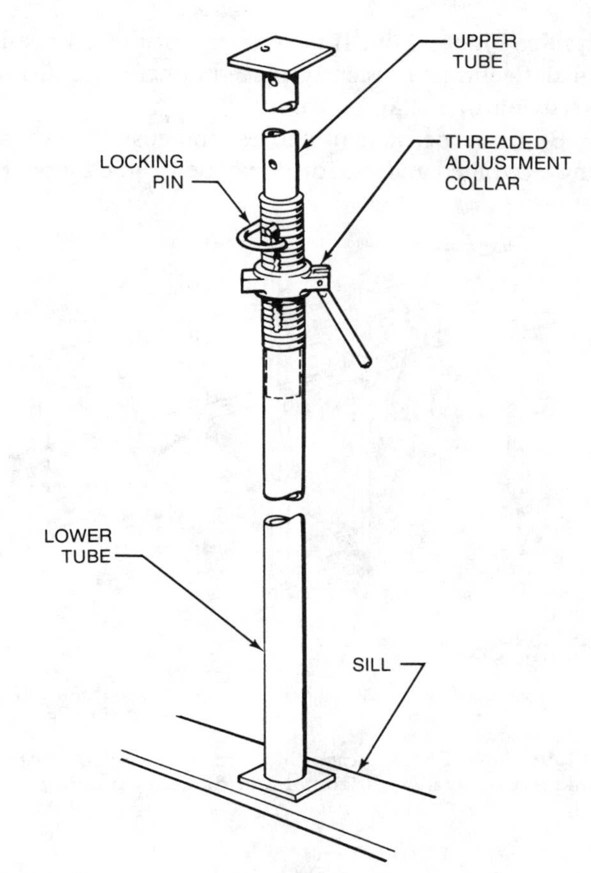

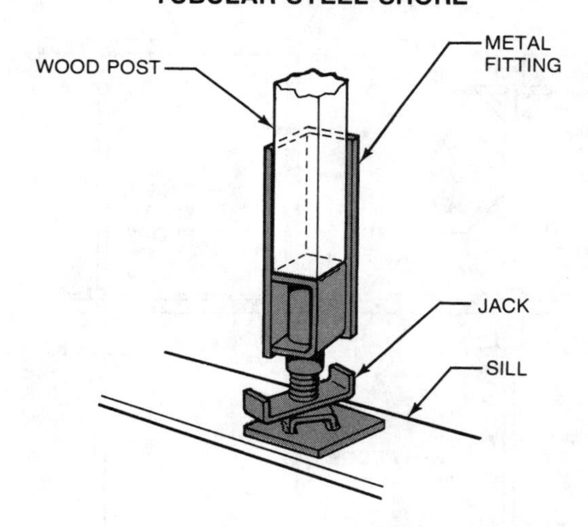

Figure 5-47. Tubular steel shores support stringers and elevated formwork. Metal shore jacks are secured to wood posts to allow height adjustment without wedges.

Horizontal and Diagonal Bracing. Horizontal and diagonal bracing ensures the stability and safeness of a shoring system. Bracing reduces the possibility of form collapse resulting from overloading forms and lateral pressure caused by wind, movement of heavy weight, and disturbance of the forms caused by crane booms or other equipment.

Horizontal bracing is placed at the midpoint of the shores and extends in two directions. Shores over 10′ may require two or more rows of horizontal bracing. Diagonal bracing should also be installed in two directions. At ground level, bracing may be extended from the outside row of shores and fastened to stakes driven into the ground. See Figure 5-48.

Reshores and Permanent Shores. Reshores and permanent shores are used to temporarily reinforce recently placed concrete and formwork for structural members at a higher level. Reshores are shores placed firmly under structural members after the original shoring is removed. Permanent shores are shores that remain in place as the formwork around them is removed.

Metal shores and 4 × 4 or 6 × 6 wood shores are used for reshoring. While reshoring beneath a recently placed structural member, construction loads should not be placed on the floor level above. Reshores are placed directly above one another because they may be required to stay in place until the entire structure has been completed. Reshores should not be wedged or jacked up to lift or crack the concrete above. In order to keep load distribution from changing, reshores are tightened in position uniformly and only enough to hold them securely. See Figure 5-49. Improper reshoring can result in deflection of recently placed concrete, causing cracking or collapse.

By using permanent shores, the cost of reshoring is avoided and the forms can be stripped sooner.

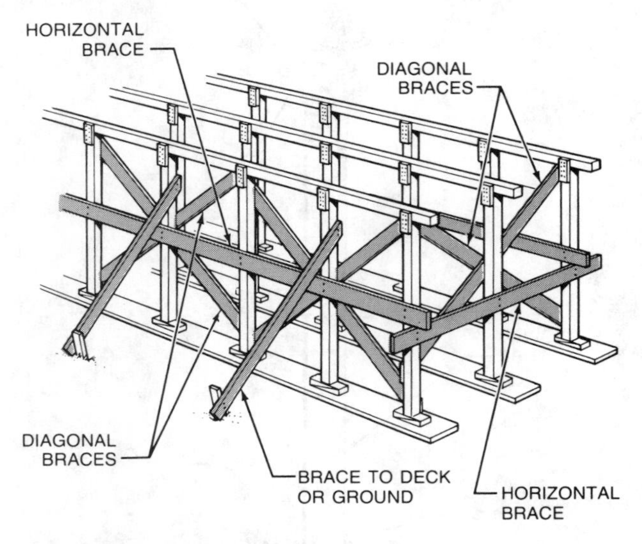

Figure 5-48. Shores supporting formwork for floor systems require extensive horizontal and diagonal bracing.

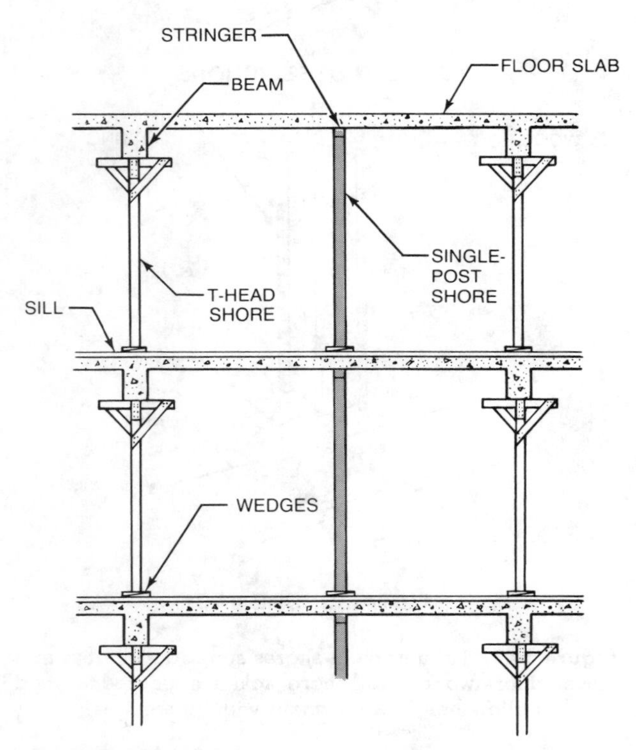

Figure 5-49. Reshores are single-post shores placed under structural members after the original shoring is removed. Reshores are placed directly above one another in successive levels.

Permanent shores are erected with the top of the stringers flush with the top of the floor slab sheathing. Ledgers are held down the thickness of the floor slab sheathing from the tops of the stringers and bolted into position. The floor slab sheathing is then placed on top of the ledger. After the concrete has set, the intermediate shores, floor slab sheathing, and ledgers are removed, leaving the permanent shores and stringers in place. See Figure 5-50.

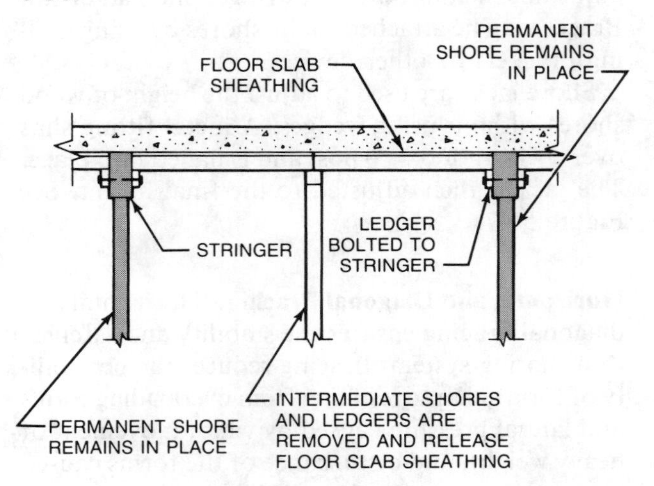

Figure 5-50. Permanent shores remain in place after the floor slab forms have been removed.

As each floor level has gained sufficient strength to carry loads, the reshores or permanent shores are removed. The removal begins at the upper floor level and proceeds toward the lower levels of the building.

Scaffold Shoring. In scaffold shoring, sections of tubular steel frames are assembled to different heights. Scaffold shoring is used to support slab and beam, flat plate and flat slab floor forms. Scaffold shoring is also used in structures with high beam and slab soffits, and bridge and overpass construction. See Figure 5-51.

Scaffold shoring is erected according to a layout plan. Safe working loads for scaffold shoring range from 4000 to 25,000 pounds per leg, depending on the type of frame, bracing, and height of the scaffold. Greater working loads are possible with specially designed towers. Typical frame sections for scaffold shoring are 2′ to 4′ wide and 4′ to 6′ high.

Individual frames are made up of two tubular steel uprights joined by horizontal members. Cross bracing extending across the uprights of individual

Symons Corporation

Figure 5-51. Scaffold shoring is used to support flat plate and flat slab floor forms.

frames provide lateral stability. The individual frames are mounted on top of each other and secured with coupling insert pins. Opposite frames are fastened to each other with diagonal cross bracing. The assembled scaffold shoring is supported by adjustable swivel screw jacks that rest on metal base plates. The plates rest on sills similar to the types used for wood shores.

Scaffold shoring supporting heavy loads over soil is placed on thick timbers to distribute the load over a large area. A *beam clamp* (adjustable U-shaped device) is installed at the top of the shoring to hold and support the stringers. See Figure 5-52.

CONCRETE STAIRWAYS

Concrete stairways are classified as exterior or interior stairways. Exterior stairways are built over sloping soil and receive the main support from the earth. Interior stairways are built over an open area and receive the main support from the reinforced soffit below the steps.

Interior stairways are further classified as *open* and *closed*. Open stairways are not enclosed by walls and rely entirely on built-in-place forms and shoring to support the concrete. Closed stairways are enclosed by walls on two sides. The walls are used to support the skirt boards and riser form boards. A stairway constructed with a wall on one side is a combination of the two designs. See Figure 5-53.

Most concrete stairways are variations of a *straight flight* design. A straight flight stairway extends from one floor to the next without change in direction. *U-shaped* and *L-shaped* stairways have landings, which change the direction of the stairway.

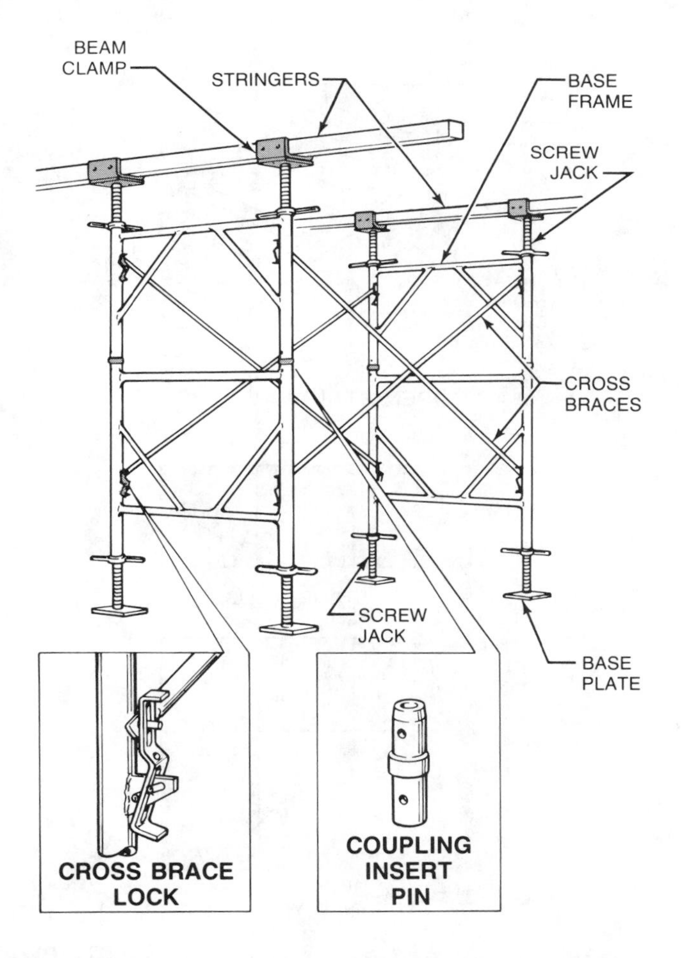

Figure 5-52. Tubular steel scaffold shoring consists of a frame secured together with insert coupling pins. A cross brace lock secures cross braces extending from the frames.

Riser and Tread Layout

Riser and tread layout establishes the height of the individual risers (vertical surface) and the depth of the individual treads (horizontal surface). The riser height (unit rise) is determined by dividing the *total rise* of the stairway by the number of risers. Total rise is the vertical distance from one floor to the floor above. The tread depth (unit run) is determined by dividing the *total run* by the number of treads. Total run is the horizontal length of a stairway measured from the foot of the stairway to a point plumbed down from where the stairway ends at a floor or landing above.

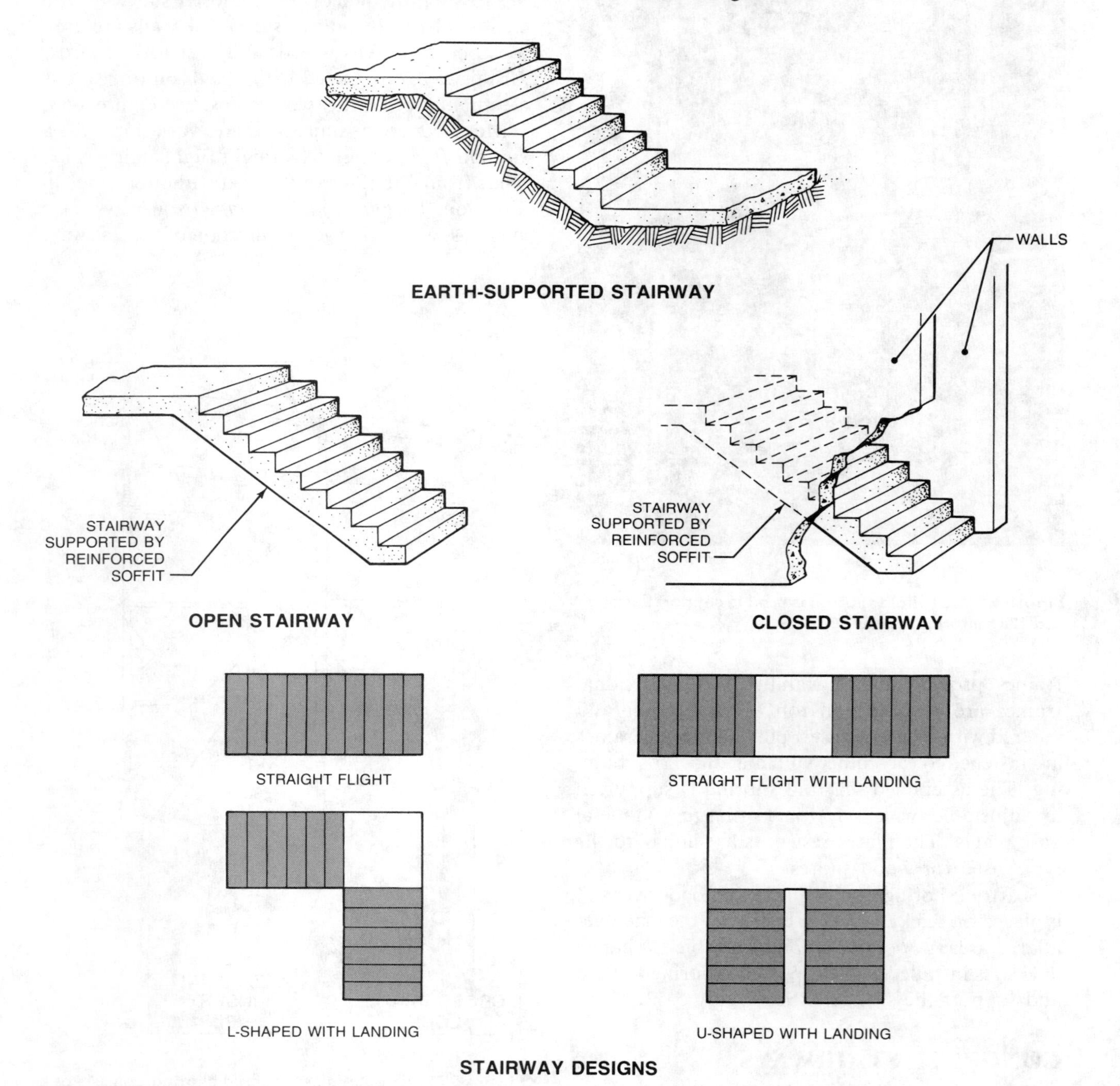

Figure 5-53. Stairways are earth-supported or supported by reinforced soffits. Stairway designs may incorporate landings to change the direction of the stairway.

Calculating Unit Rise and Unit Run. Many codes specify minimum and maximum riser heights and minimum tread depth. The Building Officials and Code Administrators International (BOCA) recommends a minimum riser height of 4″ and a maximum of 7″. Minimum recommended tread depth is 11″. The BOCA code also allows a $\frac{3}{16}$″ variation in the depth of adjacent treads or height in adjacent risers. The Uniform Building Code (UBC) recommends a minimum rise of $4\frac{1}{2}$″ and a minimum run of 10″ for every step in the stairway. The code having jurisdiction in the area should be consulted before determining unit rise and unit run.

Unit rise and unit run are calculated by converting the total rise and total run to inches and dividing by the number of risers and treads. For example, the unit rise for a stairway with a total rise of 10′-2″ is determined as follows.

Step 1: Convert total rise to inches.

Total rise (inches)
= (*no. of feet* × 12) + *no. of inches*
= (10′ × 12) + 2″
= 120 + 2
Total rise = 122″

Step 2: Determine the number of risers by dividing the total rise by the minimum desired riser height.

No. of risers = *total rise* ÷ *minimum riser height*
= 122 ÷ 7
No. of risers = 17.428

If the answer contains a decimal value, a fraction of an inch is added to each 7″ riser to ensure that the 17 risers are equal height.

Step 3: Determine the exact riser height by dividing the total rise by the number of risers. Calculate the answer to three decimal places.

Riser height = *total rise* ÷ *no. of risers*
= 122″ ÷ 17
Riser height = 7.176″

Step 4: Convert the decimal value to sixteenths of an inch by multiplying the remainder by 16.

Fractional equivalent (16ths)
= *decimal value* × 16
= .176 × 16
Fractional equivalent (16ths) = 2.816

The whole number to the left of the decimal point (2) indicates the number of sixteenths ($\frac{2}{16}$″ = $\frac{1}{8}$″). If the value to the right of the decimal point is .5 or above, add $\frac{1}{16}$″ to the value at the left of the decimal point. If the value is less than .5, disregard the value. In this example, the value 2.816 is converted to $\frac{3}{16}$″. The whole number in step 3 (7″) is added to the fractional equivalent obtained in step 4 ($\frac{3}{16}$″) to equal the exact riser height ($7\frac{3}{16}$″).

When calculating the tread depth, the total run is converted to inches and divided by the total number of treads. The total number of treads is always one less than the number of risers. For example, the tread depth of a stairway with 17 risers and a total run of 16′-0″ is determined as follows.

Step 1: Convert the total run to inches.

Total run (inches)
= (*no. of feet* × 12) + *no. of inches*
= (16′ × 12) + 0″
Total run = 192″

Step 2: Determine the tread depth by dividing the total run by the number of treads.

Tread depth = *total run* ÷ *no. of treads*
= 192″ ÷ 16
Tread depth = 12″

If the answer has a decimal value, use the same procedure used for determining fractions when calculating riser height.

The riser and tread dimensions are laid out on a pair of *skirt boards* using a steel square. The skirt boards are part of the stair form to which the riser form boards are fastened. The riser height is marked on the tongue of a steel square and the tread depth is marked on the blade. Square gauges may be used to hold the square in position along the edge of a skirt board. The steel square is positioned along the top edge of the skirt board and the steps are laid out. See Figure 5-54.

Constructing Stairway Forms

Concrete stairway forms require accurate layout to ensure accurate finish dimensions for the stairway. Interior or exterior stairways are reinforced with rebars that tie into the floors and landings.

Concrete stairways are formed monolithically with floor slabs by framing the stairway forms into the floor slab forms. Concrete stairway forms may also be constructed after the concrete for the floor slabs has set. Concrete stairways formed after the floor slab has set are anchored to a wall or beam by tying the stairway rebars to rebars projecting from the wall or beam, or by providing a keyway in the wall or beam. Step supports are commonly placed monolithically with the exterior wall. Rebars projecting from the walls may also be used to tie

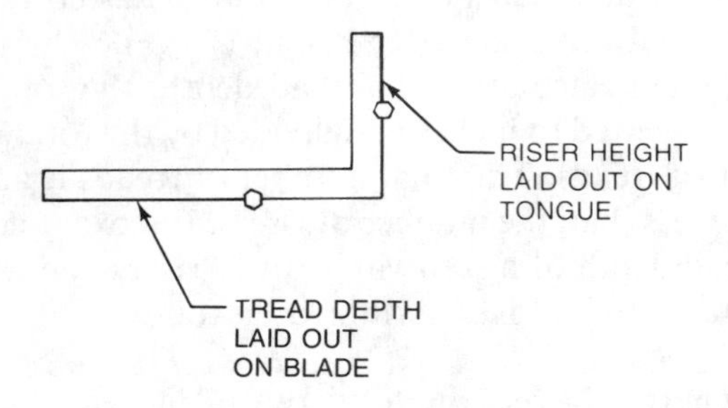

1. Clamp square gauges on the tongue of the steel square at the riser height and on the blade at the tread depth.

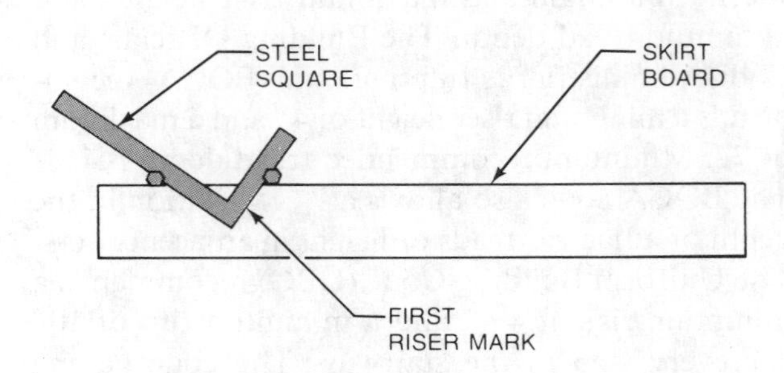

2. Position the steel square along the top edge of the skirt board and lay out the first riser.

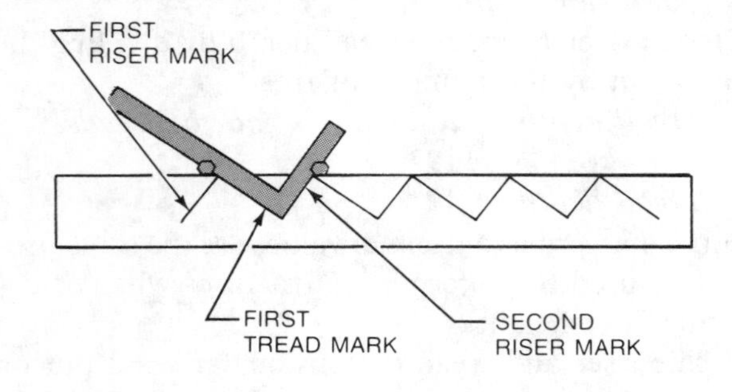

3. Slide the steel square until the tread square gauge aligns with the riser mark. Lay out the first tread and second riser. Continue sliding the steel square and laying out the treads and risers.

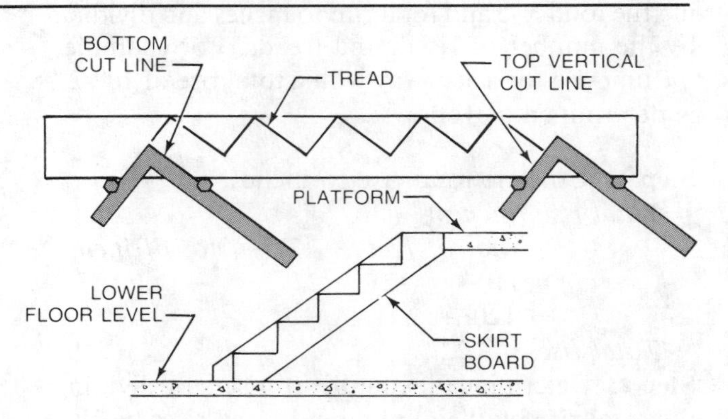

4. Place the steel square along the bottom edge of the skirt board and lay out the bottom and top cut lines.

Figure 5-54. Riser and tread dimensions are laid out on a skirt board. The skirt board is secured in position and riser form boards are fastened to it.

into rebars in the stairway form in closed stairways. See Figure 5-55.

The bottom of an interior concrete stairway may be secured by rebars projecting from the floor slab or a keyway cut in the floor slab. The area beneath the bottom step of an exterior stairway resting on the soil should be excavated to load-bearing soil. For exterior stairways, small piers are placed below the first step if a frost line or unstable soil conditions are present.

The riser of the stairway form is laid out with a ¾″ to 1″ slope to create a nosing at the front of each step. The tread slopes ⅛″ to ¼″ from front to back. See Figure 5-56. The bottom edge of the riser form boards should be beveled to facilitate troweling and finishing the treads.

Constructing Stairway Forms over Sloping Ground. Concrete stairways constructed over sloping ground receive their main support from the earth. Preliminary groundwork includes stepping the slope to prevent the fresh concrete from sliding and placing a well-compacted layer of gravel over poor draining soil.

Two skirt boards (planks or plywood) are braced and staked to the ground. Riser form boards are secured by cleats nailed to the skirt board. The riser form boards may be end nailed through the skirt boards if 1½″ form boards are used. For wide stairs, braces are placed 4′-0″ OC between the skirt boards. See Figure 5-57.

Constructing Open Stairways. Open stairways are commonly constructed inside a structure to provide access from one level to another. Where landings are required, a landing form is integrally constructed with the stairway form. An open stairway is constructed by setting up temporary panels along the

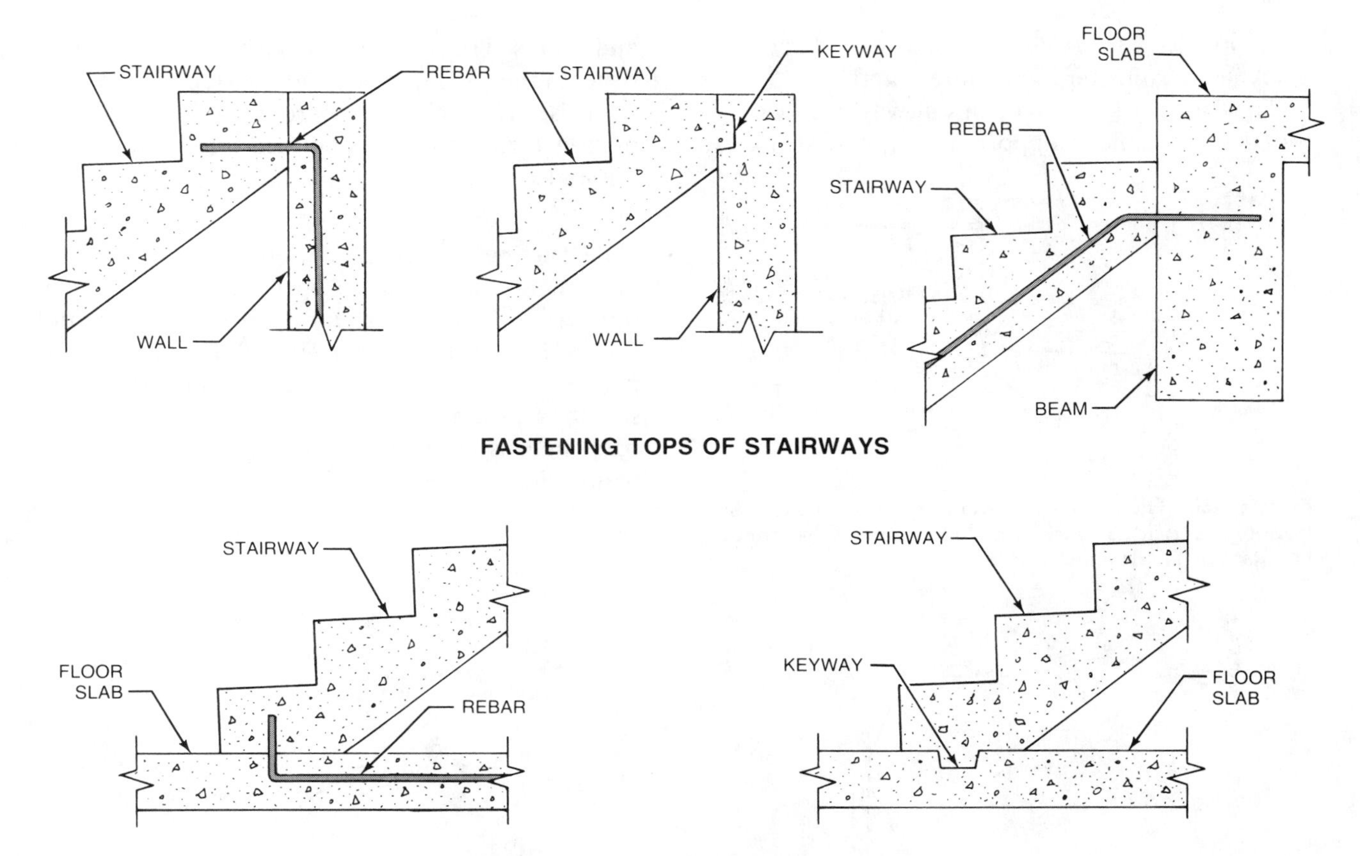

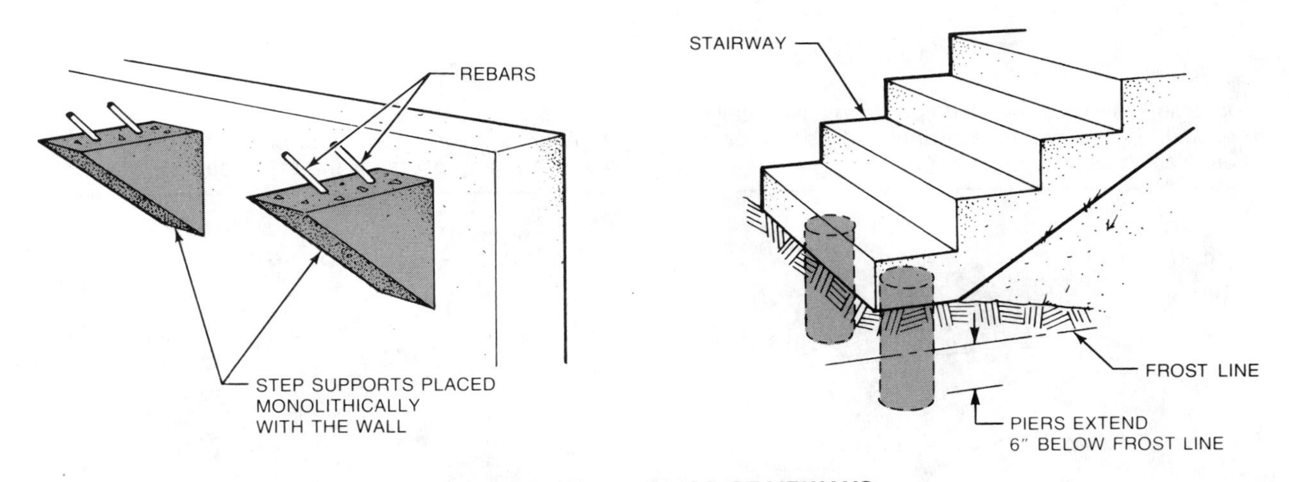

Figure 5-55. The tops and bottoms of concrete stairways are secured to walls or beams and floor slabs using rebars or keyways. Exterior stairways may require step supports or piers.

form construction area. The treads and risers and slab thickness are laid out on the panels. The slab thickness is laid out perpendicular to the stairway angle. The thickness of the soffit panel sheathing, width of the joists, and width of the stringers are laid out. The width of the skirt board is determined by measuring from the end of a tread to the top of the joists. The length of the shores is determined by subtracting 3″ (1½″ sill thickness + 1½″ wedge thickness) from the total shore length.

The formwork for an open stairway includes plywood soffit panels supported by joists, stringers, and shores. The stringers are positioned and braced beneath the soffit panels, and the steps are formed with riser form boards secured by cleats. Bottoms of riser form boards are beveled to permit troweling and finishing the treads. See Figure 5-58.

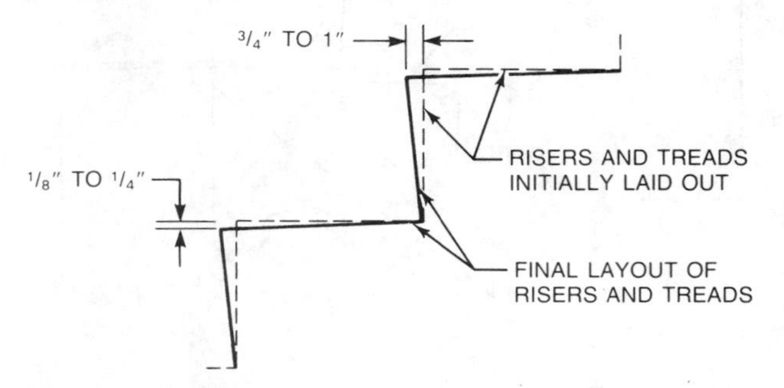

Figure 5-56. The riser is sloped to create a nosing at the front of the tread. The tread is sloped toward the front to facilitate water drainage.

Constructing Closed Stairways. Closed stairway construction is similar to open stairway construction. A soffit is supported by joists, stringers, and shores. However, the stringers and other form components are laid out on the two enclosing walls. The skirt boards are fastened to the walls and the riser form boards are secured with cleats nailed to the plank. See Figure 5-59.

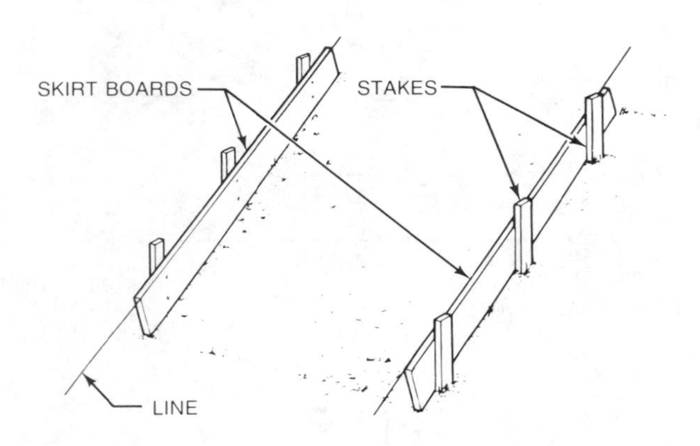

1. Drive stakes to support skirt boards to a line. Lay out the position of the top edge of the skirt boards on the stakes. Nail the skirt boards in position.

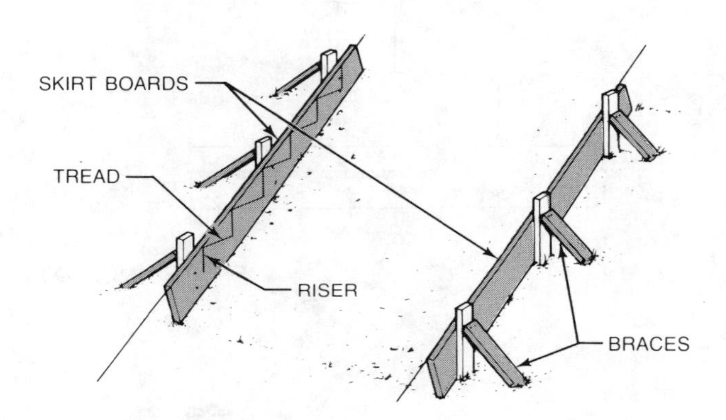

2. Align the skirt boards and brace the stakes. Lay out the treads and risers on the skirt boards.

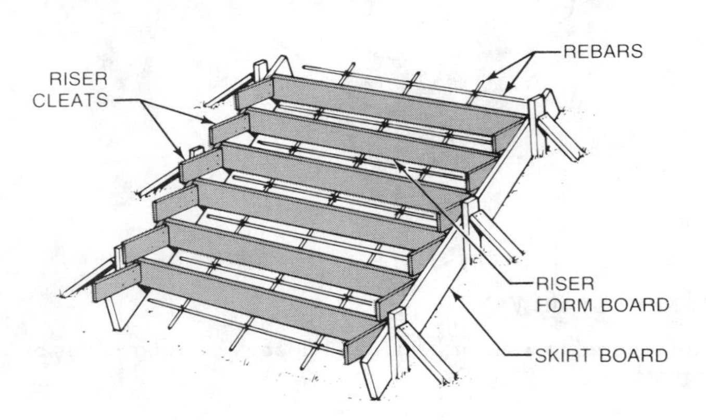

3. After rebars have been placed, nail riser cleats in position. Nail the riser form boards to the cleats.

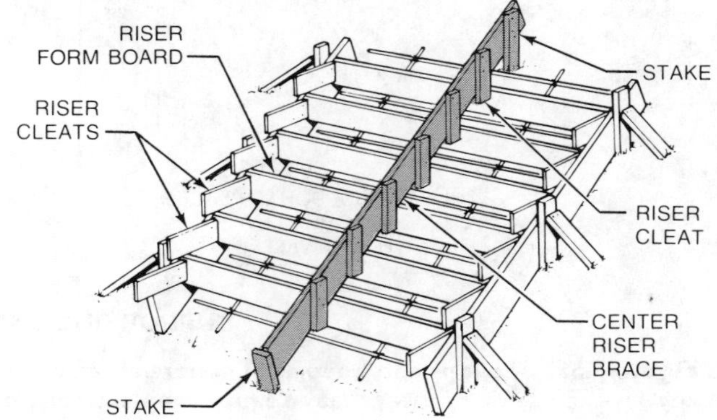

4. On wider stairways, reinforce the riser form boards with a center riser brace and cleats.

Figure 5-57. Two skirt boards are required to form a stairway over sloping ground. For wide stairways, a center riser brace and cleats are used to prevent distortion.

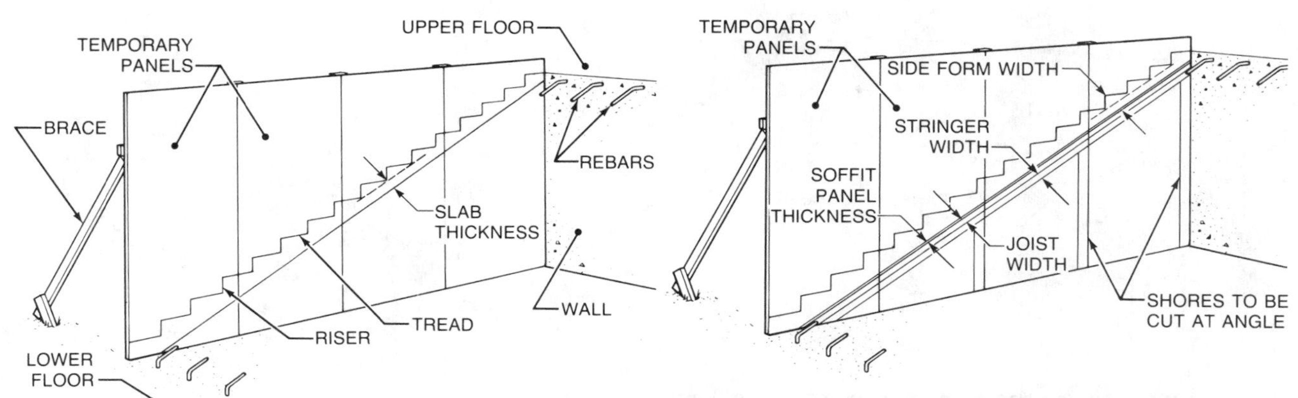

1. Set up and brace temporary panels along the stairway construction area. Lay out treads and risers on the panels. Measure the slab thickness at a right angle to the slope of the stairway and snap a line.

2. Lay out the soffit panel thickness, joist width, and stringer width and snap lines. Determine the shore length by measuring the distance from the lower stringer line to the floor and subtracting sill and wedge thickness. Determine the side form width by measuring at a right angle to the slope of the stairway.

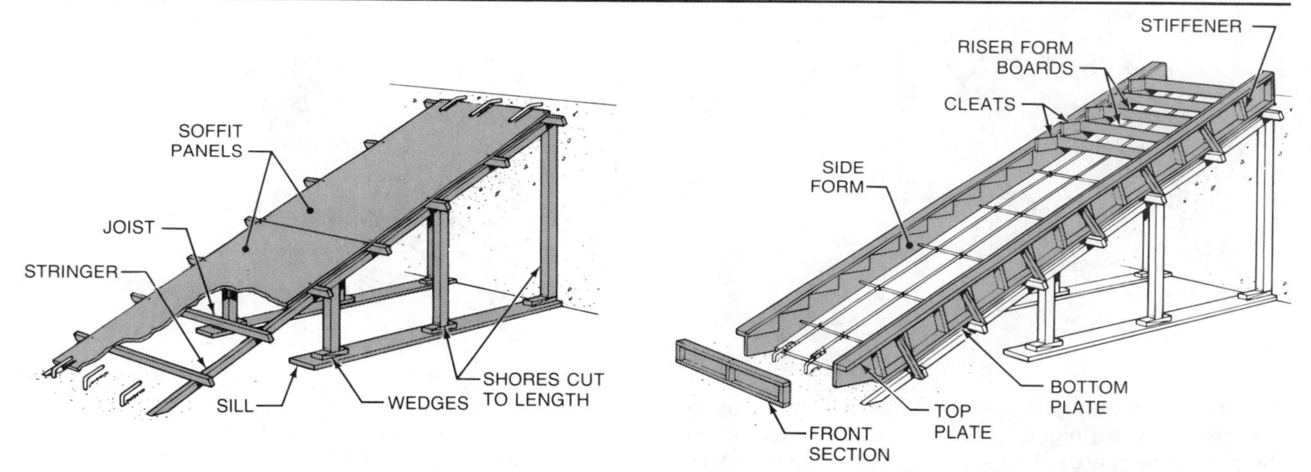

3. Cut shores to length and secure in position. Nail stringers to the tops of the shores. Nail joists to the tops of the stringers. Nail the soffit panels in position. Remove the temporary panels.

4. Lay out treads and risers on the side form. Nail top and bottom plates and stiffeners through the side forms. Fasten side forms to the top of the joists. Align and brace side forms. After the rebars have been placed, fasten cleats and riser form boards to side forms. Nail the front section into place.

Figure 5-58. Temporary panels are set up along the stairway construction area for the layout of open stairways. After the soffit panels are positioned, side forms are secured in position and riser form boards are nailed to the side forms.

HIGHWAY CONSTRUCTION

Highway construction includes the construction and maintenance of highway systems. Most of the paving and curbing of road surfaces is accomplished with mechanical slipform paving and finishing equipment. However, form construction is required for bridges, ramps, approaches, and overpasses.

Bridges are a means to cross over natural barriers, such as rivers and canyons, and man-made barriers, such as railroads and highways. Ramps and approaches allow vehicular traffic to merge smoothly onto the highway. Overpasses are bridges that are an integral part of highway systems that provide crossovers for traffic and entrance and exit ramps at intersections. See Figure 5-60. Formwork procedures and materials, such as ties, reinforcement, bracing, and shoring used in the construction of highway bridges and overpasses are similar to other types of concrete structures.

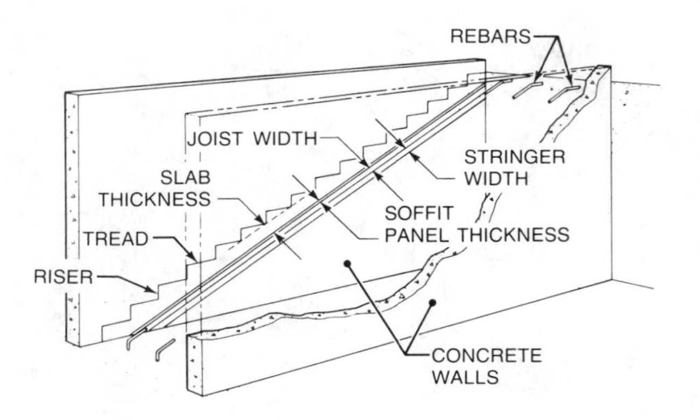

1. Lay out treads and risers on the concrete walls. Lay out and snap lines for the slab thickness, soffit panel thickness, joist width, and stringer width.

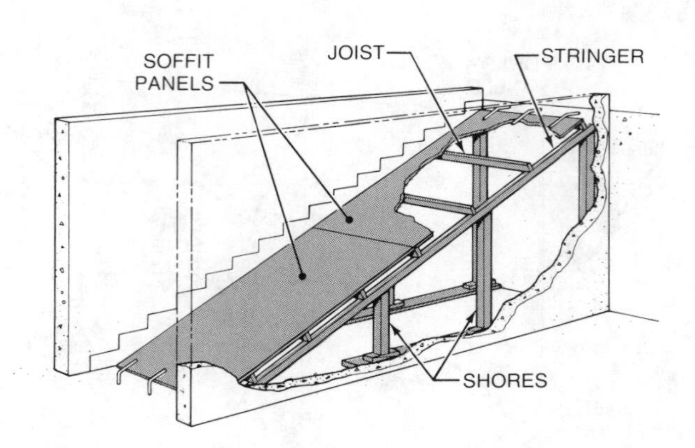

2. Lay out and cut shores to length. Construct the stairway soffit between the concrete walls.

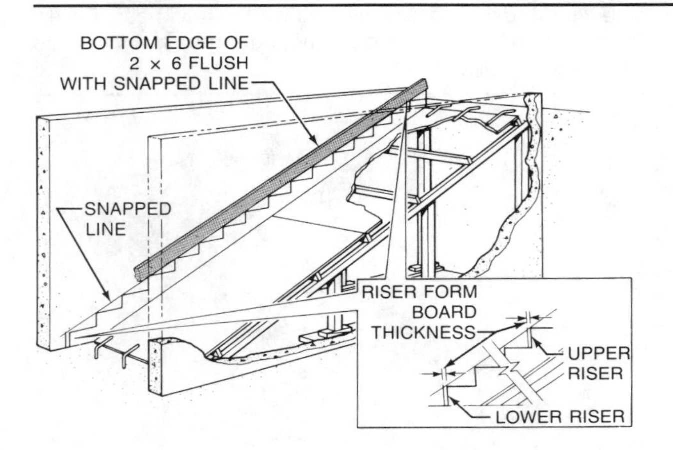

3. Lay out the riser form board thickness along the upper and lower risers. Snap a line from the outside corners of the riser form boards. Nail a 2 × 6 to the concrete walls with the bottom edge flush with the line.

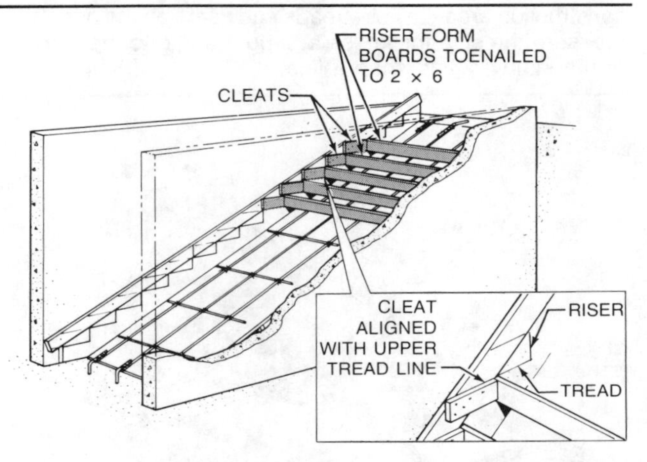

4. Mark the outer corners of the risers on the 2 × 6 and extend tread lines across. Nail form board cleats to the riser and tread marks on the 2 × 6. After the rebars have been placed, toenail the riser form boards to the 2 × 6 and against the cleats.

Figure 5-59. Risers and treads are laid out on the enclosing walls when laying out a closed stairway.

Portland Cement Association

Figure 5-60. Bridges, ramps, and overpasses are an integral part of a highway system.

The structural components of bridges and overpasses are classified as part of the *substructure* or *superstructure*. The substructure is the footings, piers, pier caps, and abutments that support bridges, ramps, or overpass decks. The superstructure is the bridge deck, sidewalks, and parapets (low walls formed along the edges of the deck). See Figure 5-61.

Precast concrete members are commonly used in highway construction. Precast members are prefabricated in casting yards and delivered by truck to the job site. They are lifted in position by crane and tied together with steel dowels and/or welding plates.

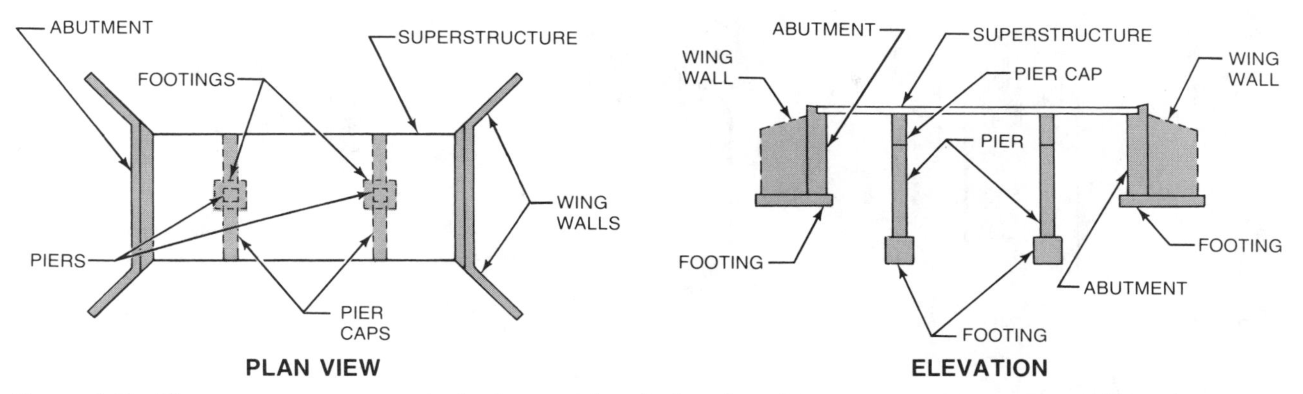

Figure 5-61. The major components of a bridge are classified as the substructure or superstructure. The substructure includes footings, piers and pier caps, abutments, and wing walls.

Substructure

The footings of a substructure are located beneath the piers and *abutments* and provide a solid foundation for the bridge. An abutment is the end structure that supports the beams, girders, and deck of a bridge or arch. Piers extend from the footings and support the *pier cap* and superstructure. The pier cap directly supports and provides a larger bearing surface for the superstructure. See Figure 5-62.

Wing walls are commonly placed monolithically with the abutments. A wing wall is a short section of wall at an angle to the abutment used as a retaining wall and to stabilize the abutment. Grouped piles or friction piles are driven into solid bedrock where poor soil conditions exist. A *pile cap* is then constructed over the piles to transmit loads to the piles. Rebars extending from the piles are tied to the rebars in the pile cap.

Cofferdams are constructed to restrain water when constructing footing forms in rivers, lakes, and other bodies of water. A cofferdam is a large, rectangular, watertight enclosure constructed of interlocking sheet piling. The interlocking sheet piling is driven into the river bottom around the work area, and water is pumped out of the enclosure to permit access for the formwork.

Economy Forms Corporation

Figure 5-62. Piers extending from the footings support pier caps.

Pier forms are constructed over footings after the concrete for the footings has set. Pier forms consisting of sheathing, studs, and/or walers secured with ties are similar to column forms for buildings. Tall piers may require climbing forms that move vertically for successive lifts. Single or multiple piers may be used to support the pier caps. Some pier designs extend the the full width of the superstructure. Pier designs include round, square, rectangular, battered on two or four sides, and inverted-batter. See Figure 5-63. Some pier designs incorporate a *tie beam* (tie strut) midway between the footing and pier cap to tie two piers together. Piers may rest on and be joined to *crash walls*. Crash walls prevent structural damage caused by moving vehicles.

The type of forms used to form piers depends on the shape and height of the piers, and the number of similar piers to be formed. *Custom-made forms* are used to form multiple piers with the same design. Custom-made forms are constructed by form manufacturers to the exact dimensions of the piers. Custom-made forms are constructed of steel, metal-framed plywood panels, or all-wood forms, and may be reused many times. See Figure 5-64.

Piers and footings provide the main support for the bridge superstructure. Abutments and wing walls acting as earth-retaining walls support the ends of bridges. Abutments and wing walls are commonly constructed over large spread footings. The formwork for abutment and wing walls is similar to forming walls for a building. The concrete for an

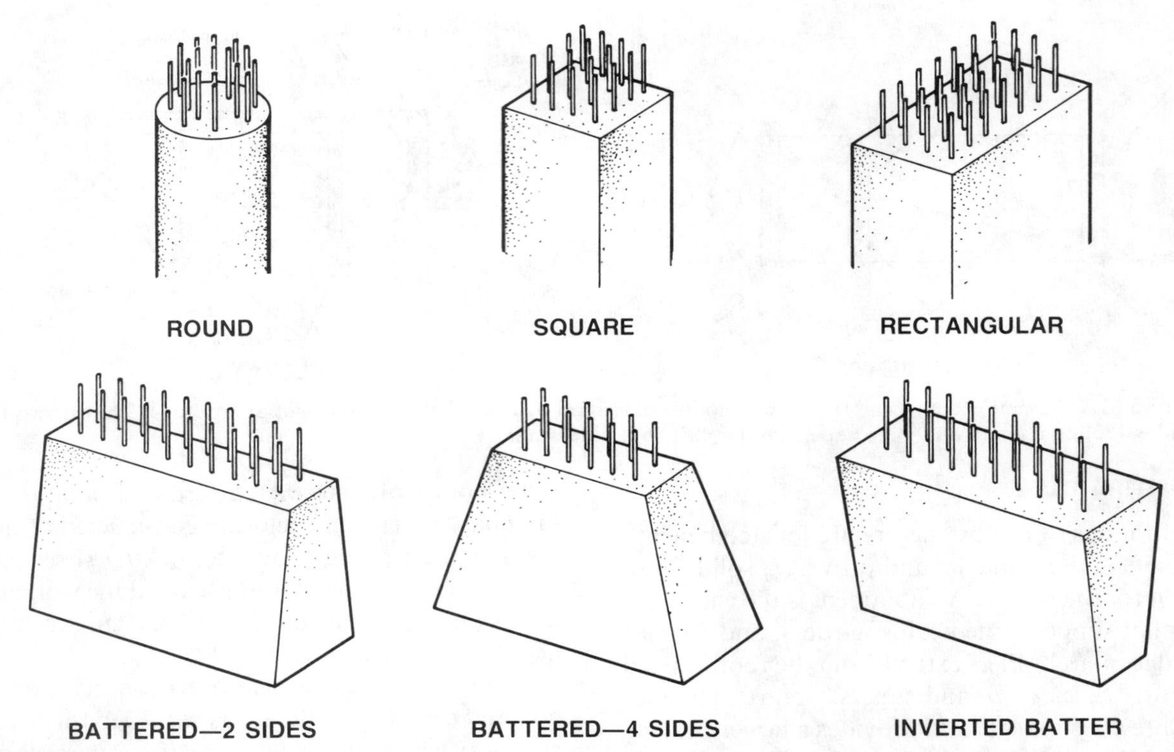

Figure 5-63. Common bridge pier designs are round, square, rectangular, battered on two or four sides, and inverted batter. Rebars extending from the top of the pier tie into the pier cap.

Economy Forms Corporation   Economy Forms Corporation

Figure 5-64. Custom-made forms are used to form multiple piers with the same dimensions.

abutment is placed in two lifts. The first lift forms the *stem wall*, and the second lift forms the *head wall* (back wall). See Figure 5-65. After the stem and head walls are constructed, concrete *bridge seats* are placed on top of the stem wall to support the superstructure girders.

Pier cap forms may be constructed over the completed piers or the concrete for the pier caps can be placed monolithically with the piers. Pier caps support the full width of the bridge superstructure. The *hammerhead pier cap* (T-cap) rests on a single round or rectangular pier. Pier caps serve as tie beams when they are supported by two or more piers. See Figure 5-66.

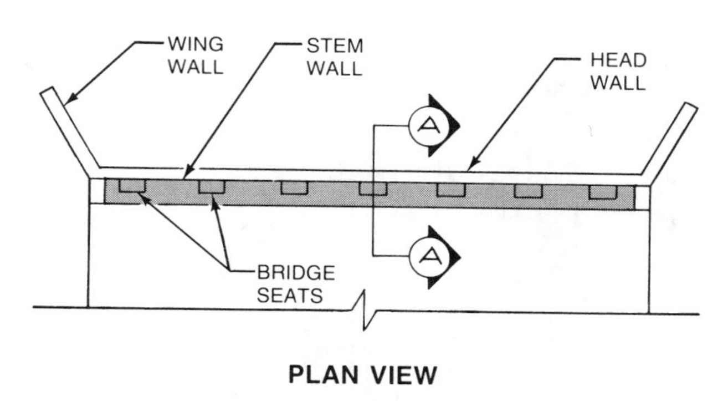

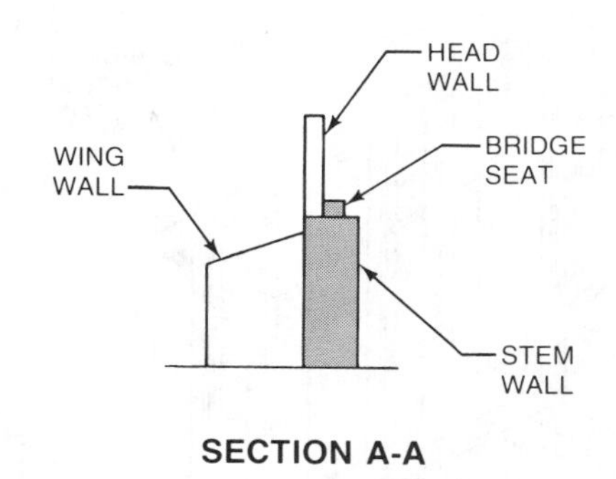

Figure 5-65. An abutment consists of a head wall and a back wall. Abutments and wing walls are earth-retaining walls that support the ends of bridges.

Pier cap forms consist of two sides, a soffit, and bulkheads at the ends. The formwork is reinforced with studs and/or walers with internal disconnecting ties extending between the opposite walls. Lift plates are secured to the top of the forms for the crane attachment. Large pier cap forms are fitted with metal walkway brackets to support a work platform. Wood uprights are inserted into the ends of the walkway brackets and a safety railing is nailed to the uprights.

Friction clamps or *screw jack support brackets* support pier cap formwork over the piers. Friction clamps support small pier cap formwork placed over round columns. A screw jack support bracket supports pier cap formwork for a rectangular pier. Pier cap formwork may also be supported by the existing formwork on the pier. See Figure 5-67. Heavy pier cap formwork is supported by steel beams secured to wide flange beam inserts that are embedded in the pier. Scaffold or timber shoring may also be required for additional support.

Superstructure

The superstructure of a bridge includes the bridge deck, curbs, sidewalks, and parapets. The deck is the concrete surface that supports the traffic load. Curbs form the edge of the deck and are used to direct the flow of rainwater. Sidewalks commonly border the curbs and are used to support pedestrian traffic. Parapets are short walls that act as a safety barrier along the edge of the superstructure.

A bridge deck for short spans consists of a concrete slab resting directly on the pier caps. Long spans may require slabs to be reinforced by concrete or steel girders. Bridge decks are cast-in-place or precast concrete and are supported by steel or precast concrete girders.

PIER CAP FORM

PIER CAP FORM SECURED TO TWO PIERS

Symons Corporation

Figure 5-66. Pier caps are secured to the top of the piers and support the full width of the bridge superstructure.

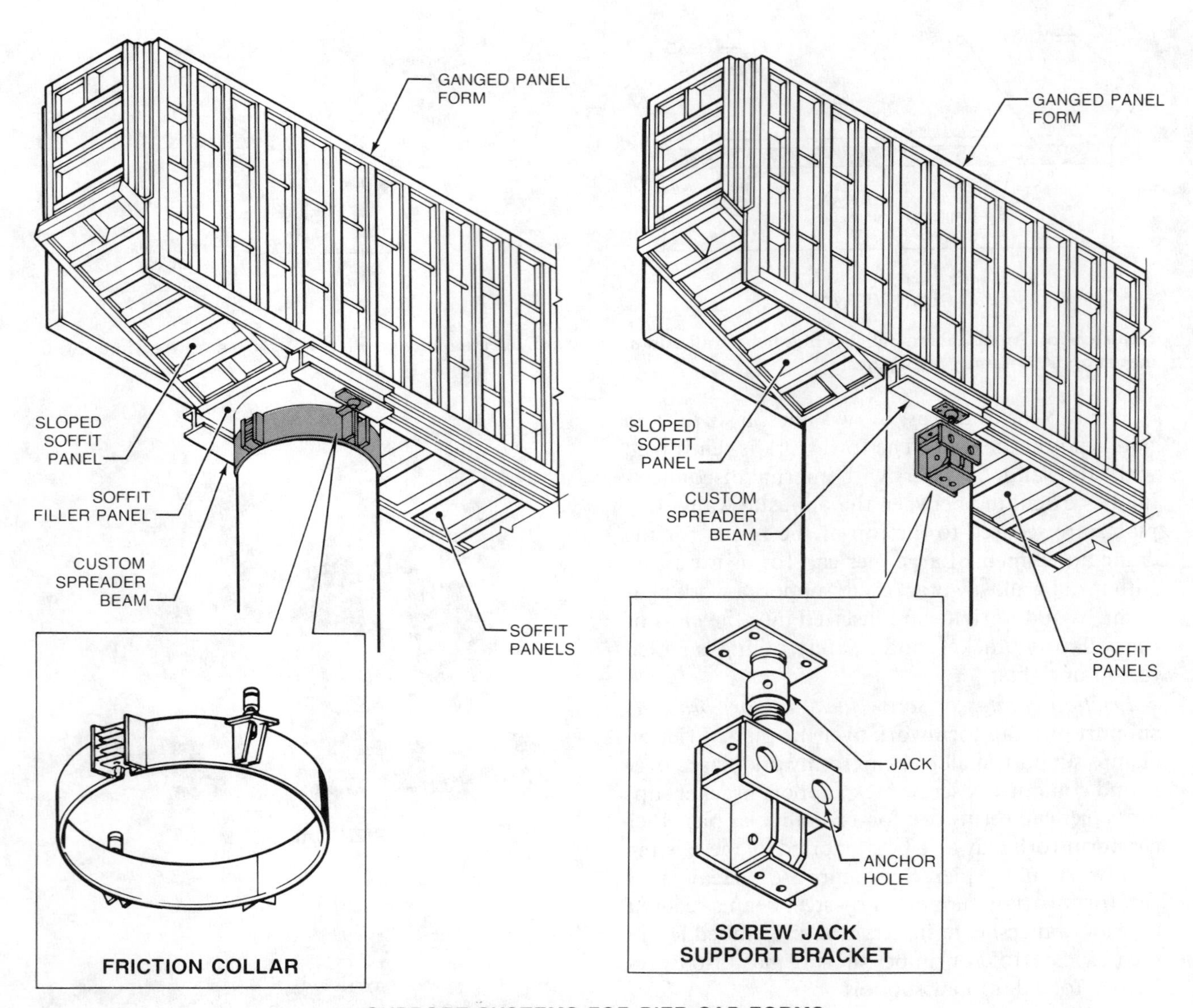

SUPPORT SYSTEMS FOR PIER CAP FORMS

Symons Corporation

PIER CAP FORMWORK SUPPORTED BY EXISTING PIER FORMWORK

Figure 5-67. Friction collars, screw jack support brackets, and existing pier formwork are used to support pier cap forms.

Cast-in-place bridge deck construction requires formwork attached to the pier caps and/or girders. Forms for bridge decks may be built-in-place in a similar manner to constructing formwork for a floor slab. Plywood or metal panels are used for deck sheathing. The deck sheathing is supported by joists and stringers secured in place by wood or metal shores. The joists and deck sheathing must extend beyond the deck width to allow for edge forms, bracing, walkways, and safety railings. See Figure 5-68. The edge form should be at the same elevation as the top of the slab if low parapets are constructed with slipforming machines. Wall forms are not required when using slipforming equipment and the concrete is placed over rebars extending from the bridge deck. When constructing high parapets an outside wall form is used as the edge form.

Shoring for bridge deck formwork is similar to shoring floor slabs for upper levels in concrete buildings. In many cases, bridge deck shoring extends to greater heights and must be designed for traffic movement below. Wood and metal scaffold shoring placed over sills is used to support high bridge formwork. The shoring is braced horizontally and diagonally in both directions at every tier. See Figure 5-69.

Cast-in-place bridge decks placed over steel or precast girders are commonly used to erect bridges, ramps, and overpasses. The girders are spaced closely together and rest on top of the pier caps. Hangers are placed over the girders and bolts supporting ledgers and/or joists are placed between the girders. The bridge deck is then formed with plywood panels placed over the joists. See Figure 5-70.

Patented steel forming systems are also used for cast-in-place bridge decks with short spans between the piers. Sheathing and edge forms supported by trusses are attached to the piers between spans. Additional shoring is not required. After the concrete sets, the entire forming system is stripped and moved to form the bridge deck between the next span of piers.

DECK FORMWORK SUPPORTED BY STEEL SCAFFOLD SHORING

CONCRETE DECK FINISHED OFF WITH MECHANICAL FINISHING EQUIPMENT

Gomaco Corporation

Figure 5-68. Bridge deck formwork is supported by pier caps or girders. Bridge deck sheathing must extend beyond the deck width to support walkways and safety railings.

Figure 5-69. High bridge shoring is braced horizontally and diagonally in both directions at every level.

DECK SHEATHING PLACED OVER JOISTS AND STRINGERS

CONCRETE PLACED IN CAVITIES TIES DECK TO PIER CAP

Figure 5-70. Bridge deck formwork is placed over steel or precast concrete girders. Hangers fit over the girders to secure the joists supporting the bridge deck formwork.

Chapter 5—Review Questions

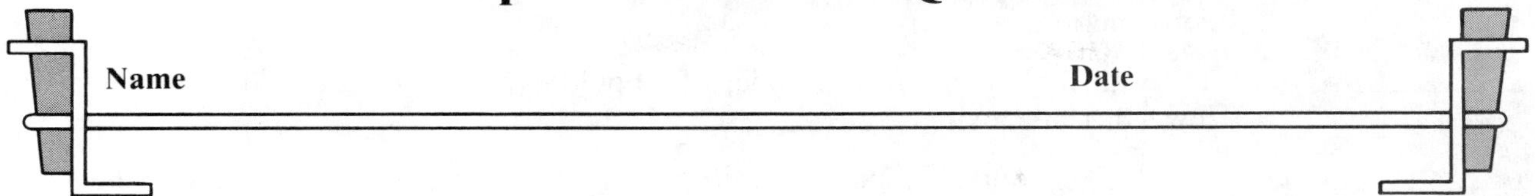

Completion

_______ 1. T-foundations and _______ beam and pile foundations support heavy concrete structures.

_______ 2. Mat and raft foundations are types of _______ foundations.

_______ 3. A(n) _______ beam is a reinforced beam running along the surface of the ground that ties wall or column footings together.

_______ 4. _______ tower cranes are set in position during foundation construction and move upward as the height of the building increases.

_______ 5. A(n) _______ is a long structural member that penetrates deep into soil.

_______ 6. The tip of a concrete pile is protected from damage with a pile _______ while it is being driven into the ground.

_______ 7. A(n) _______ pile is driven completely through unstable soil layers and rests on firm load-bearing soil.

_______ 8. In the _______ method of cast-in-place piles, a metal casing is driven into the ground and remains in place while the concrete is being placed.

_______ 9. The pile _______ is the upper surface of a precast pile in its final position.

_______ 10. _______ are used in heavy construction projects where the building design and/or soil conditions make pile driving difficult or inadequate.

_______ 11. A(n) _______ caisson provides a large bearing area at the base of the caisson.

_______ 12. _______ piles are placed beneath a concrete cap and are a base for load-bearing columns.

_______ 13. A(n) _______ is the concrete placed between two horizontal construction joints.

_______ 14. _______ panel forms are constructed by bolting many small panels together.

_______ 15. Short radius curves are formed by using two or more layers of _______" plywood.

_______ 16. Large panel forms and ganged panel forms are types of _______ forms.

_______ 17. A(n) _______ joint is formed when the concrete of a wall section is placed on top of a previously placed concrete wall section.

_______ 18. A(n) _______ is constructed inside a wall form where a vertical construction joint is formed.

__________ **19.** A(n) __________ is embedded in a concrete wall to prevent water leakage at a vertical construction joint.

__________ **20.** The wall form panel for an upper lift should overlap the hardened concrete for the lower lift approximately __________″.

__________ **21.** __________ joints are shallow grooves made in a concrete wall to control cracking.

__________ **22.** __________ support beams and other bending loads.

__________ **23.** A(n) __________ wall is a light non-load-bearing section of a wall made of metal or precast lightweight concrete.

__________ **24.** The floor slab in a flat slab system is directly supported by columns and __________ panels.

__________ **25.** A(n) __________ is a flared section at the top of a column.

__________ **26.** Columns tie directly into the floor slab above without using capitals or drop panels in a flat __________ system.

__________ **27.** __________ strips are placed in the corners of column forms to produce columns with beveled edges.

__________ **28.** Round columns are formed using tubular __________ forms.

__________ **29.** A(n) __________ is a round column form constructed of a single piece of molded fiberglass.

__________ **30.** The sheathing material for beam and girder side forms is __________.

__________ **31.** A(n) __________ is nailed against the bottoms of beam and girder side forms to secure them in position.

__________ **32.** A(n) __________ tie is a snap tie with a hooked end that is driven into deck sheathing at the top of a spandrel beam form.

__________ **33.** A(n) __________ pan is secured with its flanges nailed to the top of the soffit.

__________ **34.** The head of a(n) __________ shore is centered on top of a vertical post.

__________ **35.** __________ are shores placed under structural members after the original shoring has been removed.

__________ **36.** __________ stairways are enclosed by walls on both sides.

__________ **37.** The vertical surface of a step is the __________.

__________ **38.** The __________ of a bridge includes the footings, piers, pier caps, and abutments.

__________ **39.** A(n) __________ is constructed to restrain water when constructing footing forms in rivers, lakes, and other bodies of water.

__________ **40.** A bridge __________ is the concrete surface that directly supports the traffic load.

Multiple Choice

__________ **1.** Two-way joist systems are constructed with __________ pans.
A. long
B. adjustable
C. dome
D. all of the above

__________ **2.** A __________ is placed on top of a pile head to receive the pile driver's blows and protect the head from damage.
A. driving shoe
B. pile cutoff
C. driving head
D. pile foot

__________ **3.** __________ steel piles are used as a foundation support and for shoring around deep excavations.
A. Bearing
B. H-shaped
C. Tubular
D. Friction

__________ **4.** When placing a concrete wall in two lifts, a row of tie rods is embedded __________" from the top of the lower lift.
A. 4
B. 6
C. 8
D. 10

__________ **5.** When placing a concrete wall in two lifts, a row of tie rods is embedded __________" above the bottom of the upper lift.
A. 4
B. 6
C. 8
D. 10

__________ **6.** __________ shores are placed under beam and girder forms.
A. T-head
B. Double post
C. Scaffold
D. all of the above

__________ **7.** __________ pans for one-way joist systems are nailed to the sides of the soffit.
A. Dome
B. Nail-down
C. Adjustable
D. Long

__________ **8.** __________ pans are used to form one-way joist systems.
A. Flanged adjustable
B. Long
C. Nail-down
D. all of the above

_________ **9.** A _________ is used to secure the two members of a two-piece adjustable shore.
A. cleat
B. post clamp
C. brace
D. stake

_________ **10.** Wood shores placed over poor load-bearing soil should rest on _________.
A. ¾″ plywood
B. 2 × 10 planks
C. 3 × 10 mudsills reinforced with 4 × 4 or 4 × 6 plates
D. 4 × 6 plates

_________ **11.** Splicing cleats for wood shores extend a minimum of _________″ on each side of a splice.
A. 6
B. 8
C. 10
D. 12

_________ **12.** _________ shores remain in place as formwork around them are removed.
A. Permanent
B. T-head
C. Scaffold
D. Adjustable metal

_________ **13.** In a stairway, the _________ is the vertical distance from a floor to a floor above.
A. total run
B. total rise
C. unit run
D. unit rise

_________ **14.** The _________ of a bridge includes the bridge deck, sidewalks, and parapets.
A. superstructure
B. substructure
C. abutment
D. none of the above

_________ **15.** A(n) _________ supports the ends of a bridge deck or arch.
A. pier
B. wing wall
C. abutment
D. cofferdam

_________ **16.** A _________ wall is a retaining wall used to stabilize an abutment.
A. head
B. stem
C. back
D. wing

_________ **17.** When placing concrete for an abutment, the first lift forms the _________ wall.
A. stem
B. wing
C. back
D. head

_______ **18.** _______ are low walls that act as a safety barrier along the edge of a bridge superstructure.
- A. Wing walls
- B. Parapets
- C. Head walls
- D. Bridge seats

_______ **19.** _______ support pier cap formwork for bridge piers.
- A. Friction clamps
- B. Screw jack support brackets
- C. Pier forms
- D. all of the above

_______ **20.** _______ are placed on top of the stem wall to support the bridge superstructure girders.
- A. Crash walls
- B. Head walls
- C. Bridge seats
- D. Pier caps

Identification

_______ **1.** Beam bottom form

_______ **2.** Form tie

_______ **3.** Shore head

_______ **4.** Stud

_______ **5.** Stiffener

_______ **6.** Beam side form

_______ **7.** Post

_______ **8.** Brace

_______ **9.** Chamfer strip

_______ **10.** Joist ledger

_______ **11.** Kicker

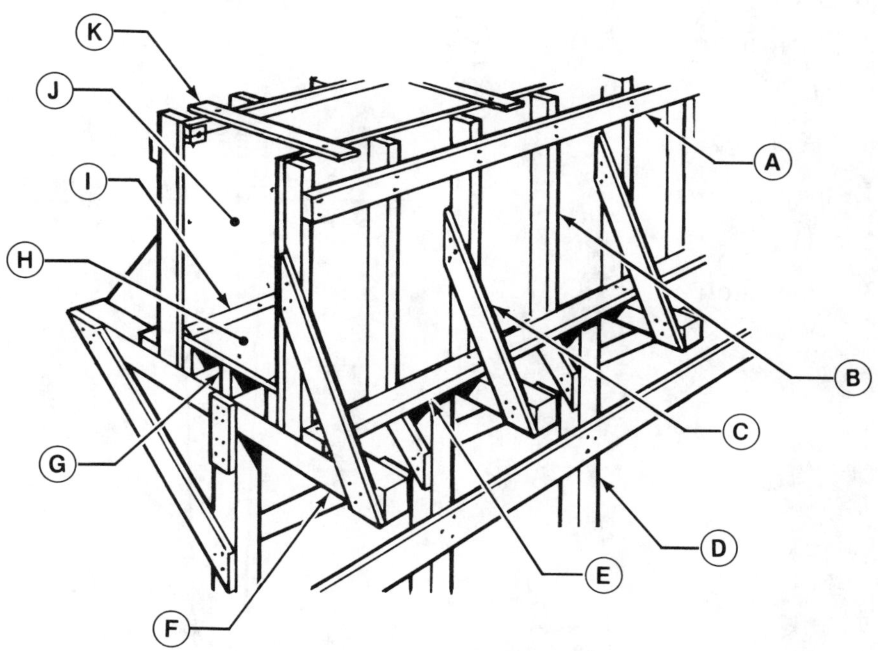

__________ **12.** Template

__________ **13.** Cleanout door

__________ **14.** Stiffener

__________ **15.** Chamfer strip

__________ **16.** Adjustable metal scissors clamp

A

B

C

D

E

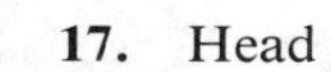

__________ **17.** Head

__________ **18.** Pile shoe

__________ **19.** Tip

__________ **20.** Foot

__________ **21.** Driving head

__________ **22.** Butt

__________ **23.** Pile cutoff

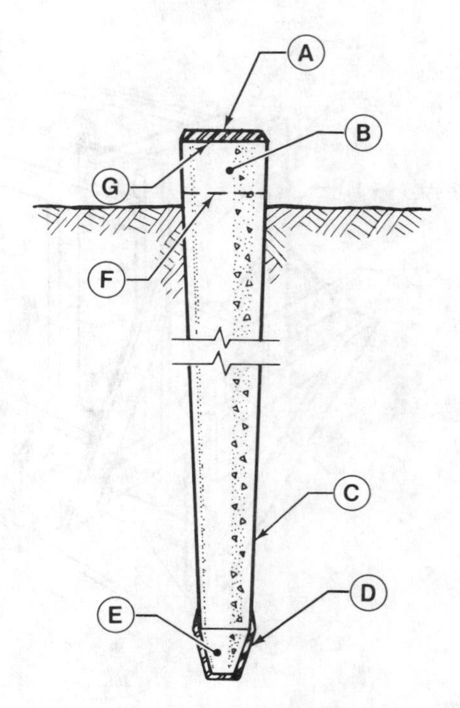

CHAPTER 6

Precast Concrete Construction

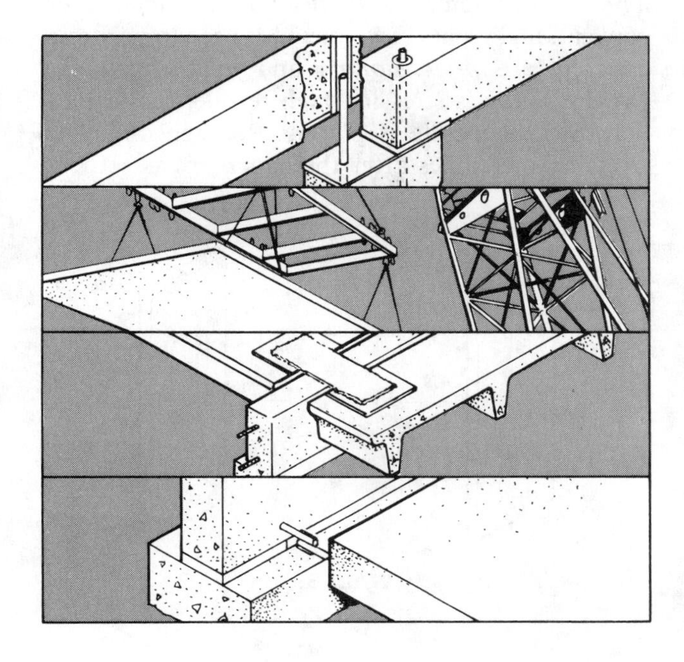

Precast concrete systems have gained wide acceptance in recent years. Many modern concrete structures are partially or completely constructed with precast concrete members. Precast concrete members are also used in the construction of bridges, tunnels, and wharves.

Precast concrete construction is advantageous in many situations. Formwork costs are reduced because fewer and simpler forms are required, less scaffolding is used to support the precast members, and production schedules are not affected as much as schedules for cast-in-place concrete.

Precast structural members are fabricated in a factory and transported to the job site by truck, or fabricated on the job site. The design and quantity of precast members, location of the casting factory, and cost of transportation are considered when determining whether to cast the precast members at a factory on the job site.

Precast members are fabricated by placing reinforcement and concrete into forms constructed over casting beds. When the concrete has set, the precast members are raised and positioned by crane. The precast members are braced in position and tied together using steel dowels and/or welding plates.

Rebars or prestressed steel cables are used to reinforce precast concrete members. Prestressed steel cables are used to reinforce precast concrete members by pretensioning or post-tensioning. Pretensioning is stressing the prestressed steel cables before the concrete is placed. Post-tensioning is stressing the steel cables after the concrete has been placed.

PRESTRESSED CONCRETE

Prestressed concrete is concrete placed in a state of compression by stressing the concrete with high tensile steel cables. Prestressed concrete members may be precast or cast-in-place. Precast concrete members for structures include walls, floor slabs, beams, girders, piles, and columns. See Figure 6-1. Precast members are reinforced with conventional rebars or *prestressed* steel cables. Prestressed steel cables are a recent development and are an improvement over conventional rebars. Two methods used to prestress concrete are *pretensioning* and *post-tensioning*. Pretensioning is stressing high tensile steel cables before the concrete is placed. Post-tensioning is stressing the cables after the concrete has set.

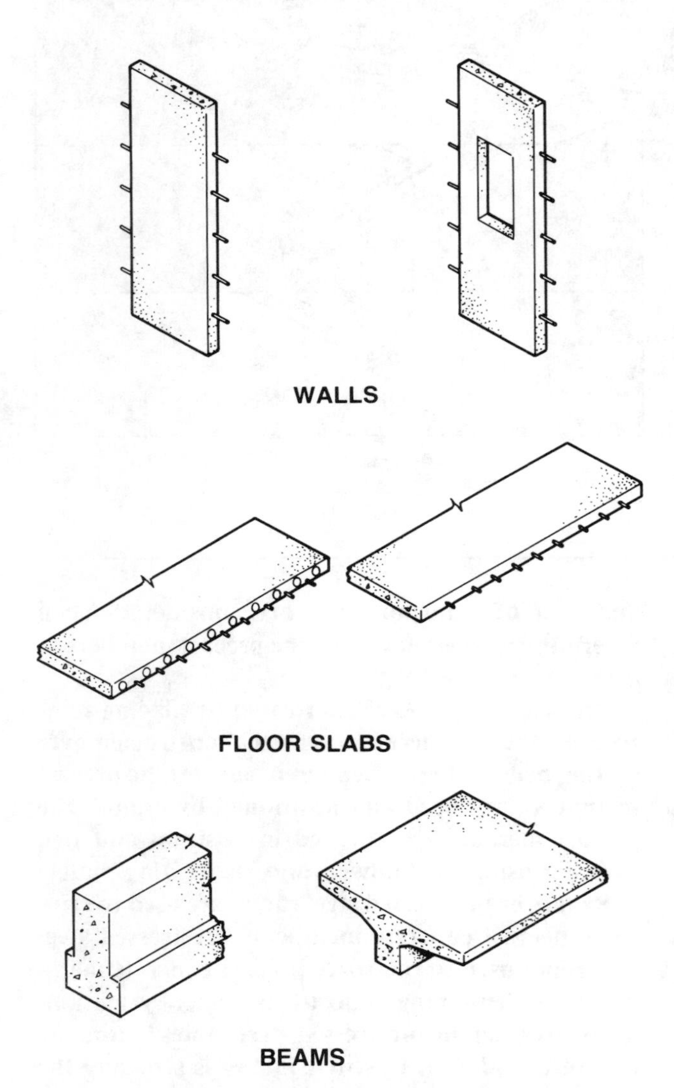

Figure 6-1. Precast members for precast concrete structures include walls, floor slabs, and beams.

Pretensioned concrete is commonly used for factory-produced members. Powerful jacks stretch the cables until they are under the required tension for a particular structural member. Concrete is then placed in the form. When the concrete has set to its specified strength, the tension from the jack is released. As the cables return to their original state, the concrete member is placed under compression, resulting in greater resistance to lateral loads and pressures than conventionally reinforced concrete. Prestressed members require less concrete and are lighter than members reinforced with rebars.

Post-tensioned members are commonly formed at the job site. Concrete is placed around unstressed cables that are enclosed in flexible metal or plastic ducts. The cables are stressed and anchored at both ends after the concrete has been placed and set. Post-tensioning is also used with cast-in-place concrete.

Job Site Precast Construction

Job site precast construction lowers costs and increases productivity when producing multiple precast units of a design. The time required for precast formwork is considerably less than cast-in-place concrete because the forms are constructed on casting beds built at ground level. Positioning of rebars, and placing, consolidating, and finishing the concrete are easier. Precasting of structural members is done during the construction of the foundation. Precast members can carry full loads, thereby eliminating the need for temporary shoring. See Figure 6-2.

Job site precasting is used when there is adequate space for a casting yard within or adjacent to the job site, and when lifting equipment (cranes or hoists) is available. A casting yard must be well-organized for efficient assembly line production. Materials used to build the forms, the reinforcing steel, and an area for stockpiling completed precast members should be nearby. If concrete is mixed on the job site, an adjacent area for storing cement, aggregates, and water is required.

Casting Beds and Forms

Casting beds are a base and support for the forms. The surface of a casting bed is smooth, level, and free from defects. Casting beds must be rigidly supported to prevent deflection from the weight of the concrete. A casting bed is constructed of plywood panels supported by timbers or a concrete slab.

CONCRETE PLACED IN CASTING BEDS

PANELS RAISED BY CRANE

PRECAST PANELS SET IN PLACE

Portland Cement Association

Figure 6-2. Entire structures may be constructed of precast wall panels and floor slabs that are cast at the job site. When the concrete has set, they are raised into position with a crane.

Precast wall and floor panels require edge forms around the perimeter of the casting bed. Wood casting bed forms are commonly lined with plastic or hardboard to facilitate stripping and increase the life of the form.

Prefabricated metal and plastic forms are also available for standard structural units, or they can be custom-made to different dimensions. Prefabricated metal and plastic forms have a long life expectancy and can be reused many times. See Figure 6-3.

Precast forms are designed for easy stripping. The side sections are hinged or bolted so they can be folded down or removed easily. The forms must also be oiled or treated with some type of release agent

REBARS POSITIONED PRIOR TO PLACING CONCRETE

FINISHING CONCRETE—BOLTS POSITIONED IN INSERTS

Symons Corporation

Figure 6-3. Prefabricated casting beds are commonly used to form precast concrete members. Reinforcement is placed in the beds prior to the the concrete placement.

to facilitate stripping. After the concrete has set, the precast member is lifted out of the form by crane.

Inserts, anchors, or lifting units are placed in precast members for the crane attachment. See Figure 6-4. Threaded or coiled inserts are positioned below the surface of thin panels. Plastic caps are screwed into the insert while the concrete is placed. After the concrete has set, the plastic caps are removed and replaced with bolts to hold the lift plate. Anchors are used to raise the precast panels when the concrete surface is not visible. The anchors are embedded in the concrete with the opening for the crane attachment protruding from the surface. After the panel is raised the anchor is left in place. Lifting units are used on heavy panels and usually have a steel bar that is recessed in the precast member.

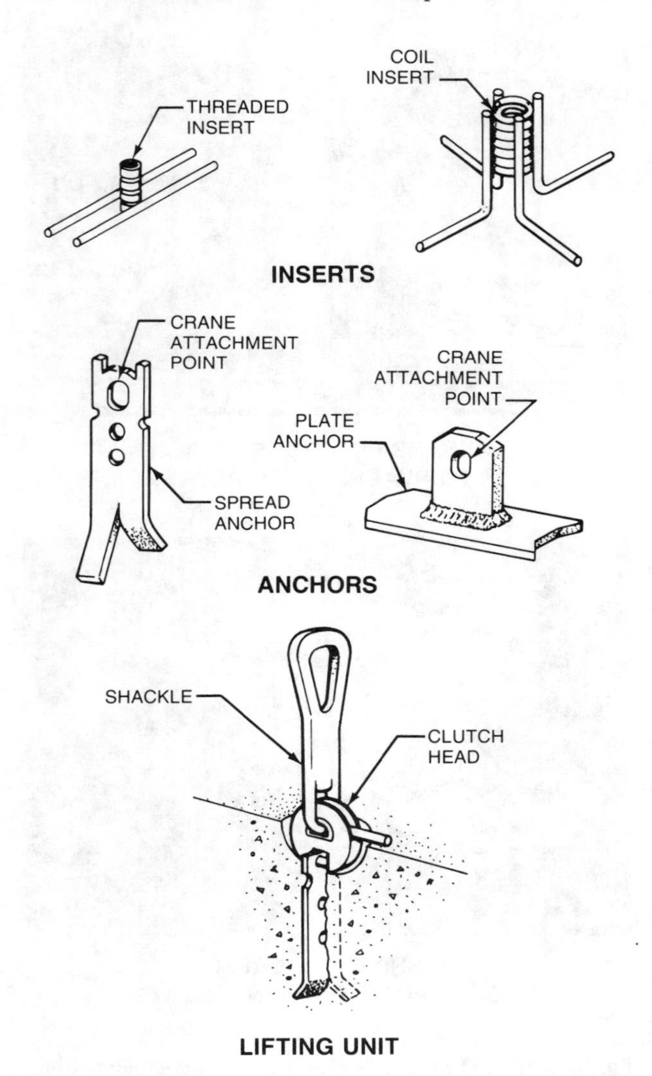

Figure 6-4. Inserts, anchors, and lifting units are embedded in precast members for the crane attachment.

The lifting unit has a flexible vinyl piece that, when removed from the precast member, creates a semicircular impression. A clutch head engages the embedded steel bar to a shackle that is hooked to the crane attachment point. After the precast member is positioned and released by the crane, the lift plates and bolts are removed, and the bolt holes are patched with a sand and cement grout mixture.

Connecting Methods. Precast members must be connected or anchored to adjoining parts of the structure after they are positioned. Steel plates or angles, dowels, tensioning bars, and bolts are methods commonly used for connections. A column resting on a footing is often secured with bolts placed in the footing. The bolts align and extend through steel plates attached to the column base. Beams resting on the top of columns are tied together by welding steel angles to tensioning bars extending from the ends of the beams. Beams are also secured with steel dowels projecting from the column below. The steel dowels slip into steel tubes embedded in the beam and the tubes are filled with grout. Precast members can also be connected with a post-tensioning bar that is stressed and secured into position with a washer after placing grout between the column and beam end. Floor slabs are commonly anchored to precast beams or walls with welding plates. Steel angles that are embedded in the precast beam are welded to the plates. The plates are welded to rebars running through the floor slab. Steel plates can also be embedded in the slabs and welded to a steel connection plate placed across the precast beam. See Figure 6-5.

Tilt-up Construction

Tilt-up construction is an on-site precast method that is commonly used in the construction of one- and two-story buildings, although higher structures may also be erected using tilt-up methods. In tilt-up construction, a concrete slab serves as the casting bed for the wall panels. The casting bed is often the ground floor slab of the building. However, when panels must be cast outside of the building, a temporary concrete slab, wood platform, or well compacted fill is used as a casting bed.

Floor slabs used as casting beds must be level and have a smooth trowel finish. The slab, as well as the compacted subgrade below, must be strong enough to carry material trucks and mobile cranes.

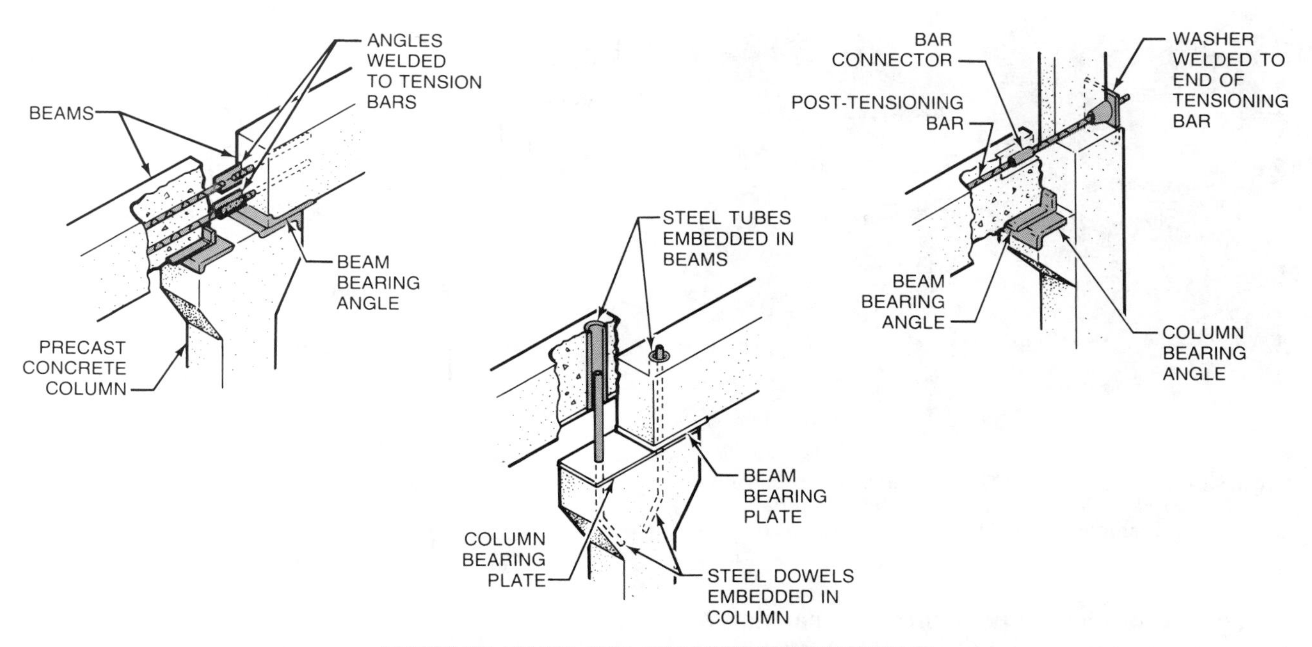

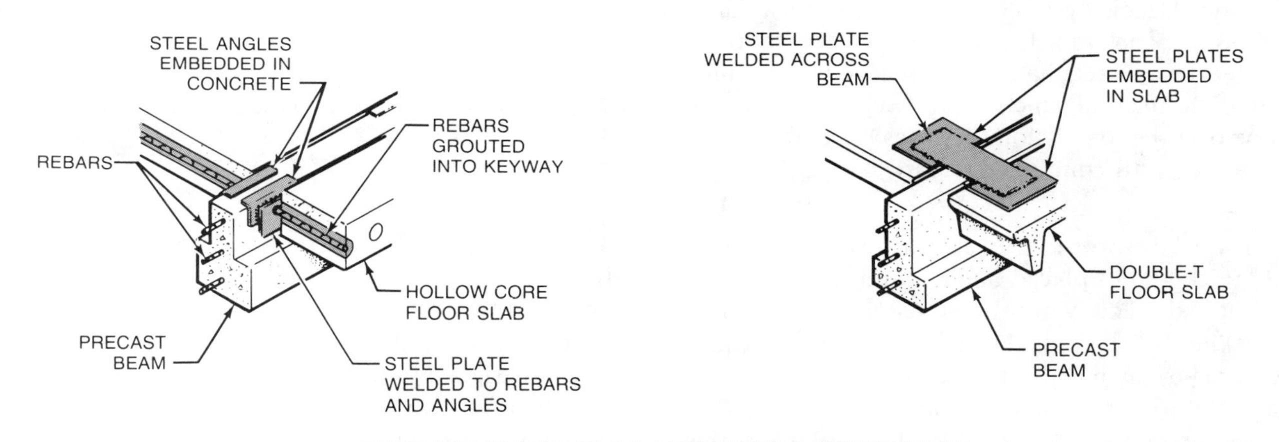

Figure 6-5. Steel welding plates, angles, dowels, and tensioning bars are used to connect precast members.

Pipe openings and other utility openings should be temporarily filled with sand and topped off with a $^3/_4$" layer of concrete. See Figure 6-6.

A bond breaking agent is sprayed on the floor slab before the wall panels are constructed to ensure a clean lift when they are raised. Chemical, resin, and wax-base compounds that act as a bond breaking agent and also as a curing compound are commonly used. A second coat of compound is applied after the edge forms have been constructed and before rebars and inserts are placed.

After the edge forms for the casting beds have been built and secured to the floor slab, window and door bucks are fastened into position. Rebars are then positioned. The size and spacing of the rebars are based on the dimensions of the wall and the anticipated vertical and lateral loads. The areas around door and window openings and the edges of the wall are more heavily reinforced than other parts of the wall panel. Wall panels containing larger openings also require additional reinforcement to withstand the strain at the time of lift. Electrical conduit, outlet boxes, and inserts for crane attachments are positioned after the rebars are positioned.

The Burke Company

Figure 6-6. Forms for tilt-up wall panels are constructed on the floor slab. Window and door bucks, rebars, and inserts are positioned prior to placing the concrete.

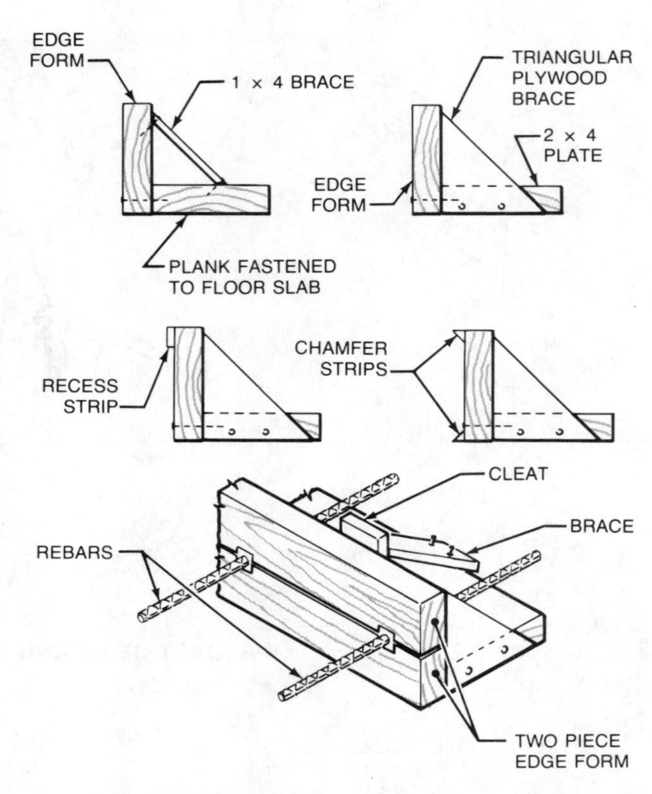

Figure 6-7. Edge forms for tilt-up construction are constructed with 2″ thick planks. Offset and beveled edges are formed by using recess or chamfer strips.

Avoid disturbing the bond breaking materials on the floor slab during the vibrating and working of the concrete. To break the impact of the concrete during placement, the concrete is placed on a slanted plywood panel, rather than directly on the floor slab. After the concrete has set and reached the required strength, the wall panels are raised and set into position over the foundation footings. They are then tied together with columns or chord bars.

Tilt-up Formwork. Tilt-up formwork consists of 2″ thick planks placed on edge and fastened to the floor slab. The width of the planks vary with the thickness of the wall. The edge form planks can be secured by laying other planks behind the edge form and bolting or pinning the flat planks to the floor slab. The bottom of the edge form plank is then nailed to the flat plank, and the top of the edge form is secured with short 1 × 4 braces. Another method is to brace the edge forms with triangular plywood braces nailed to short 2 × 4 pads that are secured to the floor slab. Metal brackets are also available for securing edge forms. See Figure 6-7.

Chamfer strips nailed to the edge form create beveled corners at the vertical ends of walls where they butt against columns. A recess strip is used to produce an offset where the vertical end of the wall panel fits into the recessed channel of a column. Rebars must extend a minimum of 12″ beyond the edge forms for wall panels that are tied together with cast-in-place columns. A two-piece edge form notched to accept rebars facilitates stripping the forms.

Tilt-up Inserts. Tilt-up inserts and lift plates or remote-release lift systems are required for crane lift points in the wall panel. One type of coil insert is positioned along the flat plane of the wall. Another type is designed to permit edge lifting of the panel. Inserts must also be placed in the walls to anchor the tops of wall braces. Plastic caps are temporarily screwed into the inserts to protect the threads from sand, grit, and water. The plastic cap is later removed and lift plates are bolted to the inserts. See Figure 6-8.

Remote-release systems are commonly used to raise wall panels and enable the construction workers to release the lifting units from ground level. Remote-release systems consist of an insert embedded in the wall panel and a lifting unit. An insert is supported by a base and positioned at a predetermined lift point prior to concrete placement. See Figure 6-9. A *clutch-type insert* consists of a T-bar anchor and a recess former supported by a base. The T-bar anchor provides a hook point for the lifting unit. The recess former fits over the top of the T-bar anchor and creates a void for the clutch of

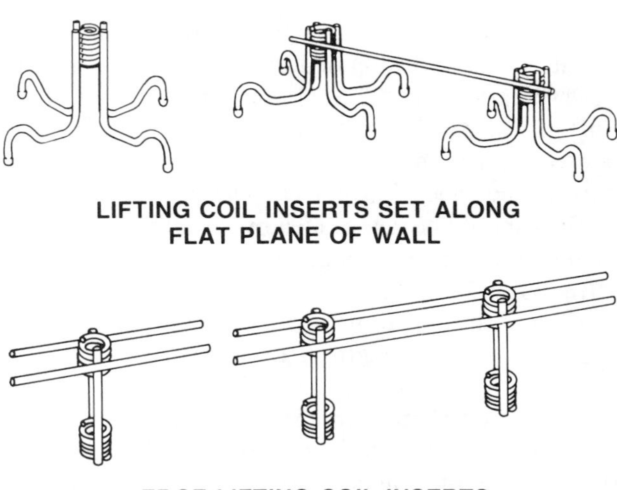

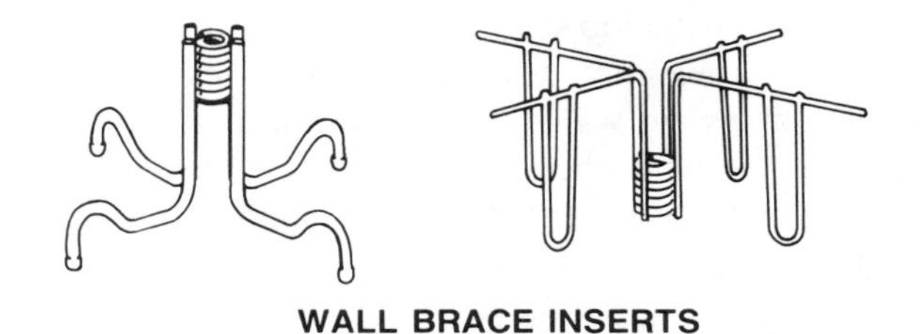

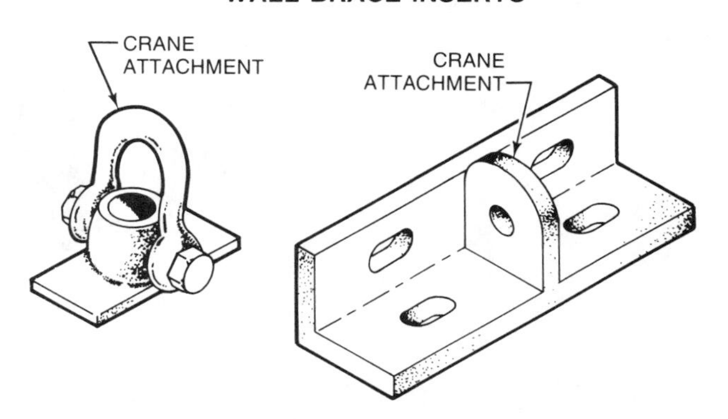

Figure 6-8. Single or double inserts are embedded in the precast panels. Lift plates are bolted to the inserts for the crane attachment.

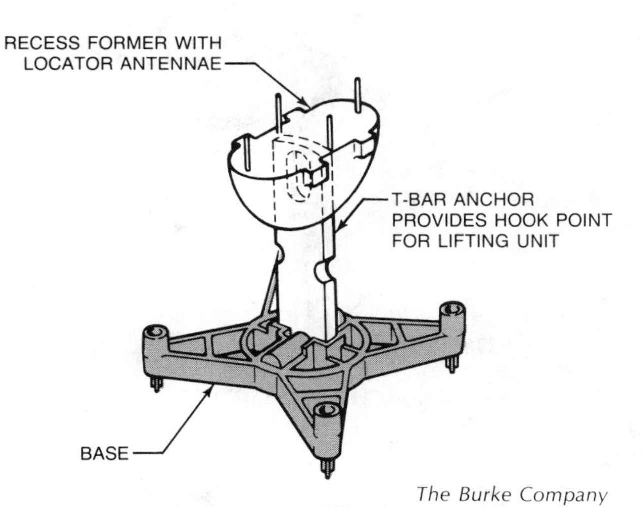

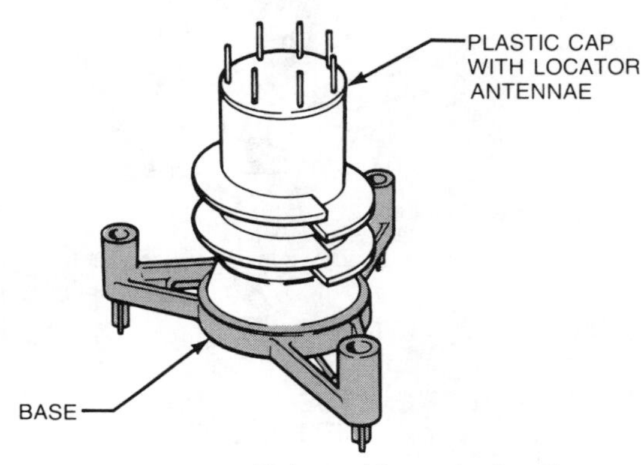

Figure 6-9. Inserts for remote-release systems are embedded in tilt-up wall panels. Lifting units are attached to the inserts and the wall panels are raised into position.

the lifting unit. Locator antennae extend from the top of the recess former to indicate the position of the insert after the concrete has been placed. An *encasement ball insert* is a one-piece insert that rests on a base. A plastic cap with locator antennae is secured in the top of the insert to prevent concrete from accumulating inside. After the concrete has set, the recess former or plastic cap is removed and the lifting unit is attached.

Lifting units for remote-release systems are attached to inserts after the concrete has obtained its required strength. See Figure 6-10. The lifting unit for a clutch-type insert consists of a clutch ring mechanism and a shackle. The clutch ring is lowered into the preformed void and attached to the T-bar anchor by pushing the clutch bar against the wall panel. The shackle is slipped over a crane hook and the wall panel is raised into position. After the wall panel is braced, a lanyard attached to the clutch bar is pulled and the lifting unit is released. Lifting units for encasement ball inserts consist of a shaft with encasement balls and an adjustment mechanism, two spring-loaded plungers, and a shackle. The encasement balls are forced against the sides of the insert as the adjustment mechanism is screwed into the lifting unit. As the lifting unit is positioned in the insert, the spring-loaded plungers are depressed. A safety stop key is placed against the adjustment

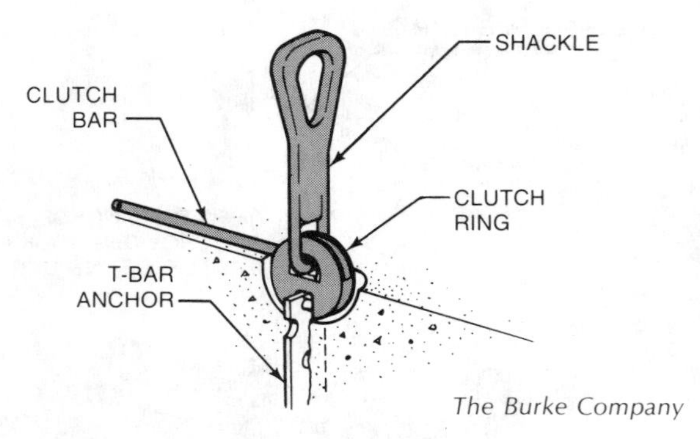

The Burke Company

CLUTCH-TYPE LIFTING UNIT

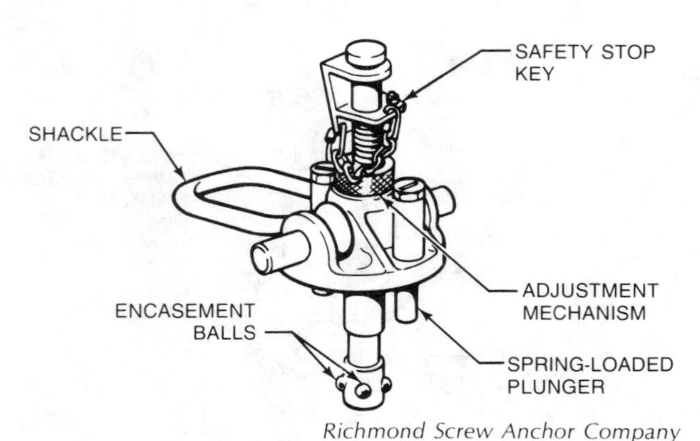

Richmond Screw Anchor Company

ENCASEMENT BALL LIFTING UNIT

Figure 6-10. Lifting units for remote-release systems are fastened to the embedded inserts. The lifting units are released from ground level.

mechanism to secure the encasement balls in position. The shackle is slipped over a crane hook and the wall panel is raised into position. After the wall panel is braced, a lanyard attached to the safety stop key is pulled to release the pressure between the encasement balls and the insert. The spring-loaded plungers eject the lifting unit from the insert.

Raising and Bracing Wall Panels. When precasting a wall panel on a floor slab, the formed panels are cast next to each other in a row. After the concrete has set and gained sufficient strength, the wall panels are raised and set into place in one continuous operation. Wall panels may also be *stack cast*. Stack casting consists of panels cast on top of each other and then raised and placed in position.

Cranes used for lifting tilt-up wall panels are equipped with horizontal spreader bars. Steel cables are attached to the lift plates or lifting units of the wall panels and threaded over pulleys fastened to the spreader bars. The *lift points* (placement of the lift plates or lifting units) must be positioned carefully to equalize the lifting force when the wall panels are raised. See Figure 6-11. The layout of the lift points should be determined by a qualified engineer. An engineer often utilizes a computer to detail the lift points based on

1. weight and dimensions of wall panel,
2. concrete strength at time of lift,
3. type of concrete used,
4. location and dimensions of openings,
5. preferred rigging configuration.

After the lift points are determined, the crane size is selected based on the lift point position, position of panel on the job site, and potential obstructions. Wood timbers may be used as strongbacks for thin wall panels or wall panels containing numerous openings. The timbers are temporarily bolted to the wall panels to prevent structural damage from lifting stress.

The Burke Company

Figure 6-11. The lift points for tilt-up wall panels must be positioned to equalize the lifting force when raising them into place.

A wall panel must be temporarily braced after it is raised in position. The braces must be able to resist all lateral stress, including wind stress. Telescoping steel braces are commonly used to resist lateral stress. The brace shoe is bolted to an insert in the wall. A telescoping brace is adjusted to rough size and the panel is erected. The bottom of the brace shoe is attached to an insert in the floor. A screw jack located at the lower end of the brace is used for final adjustment. Coil or threaded inserts for the braces must be laid out accurately so the wall inserts align with inserts embedded in the floor slab. See Figure 6-12.

Braces are commonly anchored by placing the anchoring devices after the floor slab has set. After drilling a hole completely through the concrete floor slab, a wedge bolt with a bottom lip is placed into the hole. A wedge is then driven next to the bolt to secure it in position and engage the bolt lip to the bottom of the slab. A brace shoe is then fastened to the bolt. See Figure 6-13. When the bracing is no longer required, the wedges are pried up and the bolts are removed. The holes are then filled with a sand and cement grout mixture. Expansion bolts are not recommended for anchoring braces as they may not withstand the tension and shear forces exerted by the wall.

The Burke Company

Figure 6-12. Telescoping braces secure precast wall panels to resist anticipated lateral stress.

Securing Wall Panels. After wall panels have been raised, they are permanently fastened into position. Wall panels can be fastened by placing the panels directly on grout pads placed on top of the foundation footing. Rebars extending from the wall are welded or tied to rebars extending from the floor slab. The area between the bottom of the wall panels and the foundation is then filled with grout. Concrete is placed between the wall and the floor slab. See Figure 6-14. Other methods include placing the bottom of the wall in a slot at the top of the foundation wall; slipping the walls over steel dowels extending vertically from the foundation; and securing the wall panel bottoms with steel welding plates.

Various methods are used to tie vertical edges of wall panels together. Cast-in-place columns may be formed between the wall panels. Recent tilt-up designs feature independent precast columns positioned before the wall panels are placed, or columns cast monolithically with the wall panels. Independent precast columns are formed with oversize recesses that accommodate the edges of the wall panel. The wall panels are secured to the columns with steel welding plates. When columns and wall panels are cast monolithically, one-half of a column is formed at both ends of the wall panel. The half column is tied to a half column extending from an adjoining wall panel. Another tilt-up method utilizes heavy *chord bars* that extend horizontally through the wall panels. Chord bars are heavy rebars that resist lateral pressure exerted on the wall panels, and tie the wall panels together. When forming the wall panels a small pocket around the ends of the chord bars is blocked out. After the wall panels have been raised and positioned, the exposed bars are welded together and the pocket is filled with concrete. See Figure 6-15.

The exterior wall panels for tilt-up structures that are at least two stories extend the entire height of the structure. The floor slabs are suspended and cast-in-place. The floor slab forms are supported by stringers held in position by wood or metal scaffold shoring or by wood or metal joists or trusses. The

1. Insert the bolt through a predrilled hole in the floor slab. The bolt lip must be opposite the brace shoe and engage the bottom of the slab.

2. Insert the wedge opposite the bolt lip and drive it into the slab until the removal handle touches the slab.

3. Slide the brace shoe over the bolt. Tighten a hex nut on top of the brace shoe and secure with a nut lock.

The Burke Company

Figure 6-13. Wedge bolts are used to fasten the brace shoe to the floor slab.

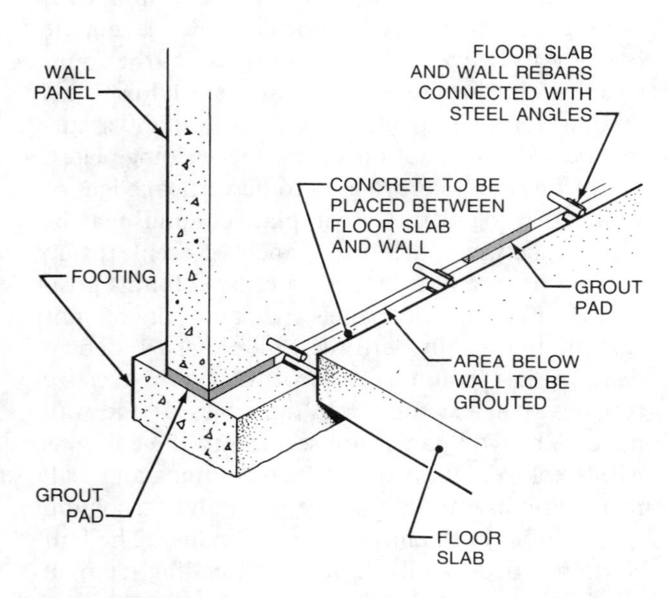

Figure 6-14. Tilt-up wall panels are placed on top of the footing. Rebars extending from the floor slab and wall panel are tied together and concrete is placed between the wall and slab.

joists or trusses are secured into position with metal brackets or steel angles that are bolted to the exterior wall panels. The floor slab sheathing is placed over the joists or trusses and the concrete is placed in the forms using buggies or pumps.

Roofs for tilt-up structures are commonly constructed with *glulam timbers*. Glulam timbers are several pieces of lumber that are glued together with a strong adhesive resulting in a heavier laminated member. The glulam timbers are framed with purlins and covered with plywood sheathing.

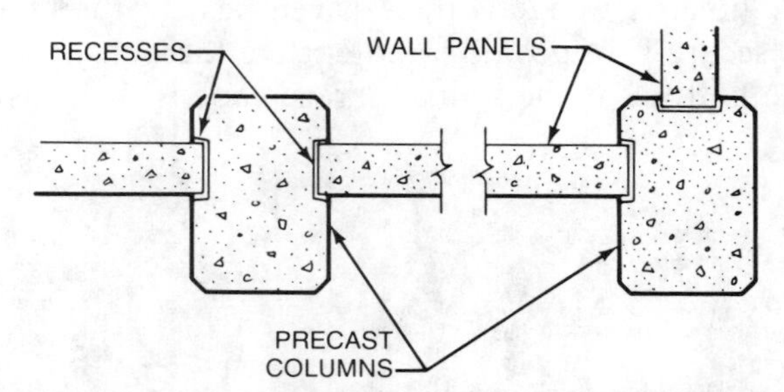

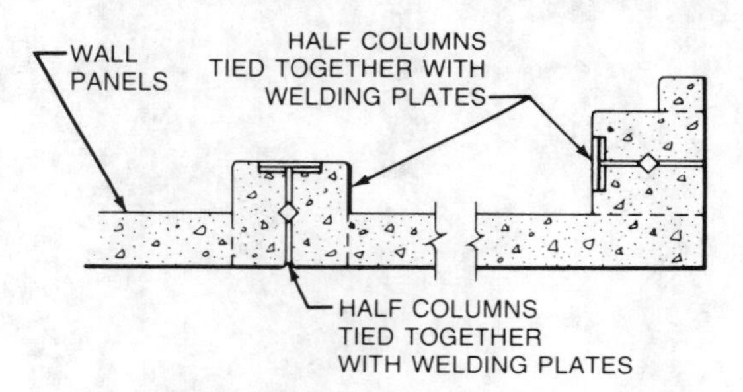

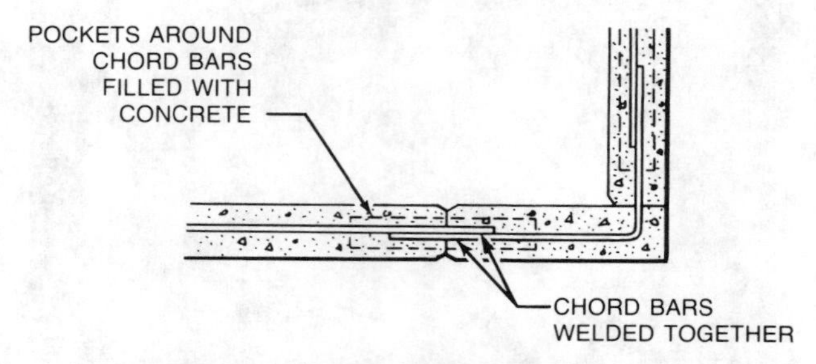

Figure 6-15. Tilt-up wall panels are tied together with precast columns or chord bars.

Chapter 6—Review Questions

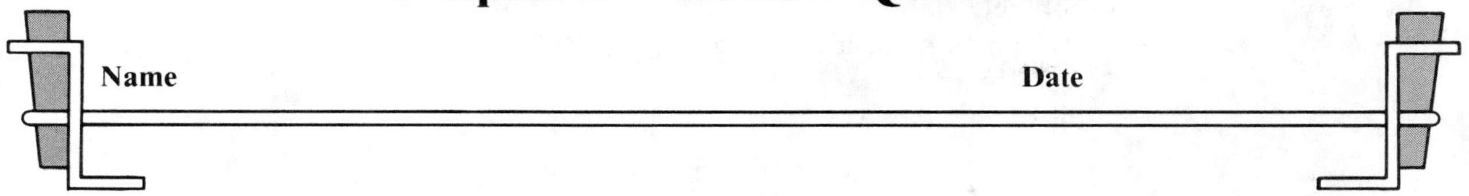

Completion

_______ 1. Precast members are reinforced with rebars or _______ steel cables.

_______ 2. Two methods used to prestress concrete are _______ and _______.

____

_______ 3. _______ beds act as a base and support for precast forms.

_______ 4. Casting beds must be _______ supported to avoid deflection.

_______ 5. _______ concrete is usually associated with job site precasting methods.

_______ 6. High tensile _______ cables are placed in the form when pretensioning concrete.

_______ 7. Concrete is placed under _______ when pretensioning occurs.

_______ 8. Advantages of job site precasting include _______ costs and greater _______.

____

_______ 9. In addition to wood, _______ and _______ may be used for precast forms.

____

_______ 10. Forms should be _______ before the concrete is placed to facilitate stripping.

_______ 11. Precast wall and floor sections only require _______ forms over the casting bed.

_______ 12. Inserts or anchors are set in precast members for the _______ attachments.

_______ 13. _______ plates are secured to the precast member by using bolts.

_______ 14. Steel plates and dowels are used to _______ adjoining precast members.

_______ 15. _______ plates are often used to fasten precast floor slab sections to precast beams.

Multiple Choice

_______ 1. Tilt-up construction is _______.
- A. a factory precast method
- B. a method for constructing high rise buildings
- C. primarily used in the construction of one and two story buildings
- D. not a widely accepted method of construction

_____ 2. In tilt-up construction, a _____ may be used as a casting bed.
A. concrete slab
B. ground floor of a building
C. wood platform
D. all of the above

_____ 3. To assure a clean lift when raising a tilt-up wall panel, _____.
A. paper should be put down before placing the concrete
B. a bond breaking material should be sprayed on the casting bed
C. the floor slab must be free of any dirt and sediment
D. the wall panel must be raised very carefully

_____ 4. When placing rebars for tilt-up wall panels, the areas around the door and window openings require _____ reinforcement than other parts of the wall.
A. less
B. the same amount of
C. more
D. no reinforcement

_____ 5. When placing concrete into the casting panels, _____ to avoid disturbing the bond breaking material.
A. discharge the concrete against a slanted plywood panel
B. allow the concrete to free-fall
C. place the concrete directly on the floor
D. none of the above

_____ 6. A _____ strip is a wood piece nailed to the edge forms to provide beveled corners at the vertical ends of the wall panel.
A. recess
B. angle
C. chamfer
D. vertical

_____ 7. Inserts for tilt-up wall panels are _____.
A. placed along the flat plane of the wall
B. placed for edge lifting
C. protected by plastic caps during the placement of concrete
D. all of the above

_____ 8. Stack casting is a procedure _____.
A. commonly utilized when casting panels on the floor slab of the building
B. in which a series of panels are cast on top of each other and then lifted into place
C. used most often in tilt-up construction
D. none of the above

_____ 9. After raising a wall panel, the first step is to _____.
A. tie it to a column
B. securely brace it
C. remove the lift plates
D. patch the insert holes

_____ 10. Grout pads are used to _____.
A. level the bottom of the wall panel
B. fill in the space between the wall bottom and the foundation
C. secure the wall to the foundation
D. waterproof the joint between the wall and the foundation

CHAPTER 7
Concrete Mix and Placement

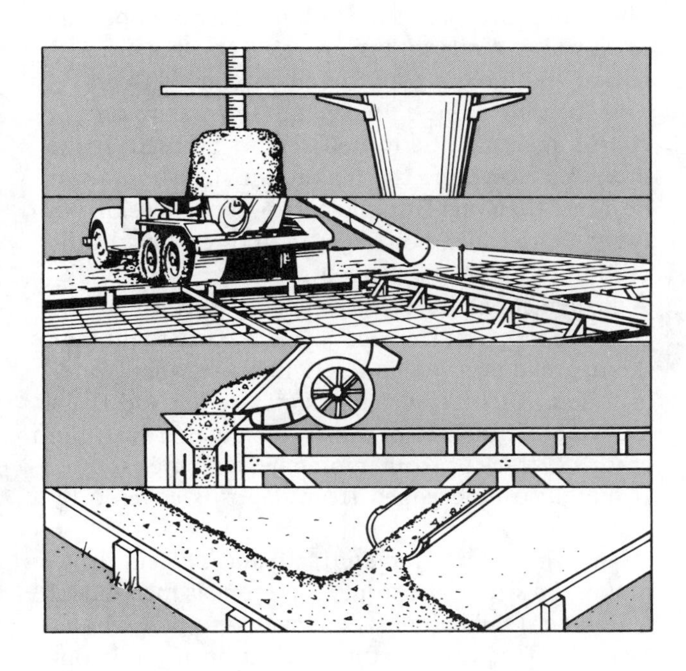

Concrete is one of the most durable materials used in modern construction. Concrete is strong, fireproof, and resists decay. Concrete is often the principle construction material used in buildings, highways, and other heavy construction projects.

Concrete is composed of portland cement, coarse and fine aggregate, and water. When water is added to these ingredients, concrete in a plastic state is produced. Hydration, a chemical reaction between cement and water, occurs and the concrete begins to set and solidify.

Modern techniques used to manufacture portland cement have been used for over 160 years with minor modifications. An Englishman, Joseph Aspdin, obtained a patent for the manufacturing process in 1824. Portland cement refers to rocks and limestone quarried from deposits on the island of Portland, off the coast of England.

Quality control of concrete and its placement is essential to ensure its final strength and appearance. Proper mixing techniques and proportions must be utilized to produce concrete with the proper amount of slump and the desired compressive strength. Proper mixing, transportation, and placement methods must be utilized to prevent segregation of the concrete.

The forms are removed or stripped from the concrete member after the concrete has set and it is hard enough to resist damage. Built-in-place forms are removed from the concrete in the opposite manner that they were erected. Large panel forms and ganged panel forms are stripped using a crane.

COMPOSITION OF CONCRETE

Concrete is composed of fine and coarse aggregate, cement, and water. Fine and coarse aggregate make up the greatest part of the concrete mixture. Fine aggregate is sand and coarse aggregate is gravel or crushed stone. Coarse aggregate ranges in size from ¼" to 1½" in diameter.

The size of coarse aggregate chosen for a concrete mixture is based on the spacing between the rebars placed in the form and the distance between opposite form walls. Walls with many rebars or that are narrow in width require a concrete mixture with gravel or stone that is small. In general, the maximum size coarse aggregate permitted should be used in the mixture. However, the maximum size should not be larger than one-fifth the narrowest dimension between form walls, nor larger than three-fourths the minimum distance between rebars.

Cement constitutes the smallest portion of the concrete mixture. Cement acts as a paste that binds the aggregate in the concrete mixture when water is added. Most cement products are derived from limestone, which is obtained by digging into the earth's surface or from mining beneath the ground. It may also be dredged from deposits covered by water.

The first step in manufacturing cement is to reduce the size of the stone. The stone is then mixed with other raw materials and ground to powder and blended. This raw mixture is then burnt and converted to cement clinkers. Gypsum is combined with the clinkers and ground up to complete the process. See Figure 7-1.

When water is added to cement, a chemical reaction called *hydration* occurs. Hydration is a chemical reaction necessary to produce the hardened concrete. The rate and degree of hydration directly affect the final strength of concrete. The water used in a concrete mixture should be clear and free of oils, alkalis, and acids. The amount of water combined with the cement and aggregate is important to the strength of the concrete. Too much water dilutes the cement and causes the aggregate to separate, producing weak concrete. Too little water results in poor mixing action of the cement and aggregate and also produces weak concrete.

The Concrete Mix

The concrete mix is the proportion of cement, fine and coarse aggregate, and water in a batch of concrete. As the size of the coarse aggregate increases, the proportion for coarse aggregate increases also. See Figure 7-2. Information about the structure being built must be determined before the concrete mix is proportioned, including shape and size of structural members (walls, slabs, beams, columns, etc.) and their required design strength. Exposure of concrete members to weather conditions and other environmental factors must also be considered.

The concrete mix affects the compressive strength of concrete. Compressive strength is the amount of force the concrete can withstand 28 days after it has been placed. Twenty-eight days is the average period of time required for concrete to gain its full strength. Compressive strength is measured in pounds per square inch (PSI).

Water-Cement Ratio. The water-cement (w/c) ratio of a concrete mix is the major factor affecting compressive strength of concrete. The water-cement ratio is the amount of water used in the mix in relation to the amount of cement. A low water-cement ratio produces strong and dense mixtures. Therefore, the water-cement ratio selected should be the lowest value possible to meet the design requirements of the structure. See Figure 7-3.

Water-cement ratio is determined by dividing the weight of the water by the weight of the cement contained in a cubic yard of concrete. For example, if the weight of the water used in 1 cubic yard of concrete is 8 pounds and the weight of the cement is 18 pounds, the water-cement ratio is .44 ($8 \div 18 = .44$).

In heavy construction work, proper proportions of the concrete mix are determined by an engineer or concrete field specialist and are included in the print specifications. Other information may include minimum cement content in relation to the lowest water-cement ratio, size and amount of coarse aggregate, and type and amount of admixtures. Local building codes may furnish the required information for an acceptable concrete mix for small projects such as residential foundations. Local batch plants that produce ready-mixed concrete may also provide additional information.

Job Site Testing

Job site testing of concrete is often required for heavy construction projects where concrete is continuously being placed. Two tests commonly conducted are the *slump test* and the *compression test*.

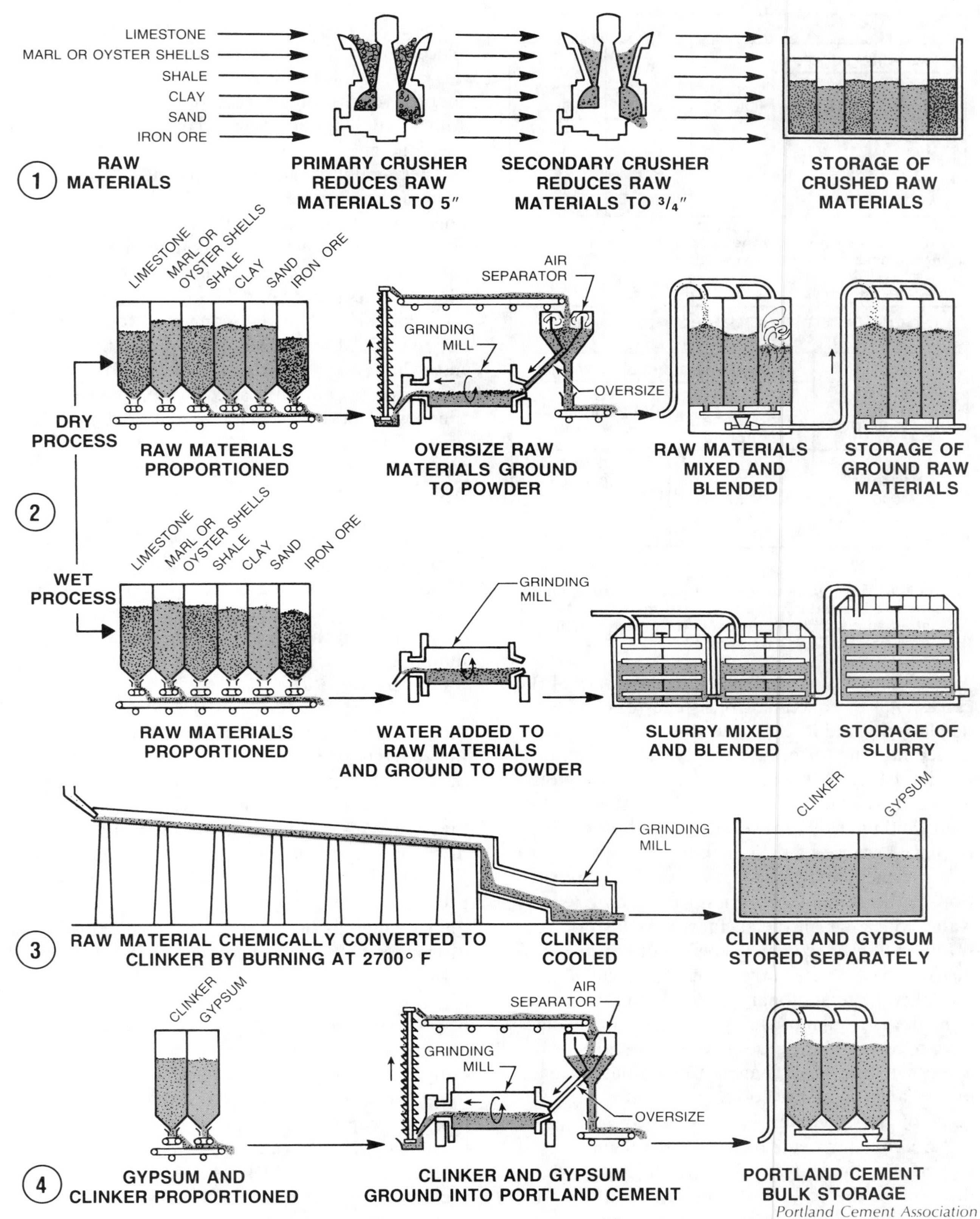

Portland Cement Association

Figure 7-1. Raw materials are crushed, blended together, burnt to partial fusion, and ground to a fine consistency to obtain portland cement.

CONCRETE PROPORTIONS				
Coarse Aggregate				
Size (Max.)	(Cu Ft)	Cement (Cu Ft)	Sand (Cu Ft)	Water (Gal.)
3/8"	1 1/2	1	2 1/2	1/2
1/2"	2	1	2 1/2	1/2
3/4"	2	1	2 1/2	1/2
1"	2 3/4	1	2 1/2	1/2
1 1/2"	3	1	2 1/2	1/2

Figure 7-2. Various concrete proportions are utilized for different sizes of coarse aggregate.

COMPARISON OF COMPRESSIVE STRENGTH TO WATER-CEMENT RATIO		
Compressive Strength at 28 Days (PSI)	Water-Cement Ratio, By Weight	
	Non-Air-Entrained Concrete	Air-Entrained Concrete
2000	0.82	0.74
3000	0.68	0.59
4000	0.57	0.48
5000	0.48	0.40
6000	0.41	—

American Concrete Institute

Figure 7-3. The compressive strength of concrete is directly related to the water-cement ratio of the concrete mix. Air-entrained concrete requires a lower water-cement ratio than non-air-entrained concrete.

The slump test is used as a rough measure of the *consistency* of concrete. Consistency of concrete is its plasticity and its ability to flow while it is being placed into the form. A stiff mix has a higher proportion of aggregate and poorer consistency than a fluid mix. The material cost of stiff mixes is less than fluid mixes. However, they are more difficult to place because of their inability to flow around reinforcement and other areas of the form. Separation of the concrete materials or *bleeding* (excess water collecting on the surface) may also occur.

Slump tests are conducted before or during placement of the concrete in the forms. Concrete samples are taken from stationary or truck mixers. Variations in the results of slump tests indicate changes have occurred in the grading or proportions of the aggregate, or in water content. Corrections are made immediately to assure correct and uniform consistency of the concrete.

A *slump cone* made of galvanized metal is used to perform slump tests. The slump cone is 8" in diameter at the bottom, 4" in diameter at the top, and 12" high. The cone is dampened immediately before use and placed on a smooth, nonabsorbent surface. See Figure 7-4. Three layers of concrete are placed in the cone and each layer is rodded with a 5/8" × 24" rod. Rodding is done with an up-and-down motion. After rodding is completed the cone is carefully lifted, without tilting or jarring the cone, from the concrete.

The amount of slump is determined by measuring from a straightedge placed on top of the inverted cone down to the top of the concrete slump pile. For example, if the top of the slump pile is 3" below the top of the cone, the concrete has a 3" slump. Slump ranges are allowed for different types of construction work. Narrow, congested forms erected for reinforced walls, beams, and columns require more fluid concrete mixtures than wide forms. These mixtures consist of small sized coarse aggregate combined with more water and cement, and produce a greater amount of slump. Wide, less congested forms such as footings and slabs allow a stiffer mix with less slump. See Figure 7-5.

The compression test is a field test that measures *compressive strength* of concrete. Compressive strength is the force the concrete can withstand after 28 days. The required compressive strength of concrete for a structure is determined by exerting force on a specimen that has set. A major factor determining the compressive strength of concrete is the water-cement ratio.

Samples of concrete must be taken at three or more regular intervals throughout the discharge of the concrete batch to conduct a compression test. Samples should not be taken at the very beginning or end of the discharge. The concrete specimens are placed in a watertight metal or nonabsorbent cylindrical mold 6" in diameter and 12" long. Each mold is filled in three layers, and each layer is rodded approximately 25 times with a 5/8" × 24" rod. When the rodding has been completed for the last layer of concrete, the surface is leveled off and covered with a glass or metal plate to prevent evaporation.

The cylinders are stored on the job site and must be protected from jarring. After 24 hours, the cylinders are taken to a laboratory where the concrete sample is removed from the cylinder and allowed to set for 28 days. At the end of this period, a thin layer of capping compound is applied to the specimen. After the cap sets, the concrete specimen is placed in a compression testing machine. Pressure is exerted until the specimen breaks. A dial on the machine indicates the pressure required to break the concrete. See Figure 7-6.

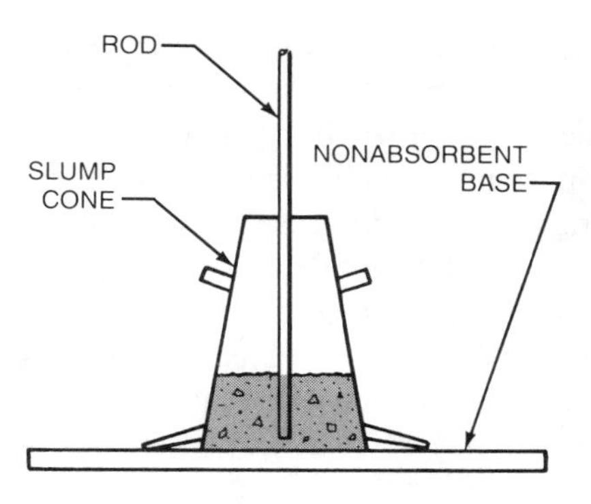

1. Fill one-third of the cone with concrete and rod 25 times.

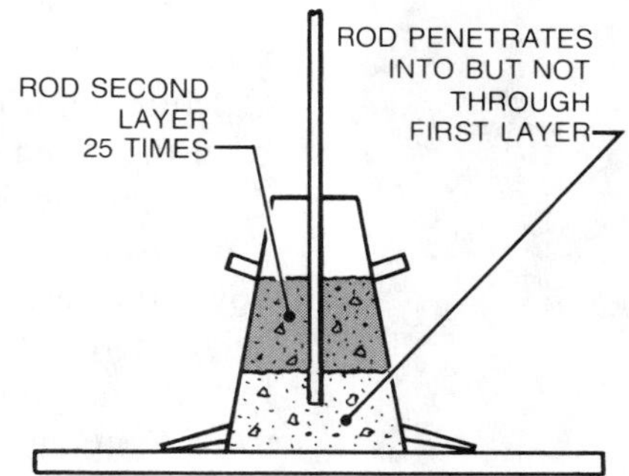

2. Fill two-thirds of the cone and rod the second layer 25 times.

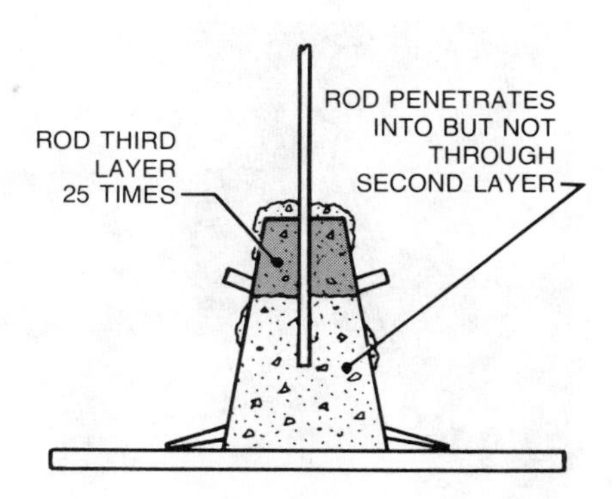

3. Fill the cone to overflow and rod 25 times.

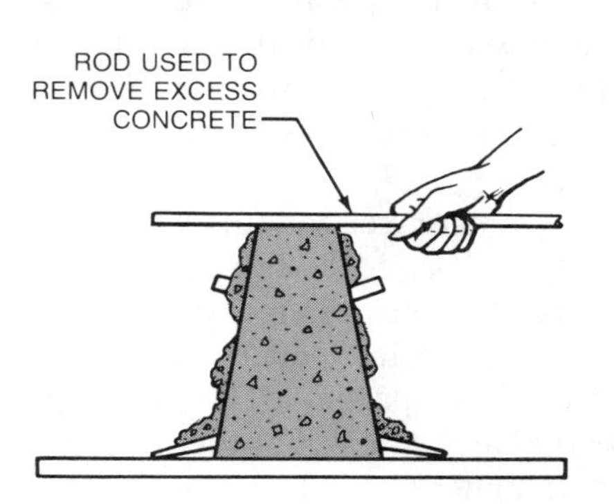

4. Remove the excess from the top of the cone and at the base.

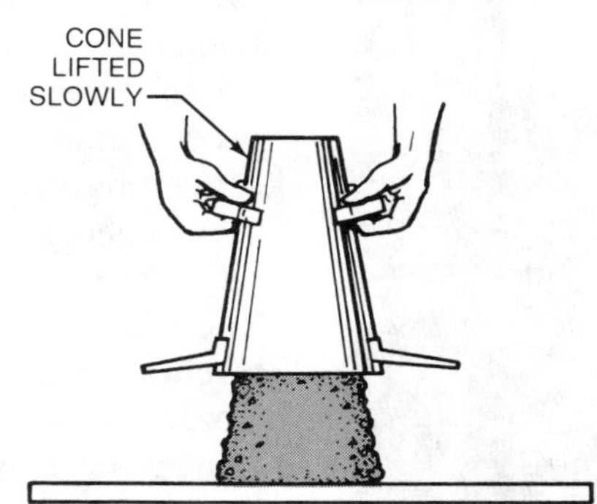

5. Lift the cone vertically with a slow, even motion.

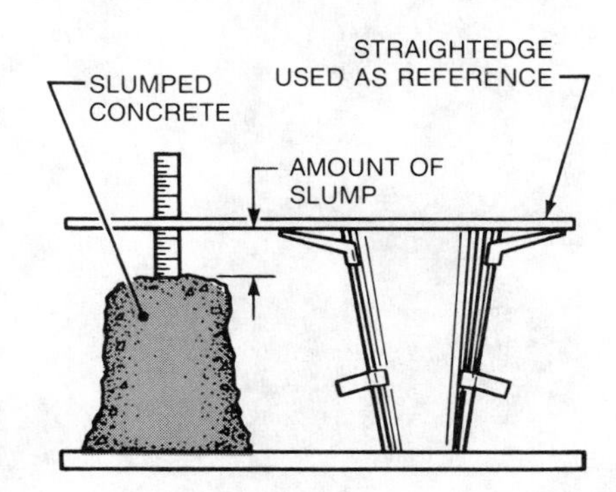

6. Invert the cone and place next to the concrete. Measure the distance from the top of the cone to the top of the concrete.

Figure 7-4. A slump test is used to measure the consistency of the concrete.

ALLOWABLE SLUMP		
Concrete Construction	**Slump**	
	Max.*	**Min.**
Reinforced foundation walls and footings	3″	1″
Plain footings, caissons, and substructure walls	3″	1″
Beams and reinforced walls	4″	1″
Building columns	4″	1″
Pavements and slabs	3″	1″
Mass concrete	2″	1″

*May be increased 1″ for consolidation by hand methods such as rodding and spading.

Portland Cement Association

Figure 7-5. Allowable slump is determined by the type of structural member.

Concrete Admixtures

A concrete admixture is a material other than cement, aggregate, and water that is added to a batch of concrete immediately before or during the mixing process. Admixtures increase the effectiveness of concrete under various conditions. The types of admixtures used most often are *air-entraining agents*, *accelerators*, *water-reducing retarders*, and *pozzolans*.

Air-entraining agents are available in liquid, ready-to-use form and are added to concrete in controlled quantities. These admixtures are a foaming material that produce tiny air bubbles in the concrete ranging in size from .01″ to .001″. Air-entrainment improves the workability of concrete and decreases bleeding and separation during the concrete placement. Air-entrained concrete requires less water per cubic yard than non-air-entrained concrete with the same amount of slump. The amount of sand in the mix may be reduced by an amount equal to the volume of the entrained air.

Portland Cement Association

Figure 7-6. A concrete sample is placed in a compression testing machine. Pressure is exerted on the sample until it breaks.

Air-entrainment increases the durability of concrete particularly in cold climates. In exposed flatwork, such as walks and driveways, air-entrainment produces concrete that is highly resistant to severe frost and the effects of salt used for snow and ice removal. Although originally designed for concrete exposed to severe weather conditions, air-entrained concrete is recommended for most concrete work because of its increased workability and durability.

Accelerators, or accelerator admixtures, speed up setting time and increase the rate of early-strength gain of concrete. Accelerators are used in cold weather construction because concrete gains early strength slowly at low temperatures. However, accelerators can be replaced by using high early-strength cement, increasing the cement content, or providing a longer curing period.

Calcium chloride is one of the most commonly used accelerators. The amount of calcium chloride used should never exceed more than 2% of the weight of cement. More than this amount may cause reinforcement corrosion and rapid stiffening. Calcium chloride is added to concrete after it has been dissolved in water to achieve better mixing.

Water-reducing retarders decrease the amount of water to produce a workable mix. They also slow down the set of concrete during hot weather and delay early stiffening of concrete placed under different conditions. Water-reducing retarders may entrain some air in the concrete. Since water-reducing retarders may have varying effects on concrete, technical advice is recommended.

Pozzolans used in concrete are manufactured from materials such as permicite, volcanic ashes, tuffs, diatomaceous earth, fly ash, calcined clays, and shale. In the presence of moisture and under normal temperature conditions, pozzolans react with other cement materials to form compounds with cementing properties.

As a general rule, concrete containing pozzolans requires more water to produce the same slump. Therefore, the set concrete has a higher rate of contraction, resulting in a greater tendency to crack. Pozzolans, when used as a partial replacement for cement, decrease the early strength of the concrete, but usually develop higher strength at a later stage. Pozzolans may be used as a partial replacement for cement in mass concrete operations, such as dam construction, where early strength development is not critical.

Hot and Cold Weather Concreting

Placing concrete during extremely hot and cold conditions creates problems that affect the hydration of concrete. The problem encountered during hot weather is the rapid rate of water evaporation. This can cause an early slump loss and rapid setting, which may result in strength loss and the possibility of cracking. The problem encountered in cold weather is the slow rate of hydration of the concrete. The ideal setting temperature for concrete is 70°F.

Concrete sets faster as the temperature increases. When the surrounding temperature reaches or exceeds 85°F, measures must be taken to control the temperature of the concrete. The most effective way to maintain a low concrete temperature is to keep the concrete ingredients as cool as possible before mixing. The water used in the mix can be cooled before it is added to the mix by refrigerating it or adding liquid nitrogen or crushed ice. When mixing concrete at the job site, aggregate should be stockpiled in a shady place before use and kept moist by sprinkling water on it. Placing equipment such as mixers, chutes, buckets, buggies, and pump lines should be shaded or covered with wet burlap. Wall forms and reinforced steel should be wetted down. The subgrade below concrete slabs should be thoroughly soaked the night before placing the concrete. The *curing* of the slabs should be started as soon as possible. Curing is the process of retaining the moisture in freshly placed concrete to ensure proper hydration.

In temperatures 0°F or below, no hydration occurs. To ensure proper hydration in concrete placed during freezing weather, the concrete mix should be heated prior to being placed in the forms. In most cases, heating of the mixing water is all that is required, provided the aggregate are free of ice and snow. If it is necessary to heat the aggregate, steam coils may be placed in the storage pile or holder bins, or steam can be injected in the pile. After the concrete has been placed, the forms can be covered with tarpaulins or sheets of plastic film to retain heat. If necessary, an enclosed area can be further warmed with gasoline-powered heaters.

Mixing and Transporting Concrete

Concrete is usually mixed at a batch plant and delivered by truck to the job site. For isolated projects located too far from a batch plant, large onsite stationary mixers are used. See Figure 7-7.

Concrete must be mixed thoroughly to ensure consistency and uniform distribution of the ingredients. Mixing time can vary; however, common manufacturer recommendations specify 1 minute of mixing for 1 cubic yard of mix plus 15 seconds for each additional cubic yard. The mixing period is measured from the time all ingredients are in the mixer. Approximately 10% of the required mixing water should be placed in the drum before the cement and aggregate are deposited. Water is then added along with the dry materials. The last 10% of water is added after all the materials are in the drum.

Ready-Mixed Concrete. Ready-mixed concrete is prepared at a batch plant and then delivered to the job site. The concrete ingredients are proportioned according to specifications for the particular job.

Kaiser Cement Corporation

BATCH PLANT

Portland Cement Association

STATIONARY MIXER

Figure 7-7. Large quantities of concrete are mixed at batch plants or in on-site stationary mixers.

Most batch plants are highly automated, ensuring fast and accurate *batching,* which is the measuring and proportioning of the concrete mix. The type of truck most often used to transport ready-mixed concrete from the batch plant to the job site is the *ready-mixed truck* or *transit mixer truck*. *Agitating trucks* are also used to transport ready-mixed concrete.

A ready-mixed truck is equipped with a large revolving drum operated by an auxiliary engine. See Figure 7-8. Truck sizes vary with drum capacities ranging from 1 to 12 cubic yards. A typical ready-mixed truck also has a water tank so water can be added en route to the job site, if necessary. A number of methods is used to mix and transport concrete with ready-mixed trucks. The concrete ingredients may be *dry batched,* a procedure in which the ingredients are placed dry then mixed with water on the way to the job. Another common method, *shrink mixing*, involves mixing the ingredients and water for approximately 30 seconds at the batch plant. The concrete is then deposited in the drum of the ready-mixed truck and the mixing is completed en route to the job.

Care must be taken with all mixing methods to avoid loss of slump and plasticity while the concrete is transported. Concrete should be delivered and discharged from the truck mixer within 1½ hours after water has been added to the cement and aggregate.

Agitating trucks can also be used to transport concrete instead of ready-mixed trucks. An agitating truck has an open-top body with a paddle mechanism. The paddle mechanism, located at the bottom of the body, maintains the proper plasticity of the concrete during transportation. Concrete delivered by agitating trucks must be premixed at the batch plant. Agitating trucks can only be used for short hauls (30 to 45 minutes) or to transport concrete from on-site concrete mixers. The concrete is discharged from the rear by tilting the body of the truck.

Special dump trucks with sealed seams may be used for short hauls in moderate weather conditions. Other transportation methods include small rail cars used on special jobs, such as concrete tunnel liners, where large quantities of concrete are required and other standard methods of transportation are not possible. Boats and barges may be used to deliver concrete to waterbound projects. Helicopters may be used to carry concrete that is placed in buckets and flown to inaccessible mountain areas.

Symons Corporation

READY-MIXED TRUCK

Portland Cement Association

AGITATING TRUCK

Figure 7-8. Ready-mixed and agitating trucks are used to transport ready-mixed concrete.

CONCRETE PLACEMENT

Concrete placement is the transfer of concrete from a concrete truck or other transporting means into the forms. Concrete must be placed properly to prevent *segregation*. Segregation is the separation of sand and cement ingredients from the coarse aggregate in the mix. Segregation results in strength loss of the hardened concrete.

Proper placing procedures are essential to guarantee uniformity of the batch of concrete. Concrete is deposited into the forms as quickly as possible after it has been transported to or mixed on the job site. A delay in placing concrete may result in slump loss, thus affecting the consistency and workability of the concrete. Segregation should be avoided when transferring concrete from a truck to chutes, buckets, and buggies.

The subgrade should be well compacted and graded to its proper elevation when placing concrete for slabs directly over a subgrade. The subgrade is dampened to a depth of 4″ to 6″ to prevent rapid loss of water from the freshly placed concrete. If the subgrade is frozen, all snow and ice should be removed from the placement area.

When placing concrete on rock or existing concrete, the surface of the rock or concrete must be clean, rough, and damp. This allows fresh concrete to firmly grip the surface. If it is necessary to cut away rock or old concrete, all cut surfaces should be horizontal and vertical (not sloping), and all loose fragments and dust should be removed.

Wall forms receiving concrete should have a final check for accurate dimensions as well as proper bracing. Sheathing joints must be tight to prevent mortar loss, and holes in the sheathing must be plugged. Sawdust, nails, and any other debris within the form walls are removed. The inside of the forms is then thoroughly wetted down with water before the concrete is placed. Rebars and inserts placed in the form should be free of rust, oil, mud, and mortar.

Placing Equipment

After concrete is delivered to or mixed on the job site, it is placed in the forms. Various methods and equipment are used and are chosen based on the type of construction project. Equipment most commonly used for transferring and placing concrete are chutes, buckets, and manual or motorized buggies. In addition, pumping concrete by hose is often used on construction jobs. See Figure 7-9.

Chutes. Chutes are used to place concrete directly from trucks into the forms if the truck can be maneuvered close enough to the location. Chutes are also used to transfer concrete to buckets or buggies that deliver the concrete to the forms. Extended chutes may be used to transfer concrete from one elevation to another. The most efficient type of chute has a round bottom constructed of metal, or one equipped with a metal liner. The sides of the chute must be high enough to avoid concrete overflow. The chute should be sloped enough to allow the concrete to flow continuously down the chute without segregation.

Buckets. Buckets are one of the more efficient and flexible methods of moving concrete on the job. Concrete is deposited from the truck or stationary mixer into a bucket ranging in capacity from ½ to 4 cubic yards. The bucket is then lifted by crane and moved over the placement area. A discharge gate is opened at the bottom of the bucket and the concrete flows into the form.

Various attachments may be used to facilitate the flow of concrete from the bucket to the form. A heavy rubber *boot* or *elephant trunk* may be attached below the discharge gate to direct the concrete into forms. Another convenient attachment is a *side dump chute.* The side dump chute provides flexibility when positioning the bucket over wall and slab forms. See Figure 7-10.

Manual and Motorized Buggies. Concrete buggies are similar in appearance to two-wheeled wheelbarrows. Concrete is deposited in the buggies by chute or bucket. Smooth, rigid runways are constructed to allow the concrete to be transported. Small amounts of concrete can be moved with buggies to points of placement that are not easily accessible by chute or bucket. Manual buggies are pushed by hand and are not as common as the motorized types. The recommended maximum delivery distance is 1000′ for motorized buggies and 200′ for manual buggies.

Pumping Concrete. Pumping concrete is a method of conveying concrete by pressure through a rigid pipe or flexible hose and depositing it at the placement area. This procedure is widely used for placing concrete for foundations, low-rise buildings, and lower levels of tall structures. Pumping concrete eliminates the need for buckets, buggies, and other placement equipment. Pumping rates vary from 10 to 90 cubic yards per hour depending on the equipment used. An effective range for pumping is from 300′ to 1000′ horizontally and 100′ to 300′ vertically.

Portland Cement Association

CHUTE

Walsh Construction Company of Illinois

BUCKET

PUMP

Portland Cement Association

MOTORIZED BUGGY

Figure 7-9. Concrete is placed using chutes, buckets, motorized buggies, and pumps.

The pumping apparatus consists of a mobile unit containing a pump, a boom and cables, and the transport lines that deposit the concrete. The pump supplies the pressure necessary to transport the concrete. The boom and cables support the transport lines, which may be rigid pipes or flexible hose. Rigid pipes made of steel, aluminum, or plastic are available in 3″ to 8″ diameters. Flexible hose made of rubber, spiral wound flexible metal, and plastic are available in 3″ to 5″ diameters.

The concrete is discharged from a concrete truck into a hopper on the pumping apparatus. The transport line supported by the boom and cables is moved in position at the placement area. The concrete is then conveyed to the forms. See Figure 7-11.

Placing and Consolidating Concrete

Concrete should be placed as close as possible to its final location within the forms to prevent excessive movement. Excessive movement of concrete within the forms results in segregation and poor consolidation. This occurs because the sand-cement paste in the mix flows ahead of the coarse aggregate materials.

The Burke Company

Figure 7-10. A bucket with a side dump chute may be used to place concrete in forms. The concrete is released by opening the discharge gate at the bottom of the bucket.

Figure 7-11. Concrete is pumped into the forms through a flexible transport line.

Concrete must be worked and compacted while it is being placed in order to consolidate each fresh layer of concrete with the layer below. Proper consolidation reduces or eliminates *rock pockets* or *honeycombs.* Rock pockets occur when coarse aggregate is visible on the surface of the concrete after it has set. Honeycombs are voids in the concrete surface caused by the mix failing to fill the spaces among the coarse aggregate. See Figure 7-12.

In small foundation forms, compaction and consolidation can be accomplished by spading the concrete with a wood rod or spading tool. The rod or spading tool must be thin enough to pass between the rebars and form walls and reach the bottom of the form. In heavy construction work, *immersion vibrators* (or *internal vibrators*) and *external vibrators* are used to consolidate the concrete. See Figure 7-13. Immersion vibrators consist of a metal head at the end of a flexible steel spring core enclosed by a rubber hose. When activated by a motor, the steel spring causes the metal head to vibrate. The vibrating head is immersed in the concrete and produces rapid consolidation. Internal vibrators are

Kaiser Cement Corporation

Figure 7-12. Surface defects, such as honeycombing, are a result of improper consolidation methods.

Portland Cement Association

Figure 7-13. Internal vibrators are used to consolidate larger masses of concrete. The vibrator is immersed vertically into the concrete at 18″ to 30″ intervals for 5 to 15 seconds.

The Burke Company

Figure 7-14. Concrete for a floor slab is placed against the face of previously placed concrete.

powered by electricity, gas, or compressed air. External vibrators are used primarily in precast concrete construction. External vibrators are attached along the outside of the casting forms at strategic positions and are vibrated using a pneumatic or electric power supply.

Floor Slabs. Concrete for floor slabs may be placed directly on a subgrade or elevated floor forms. The concrete is deposited at the far end of the slab area so that each new batch of concrete is discharged against the face of the concrete already in place. See Figure 7-14. As the concrete is placed, it is spread and consolidated. Immersion vibrators are often used to consolidate concrete for slabs covering large areas. After consolidation, cement finishers strikeoff the concrete with the screeds already set up. The screeds and screed supports are removed and the concrete is finished.

Concrete is placed and distributed in an even layer, rather than placed in piles and leveled off. When placing concrete on a sloping surface, the placement begins at the base of the slope so that compaction is increased by the added weight of each batch. A baffle should be positioned at the end of a chute when working with a sloping surface to prevent segregation and to keep the concrete on the slope. See Figure 7-15.

Walls. Concrete placed in wall forms should first be placed at the ends of the forms and progress toward the center. Working from the ends toward the center prevents water from being trapped in the corners of the form, which would result in voids in the concrete. Concrete in wall forms should always be placed in level courses from 12″ to 20″ thick. Thinner courses are recommended in forms containing heavy, closely spaced rebars, and when placing concrete from great heights.

After a course is placed, it is immediately consolidated with the course directly below. A spading tool is often used to consolidate the concrete in small and low forms and an immersion vibrator on heavy construction projects. When an immersion vibrator is used, it is held vertically and should pass through

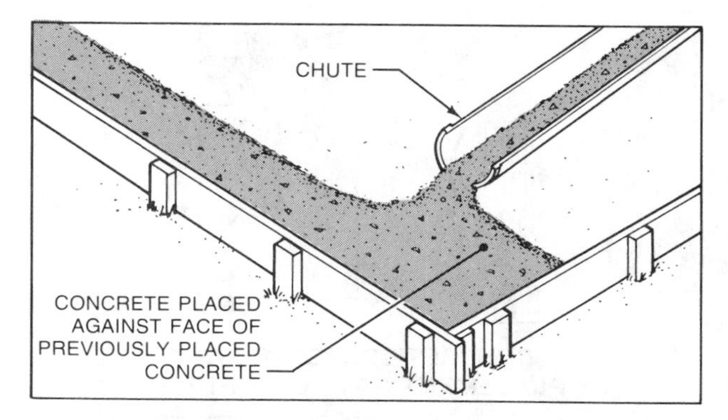

PLACING A CONCRETE FLOOR SLAB

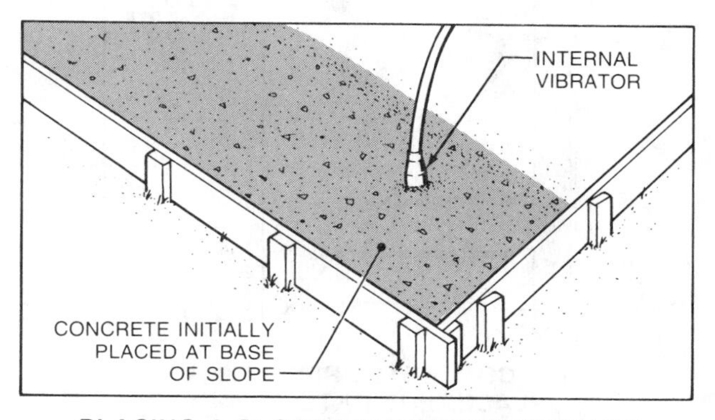

PLACING A SLOPED CONCRETE SURFACE

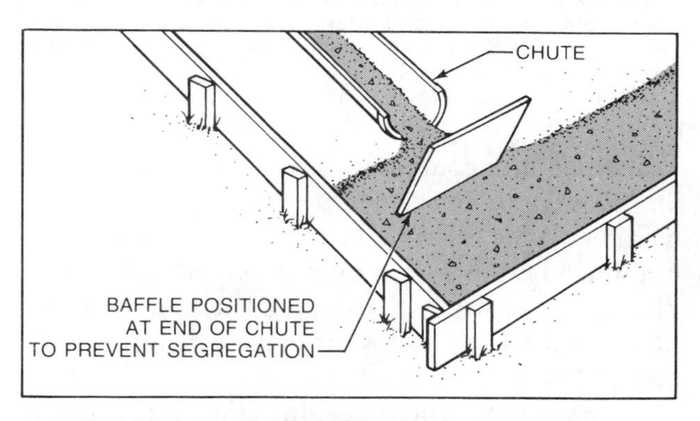

PLACING A SLOPED CONCRETE SURFACE USING A BAFFLE

Figure 7-15. Placing concrete properly ensures consolidation and uniformity of the concrete slab.

the top course, penetrating a few inches into the course below. The vibrator is immersed at regular intervals into the course for periods of 5 to 15 seconds. Overvibration should be avoided because water and paste may flow to the surface.

To avoid segregation in wall forms, the drop of the concrete must be as near to vertical as possible. Hoppers and flexible drop chutes should be used, particularly in walls with closely spaced rebars. Segregation resulting from concrete striking rebars and form walls above the placement level should be avoided. The free-fall distance of concrete within the wall form should be limited to 4′ to 6′.

In high walls and columns, openings (ports) are commonly cut at intervals in the side of the forms to eliminate long vertical drops. A pocket built along the outside of the opening will momentarily stop the concrete and allow it to flow slowly and evenly into the form. When the concrete level reaches the bottom of the openings, the openings are closed off. If concrete is placed into the form by pump, the pipe or hose should be lowered close to the point of discharge. See Figure 7-16.

Placement Rate and Form Pressure. Wall forms must be designed and constructed to withstand the *lateral pressure* of the concrete as it is being placed into the forms. Lateral pressure of concrete is pressure the concrete exerts against the forms. Lateral pressure of concrete is primarily affected by the rate and height of the placement. Other factors affecting lateral pressure of concrete are internal vibration, temperature of the concrete, weight of the concrete, type of cement, concrete slump, and admixtures.

Before concrete sets, it acts like a liquid and exerts force against the form walls in all directions. The amount of pressure at any given point in a form is directly affected by the weight and height of the concrete above that point. Wall thickness does not affect the pressure. Concrete in a plastic state normally exerts a pressure of 150 PSF (pounds per square foot) at a placement rate of 1 foot per hour. If, for example, the placement rate is 3 feet per hour, the pressure at the bottom level is 450 PSF ($3 \times 150 = 450$). A placement rate of 5 feet per hour produces a pressure at the bottom level of 750 PSF ($5 \times 150 = 750$). See Figure 7-17.

Once concrete sets, it does not exert pressure against the form walls. The time it takes concrete to set is directly affected by its temperature. Most form designs are based on an assumed air and concrete temperature of 70°F. Concrete should be placed at a temperature ranging between 60°F and 100°F. Under these conditions the concrete will set in approximately 1 hour. Concrete takes longer to set at lower temperatures; therefore, the placement rate must be decreased or the concrete heated to maintain acceptable lateral form pressure.

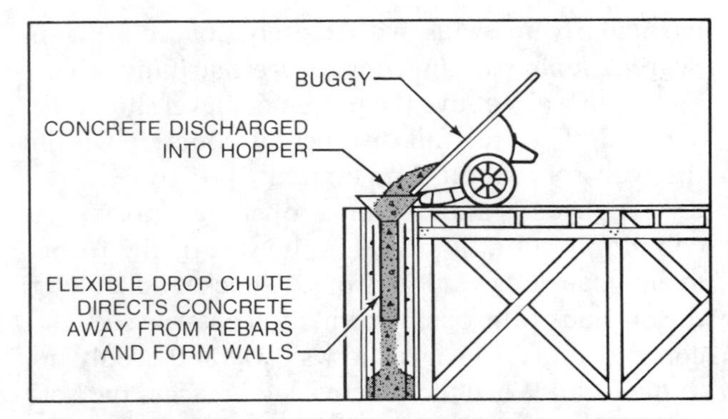

PLACING CONCRETE IN NARROW WALL FORMS

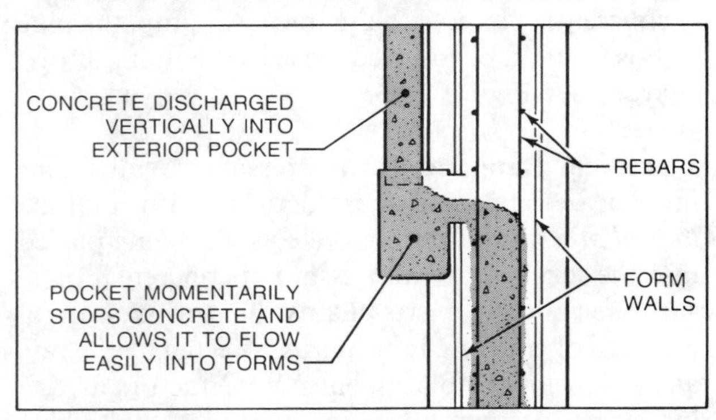

PLACING CONCRETE THROUGH WALL OPENINGS

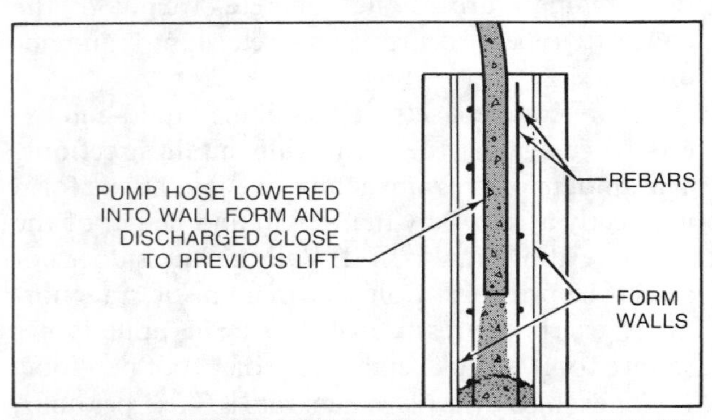

PLACING PUMPED CONCRETE INTO WALL FORMS

Figure 7-16. Considerations must be made for placement of concrete in wall forms.

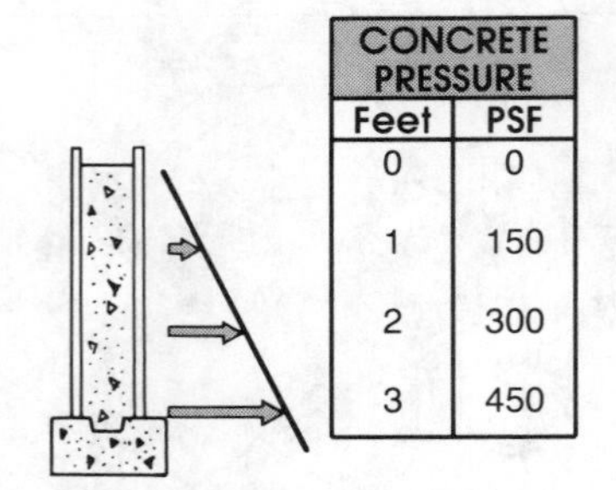

CONCRETE PRESSURE AT PLACEMENT RATE OF 3 FT/HR

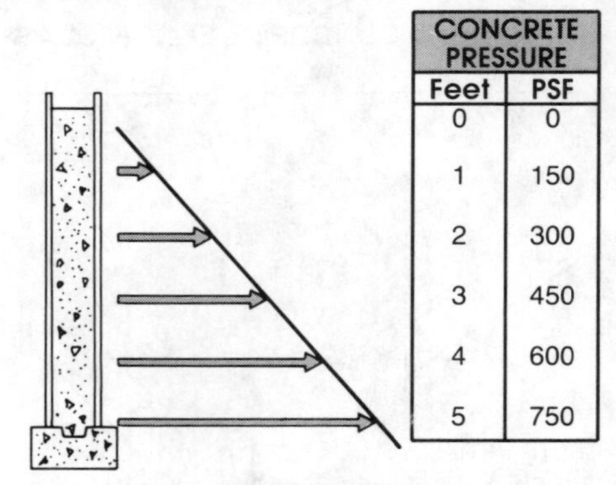

CONCRETE PRESSURE AT PLACEMENT RATE OF 5 FT/HR

Figure 7-17. Lateral concrete pressure is directly affected by the placement rate and height of the concrete in a form. Wall thickness does not affect the lateral pressure exerted on the form.

CURING CONCRETE

Curing is maintaining proper concrete moisture content and temperature long enough to allow hydration of the concrete and development of the desired properties. Hydration is the chemical reaction between water and cement, at which time the cement becomes the bonding agent of the concrete mix. Hydration begins as soon as the water and cement are combined and continues as long as there is water in the mixture and temperature conditions are favorable. If the water in the concrete mix evaporates too quickly, the hydration process will end before the concrete attains its required design strength. Rapid water loss also results in the concrete shrinking and cracking.

In the initial curing stage, the temperature of the concrete should be maintained at approximately 70°F, and the concrete should be kept thoroughly moist for a minimum of three days. This is the most critical period in concrete curing. The cement and water combine rapidly and the concrete is most vulnerable to permanent damage. Concrete attains about 70% of its strength after seven days, and about 85% of its strength after 14 days. Full strength is reached after approximately 28 days.

Curing Methods

Wall forms are kept in place during the critical drying period to facilitate the curing of the concrete.

The forms retain moisture in the walls and can be sprinkled with water and kept damp during hot, dry weather. Some types of wall forms can be loosened to allow water to run inside the forms.

A concrete slab presents a curing problem because its large exposed surface area allows a great amount of moisture to escape. However, various methods can be used to maintain moisture content. The slab can be sprayed with water by placing a pipe with a series of spray nozzles across the center of the slab. A method similar to spraying, *fogging* produces a mist-like effect.

To retain moisture, the concrete slab can be covered with waterproof paper or polyethylene film, which also protects the slab from frost, direct sun, traffic, and debris. Waterproof paper, available in 18″–96″ widths, is laid on the slab and anchored with sand or planks. White polyethylene film is recommended because the white pigment reflects heat. Water-saturated burlap material spread over the slab surface is another effective method used to retain moisture. See Figure 7-18. Chemical sealing compounds are also available and are applied by manual sprayers or automatic self-powered sprayers. The compounds are available in clear, black, and white finishes.

Concrete placed during hot weather is subject to excessive water evaporation. Because of the heat, curing should begin as soon as possible. Water should be fogged or sprayed for 12 hours before using other curing methods. A white, heat-reflective membrane or polyethylene may be applied later. The curing period should also be extended seven days or longer during hot weather.

During cold weather, freezing of the concrete must be prevented during the curing period. A minimum concrete temperature of 50°F to 70°F must be maintained at the time of placement. A number of methods may be used to maintain the temperature, including heating the concrete, covering the concrete, or providing a heated enclosure.

POLYETHYLENE FILM

WATER-SATURATED BURLAP

CHEMICAL SEALING COMPOUND

Portland Cement Association

Figure 7-18. Moisture in a concrete slab is retained by covering the slab with polyethylene film or water-saturated burlap, or spraying the slab with a chemical sealing compound.

STRIPPING FORMS

Stripping forms is the removal of forms after the concrete has set and achieved its required design strength. The concrete must also be hard enough to ensure that its surface will not be damaged when stripping the forms.

Stripping and Removal Schedules

On many building projects forms must be stripped and removed as soon as possible in order to reuse the form materials. Stripping schedules for low foundation forms can be obtained from local building codes. The period of time required before form removal on heavy construction projects may require the approval of an architect or engineer. This information is often included in the print specifications. On construction projects where stripping specifications are not given, the American Concrete Institute (ACI) recommends the following stripping schedules. These schedules apply to concrete placed under ordinary conditions.

Walls†	12 hr	
Columns†	12 hr	
Sides of beams and girders†	12 hr	
Pan joist forms‡		
30 in. wide or less	3 days	
Over 30 in. wide	4 days	
	where design live load is:	
	less than dead load	greater than dead load
Arch centers	14 days	7 days
Joist, beam or girder soffits		
Under 10 ft clear span between structural supports	7 days§	4 days
10 to 20 ft clear span between structural supports	14 days§	7 days
Over 20 ft clear span between structural supports	21 days§	14 days
One-way floor slabs		
Under 10 ft clear span between structural supports	4 days§	3 days
10 to 20 ft clear span between structural supports	7 days§	4 days
Over 20 ft clear span between structural supports	10 days§	7 days

†Where such forms also support formwork for slab or beam soffits, the removal times of the latter should govern.

‡Of the type which can be removed without disturbing forming or shoring.

§Where forms may be removed without disturbing shores, use half of values shown but not less than three days.

In the case of suspended forms (arch centers and joist, beam, or girder soffits), the forms must remain in place for a longer period where the design live load is less than the dead load. Under these conditions, a larger percentage of the design load is included in the dead load.

Stripping and Removal Methods

Forms should be designed for safe and convenient removal. Panels and other form materials must be stripped carefully to avoid damage so they can be reused. The sequence of stripping is determined by the original assembly of the form.

When stripping a wall form, the tie clamps or wedges are removed and the walers are pried off. In a panel system, the panel section and studs can be removed as a unit. In a built-in-place form, the studs and/or walers are pried off, followed by the removal of the plywood sheathing.

Column forms should be constructed so the sides can be pried and removed without disturbing the adjoining beam or girder forms. Beam or girder forms should be constructed so the side panels can be stripped before the beam bottoms. The floor soffit forms are removed by releasing the supporting shores and stringers. If it is necessary to allow a section of floor slab form to fall free, a platform or other support should be constructed to reduce the distance of drop.

Cranes, using two lines, are used to strip larger panels or ganged panel forms. One line is attached at the top of the form for the upward pull. The second line is attached at a lower point to exert an outward pull. When stripping large panel and ganged panel forms, a few ties should remain connected until the crane lines are securely attached.

Metal stripping bars should not be used to pry form panels directly from concrete surfaces, as this results in damage to the concrete. Wooden wedges are placed against the concrete surface and the stripping bar is then used to pry the form.

Chapter 7—Review Questions

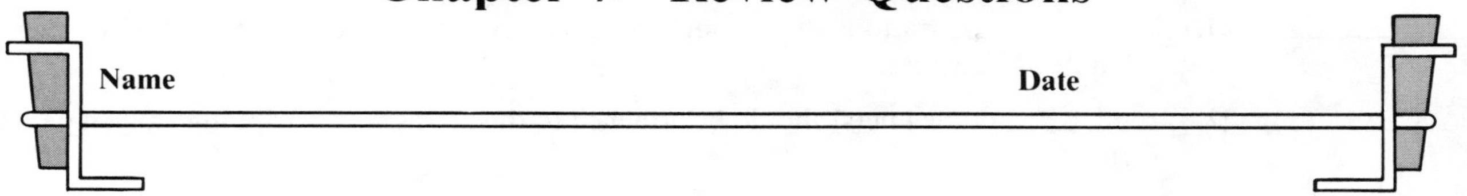

Completion

__________ __________ **1.** The basic ingredients of concrete are __________ and __________.

__________ **2.** __________ is the source of most cement products.

__________ __________ **3.** Aggregate consists of __________ and __________.

__________ **4.** __________ is the chemical reaction between water and cement that produces concrete.

__________ **5.** The smallest proportion in a concrete mix is __________.

__________ **6.** Decreased water-cement ratios produce __________ compressive strength in concrete.

__________ **7.** If the weight of water in a cubic yard of concrete is 10 pounds, and the weight of the cement is 20 pounds, the water-cement ratio is __________.

__________ **8.** A rough measure of the consistency of concrete is determined with a(n) __________ test.

__________ **9.** A compression test measures the __________ strength of concrete.

__________ **10.** Concrete gains its full strength __________ days after placement.

__________ **11.** The durability of concrete in freezing temperatures is increased by adding __________ admixtures.

__________ **12.** __________ admixtures are effective in slowing down the setting process of concrete during hot weather.

__________ **13.** Rapid water evaporation is a problem encountered when placing concrete during __________ temperatures.

__________ **14.** __________°F is the ideal setting temperature for concrete.

__________ **15.** A slow __________ rate is encountered when placing concrete during cold weather.

__________ **16.** __________ concrete is mixed at a batch plant.

__________ **17.** A revolving __________ contains the concrete in a ready-mixed truck.

__________ **18.** Concrete should be discharged from ready-mixed trucks within __________ hour(s) after water has been added to it.

__________ **19.** __________ is the separation of the sand and cement ingredients from the coarse aggregate in a concrete mix.

__________ **20.** Loss of __________ is the result of segregation during concrete placement.

__________ **21.** The __________ ratio is the amount of cement in relation to the amount of water in a batch of concrete.

__________ **22.** __________ crane line(s) should be attached to the forms when stripping ganged panel forms.

Multiple Choice

__________ **1.** A __________ is used to discharge concrete directly from a ready-mixed truck into forms.
A. buggy
B. bucket
C. chute
D. all of the above

__________ **2.** Concrete buckets are __________.
A. used when pumping concrete
B. lifted by crane and carried to the placement area
C. moved by motorized buggies
D. only used on small concrete projects

__________ **3.** The maximum recommended delivery distance for concrete using a motorized buggy is __________′.
A. 200
B. 400
C. 1000
D. 2000

__________ **4.** The effective range for pumping concrete is __________.
A. 100′ to 300′ horizontally and 300′ to 1000′ vertically
B. 250′ to 1000′ horizontally and 150′ to 300′ vertically
C. 300′ to 1000′ horizontally and 100′ to 300′ vertically
D. none of the above

__________ **5.** When placing concrete it should be __________.
A. placed and moved from one end of the form
B. allowed to free-fall 10′–12′ in the form
C. compacted as little as possible
D. placed as close as possible to its final location in the form

__________ **6.** A(n) __________ is used to consolidate concrete in heavy construction work.
A. spade
B. rod
C. immersion vibrator
D. consolidator

__________ **7.** Concrete in wall forms is placed in level lifts ranging from __________″ deep.
A. 6 to 12
B. 8 to 16
C. 12 to 20
D. 20 to 30

_________ **8.** When placing concrete for walls segregation can be avoided by _________.
A. dropping the concrete vertically
B. using hoppers and chutes
C. limiting the free-fall to 4′ to 6′
D. all of the above

_________ **9.** Lateral concrete pressure against a form wall at the time the concrete is being placed _________.
A. depends on the thickness of the wall
B. is affected only by the concrete temperature
C. is affected by the rate of pour and the concrete temperature
D. depends on the method of consolidation

_________ **10.** The most critical period in the curing and hydration process of concrete is _________.
A. during the first 3 days
B. during the first 28 days
C. after the first 3 days
D. after the first 7 days

_________ **11.** An effective method for curing a concrete slab is to _________.
A. leave the edge forms in place
B. cover the slab with boards
C. place a layer of sand over the slab
D. spread waterproof paper or plastic film over the slab

_________ **12.** A cubic yard of concrete containing 9.5 pounds of water and 22 pounds of cement has a water-cement ratio of _________.
A. .33
B. .37
C. .43
D. .46

_________ **13.** The American Concrete Institute recommends that wall forms not supporting beam soffits may be stripped after _________ hours.
A. 3
B. 6
C. 12
D. 24

_________ **14.** The maximum allowable slump for concrete used for beams and reinforced walls is _________″.
A. 1
B. 2
C. 3
D. 4

_________ **15.** _________ is one of the most commonly used accelerators.
A. Calcium chloride
B. Pozzolan
C. Sodium chloride
D. Permicite

________ 16. Hydration stops when temperature is ________ °F or below.
- A. 15
- B. 10
- C. 0
- D. −10

________ 17. The lateral pressure at the base of a wall form with a placement rate of 4 feet per hour is ________ PSF.
- A. 300
- B. 400
- C. 500
- D. 600

________ 18. Concrete attains approximately ________% of its design strength after 14 days.
- A. 55
- B. 70
- C. 85
- D. 100

________ 19. During the initial curing stage of concrete, the concrete should be kept moist for ________ days.
- A. three
- B. four
- C. five
- D. six

CHAPTER 8

Concrete Formwork Problems

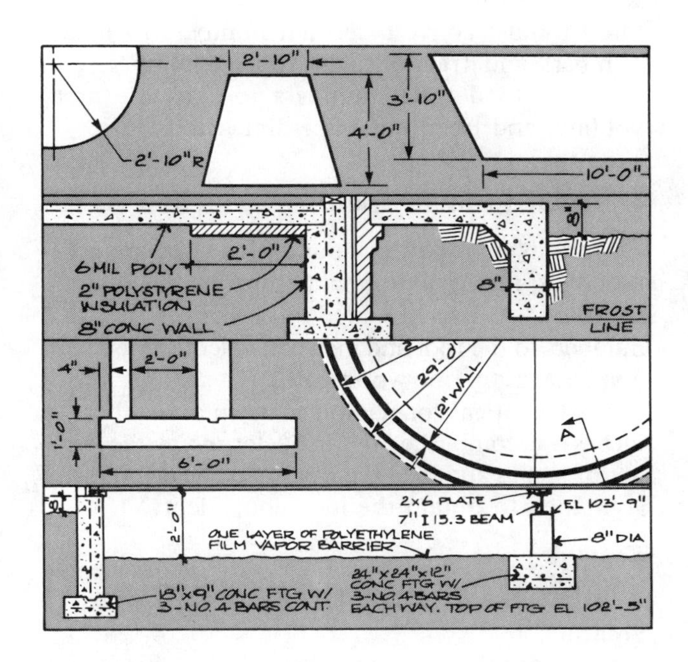

A form builder encounters applications on the job that involve math and printreading skills. The applications range from calculating area and volume to determining placement of forms for foundation walls and footings.

Section 1, Math Fundamentals, includes information regarding conversions of decimal values to foot and inch equivalents and foot and inch values to decimal equivalents, area and volume calculations, and tread and riser calculations. Conversions and area and volume calculations are used frequently to determine the surface area of a wall form or volume of concrete required for a structure. Tread and riser calculations determine the number of treads and risers and their respective dimensions.

Section 2, Printreading Exercises, include questions related to five prints: Plot Plan, Slab-on-Grade Foundation, Crawl Space Foundation, Full Basement Foundation, and Heavy Construction Foundation. The prints contain information regarding foundation construction that is presented in chapters 3 to 5 of the text.

Section 3, Estimating Form Materials and Concrete, provides a step-by-step procedure for calculating the amount of form materials and volume of concrete required for a small residential structure. The Review Questions contain exercises for estimating the amount of form materials and volume of concrete for a full basement foundation.

SECTION 1

MATH FUNDAMENTALS

A form builder must have an understanding of basic math concepts to perform estimating and formwork operations. Mixed numbers and fractions, such as $4\frac{5}{8}''$ or $\frac{1}{2}''$, are routinely used in form construction. Decimal numbers, such as .55 or 101.8′, are used when referring to ratios or elevations. A form builder must convert decimal numbers to mixed numbers and fractions, and mixed numbers and fractions to decimal numbers to calculate area, volume, and tread and riser dimensions.

Decimal Foot to Inch Equivalents

Elevations on a plot or foundation plan are commonly expressed in decimal numbers, such as 45.2′ or 10.2′. A form builder must convert decimal numbers to the foot and inch equivalent for use with a standard tape or wood rule.

Mathematical conversion of decimal numbers to inches is accomplished by first determining the number of inches and then the fraction. The answers are combined to obtain the inch equivalent.

Example
Convert .83′ to an inch equivalent.

Solution

1. Multiply the decimal value by 12 to determine the number of inches and decimal part of an inch.

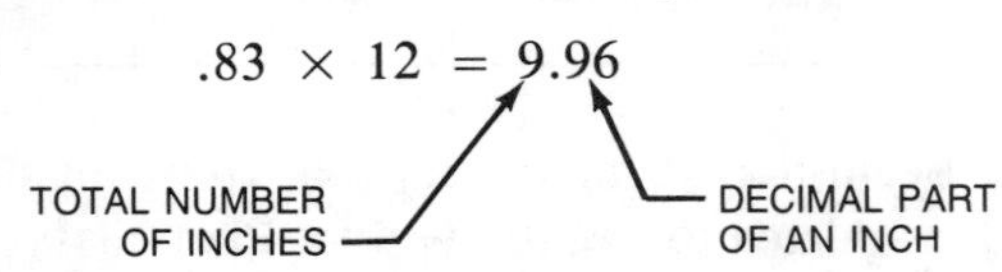

2. Multiply the decimal part of the inch by a commonly used denominator (for example 4, 8, or 16). A larger denominator value results in greater accuracy.

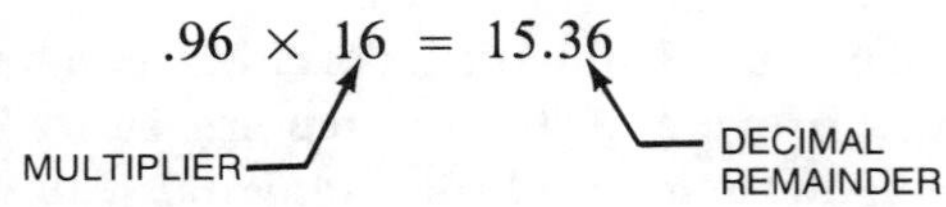

3. If the decimal remainder is .5 or greater, round the value preceding the decimal point up. If the decimal remainder is less than .5, round the value down. Use the rounded value as the numerator and the commonly used denominator in step 2 as the denominator. Reduce the fraction to lowest terms if possible.

15.36 rounded down to 15

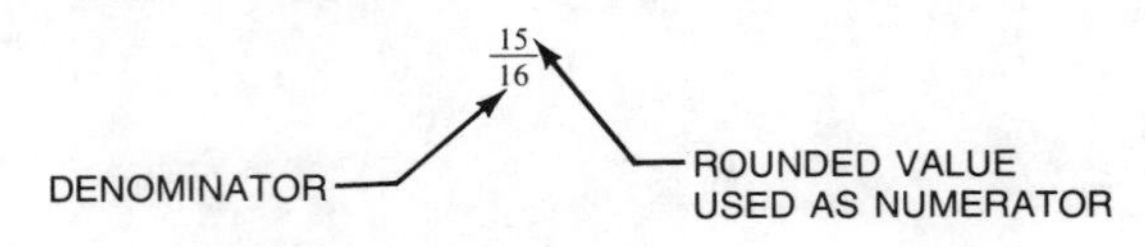

4. Combine the answers from steps 1 and 3 to obtain the inch equivalent.

$9'' + \frac{15}{16}'' = 9\frac{15}{16}''$

$.83' = 9\frac{15}{16}''$

When converting an elevation to an inch equivalent, the number preceding the decimal point is the total number of feet and the number following the decimal point is converted to inches.

Example
Convert 10.9′ to a foot and inch equivalent.

Solution

1. Multiply the value after the decimal point by 12 to determine the number of inches and decimal part of an inch.

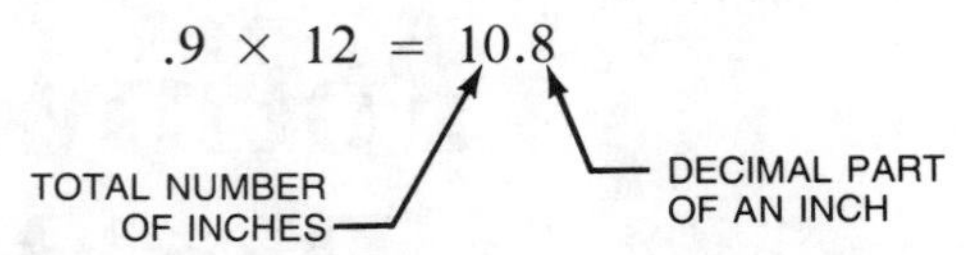

2. Multiply the decimal part of an inch by a commonly used denominator (for example 4, 8, or 16). A larger denominator value results in greater accuracy.

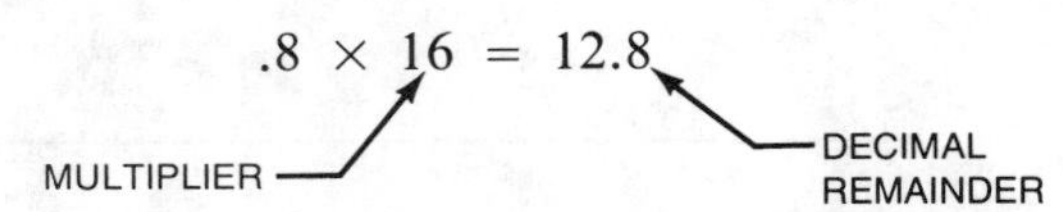

3. If the decimal remainder is .5 or greater, round the value preceding the decimal point up. If the decimal remainder is less than .5, round the value down. Use the rounded value as the numerator and the commonly used demoninator in step 2 as the denominator. Reduce the fraction to lowest terms if possible.

12.8 rounded up to 13

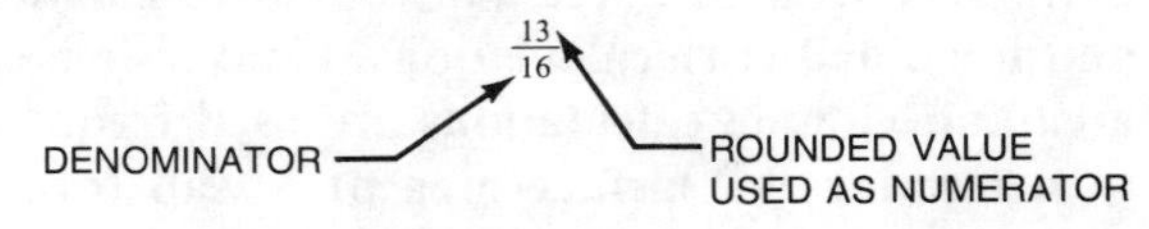

4. Combine the foot value (number preceding the decimal point in the example) with inch values.

$10' + 10'' + \frac{13}{16}'' = 10'\text{-}10\frac{13}{16}''$

$10.9' = 10'\text{-}10\frac{13}{16}''$

A conversion table is also used to convert decimal feet to inches. (See Appendix A for Conversion Table.) The decimal foot value to be converted is located in the table and the number of inches is read from the horizontal row above. The fractional part of an inch, in eighths, is read from the vertical column to the left. If a decimal foot value is not located in the conversion table, the mathematical conversion is used.

Example

Convert .22′ to an inch equivalent.

Solution

8th of an Inch	Inches 0	Inches 1	Inches 2
0	.00	.08	.17
1	.01	.09	.18
2	.02	.10	.19
3	.03	.11	.20
4	.04	.13	.21
5	.05	.14	.22
6	.06	.15	.23
7	.07	.16	.24

$.22' = 2\frac{5}{8}''$

Inch to Decimal Inch and Foot Equivalents

Fractions and mixed numbers are converted to decimal inch and foot equivalents for use in applications such as determining stair riser and tread dimensions. Decimal inch equivalents express the given value in terms of 1″, such as $\frac{3}{4}'' = .75''$. Decimal foot equivalents express the given value in terms of 12″, such as $\frac{3}{4}'' = .06'$.

A decimal inch equivalent is determined mathematically or by using a conversion table. (See Appendix A for Conversion Table.) To convert a fraction to a decimal inch equivalent, the numerator is divided by the denominator. A decimal point is placed before the equivalent.

Example

Convert $\frac{7}{8}''$ to a decimal inch equivalent.

Solution

$7 \div 8 = .875$

NUMERATOR (7)
DENOMINATOR (8)
DECIMAL INCH EQUIVALENT (.875)

$\frac{7}{8}'' = .875''$

A mixed number is converted to a decimal inch equivalent in a similar manner. The whole number remains the same and the fraction is converted to a decimal inch equivalent as in the previous example.

Example

Convert $8\frac{1}{4}''$ to a decimal inch equivalent.

Solution

$8\frac{1}{4}'' = 8.25''$

A decimal foot equivalent is determined by converting a fraction, whole number, or mixed number to a decimal inch equivalent. The decimal inch equivalent is divided by 12 to determine the decimal foot equivalent.

Example

Convert 10″ to a decimal foot equivalent.

Solution

$10 \div 12 = .833$

$10'' = .833'$

Example

Convert $8\frac{1}{2}''$ to a decimal foot equivalent.

Solution

1. Convert $8\frac{1}{2}''$ to a decimal inch equivalent.

   $8\frac{1}{2}'' = 8.5''$

2. Convert 8.5″ to a decimal foot equivalent.

   $8.5 \div 12 = .708$

   $8\frac{1}{2}'' = .708'$

Area Calculation

Area is a measurement of space that a two-dimensional plane or surface occupies. Area calculations determine the amount of form sheathing material required or the area within building or property lines. Area is expressed in units such as square feet (sq ft) or square inches (sq in.). The area for various shapes is calculated by different formulas, but in general is determined by multiplying the length by the height or width. The area of a circle is determined by multiplying π (3.14) by the radius squared or .7854 times the diameter squared.

Squares, Rectangles, and Parallelograms. The area of a horizontal or vertical square or rectangular surface, such as a building site or form wall, is determined by multiplying the two outside dimensions. The area of a horizontal surface is determined by multiplying the width by the length, and a vertical surface is determined by multiplying the length by the height.

Example
Determine the area (A) of the rectangle.

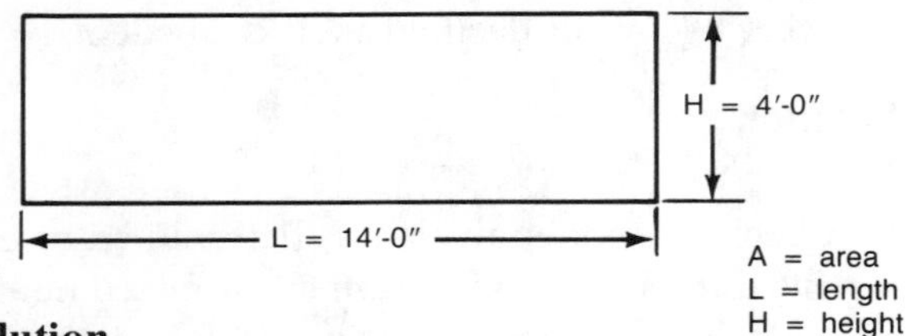

Solution

$A = L \times H$

$= 14'\text{-}0'' \times 4'\text{-}0''$

$Area = 56$ sq ft

Example
Determine the area (A) of the parallelogram.

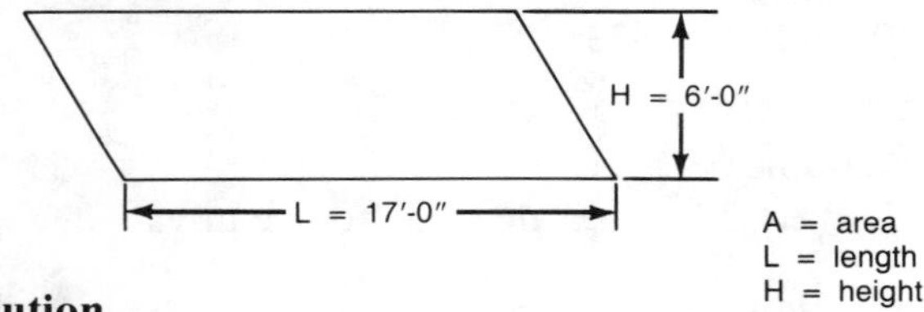

Solution

$A = L \times H$

$= 17'\text{-}0'' \times 6'\text{-}0''$

$Area = 102$ sq ft

Triangles. The area of a triangle is determined by multiplying the base dimension by the altitude and dividing by 2.

Example
Determine the area (A) of the triangle.

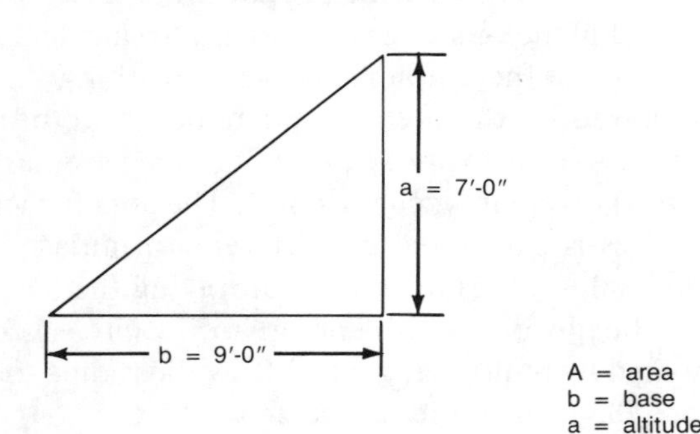

Solution

$$A = \frac{ba}{2}$$

$$= \frac{9'\text{-}0'' \times 7'\text{-}0''}{2}$$

$$= \frac{63}{2}$$

$Area = 31.5$ sq ft

Trapezoids. A trapezoid is a geometric shape having four sides in which two of the sides are parallel. Battered foundation walls or tapered pier footings are common designs incorporating a trapezoid shape. To determine the area of a trapezoid, add the length of the top and bottom and multiply the sum by the height. Divide the results by 2 to obtain the area.

Example
Determine the area (A) of the trapezoid.

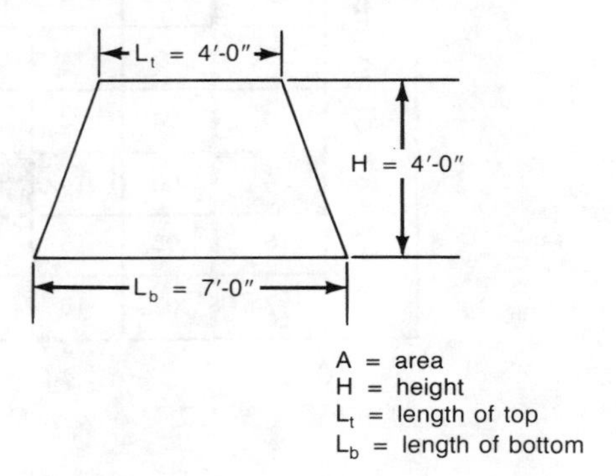

Solution

$$A = \frac{H\,(L_t + L_b)}{2}$$

$$= \frac{4'\text{-}0''\,(4'\text{-}0'' + 7'\text{-}0'')}{2}$$

$$= \frac{4'\text{-}0'' \times 11'\text{-}0''}{2}$$

$$= \frac{44}{2}$$

$Area = 22$ sq ft

Circles. The area of a circle is determined by using either the radius or the diameter of the circle. The radius is one-half the diameter and is measured from the center point to an edge of a circle. The diameter is the distance from one edge of a circle to the other and passing through the center point.

The area of a circle is calculated by multiplying π (3.14) by the radius squared (radius × radius). The area may also be determined by multiplying .7854 by the diameter squared (diameter × diameter).

Example
Determine the area (A) of the circle using the radius of the circle.

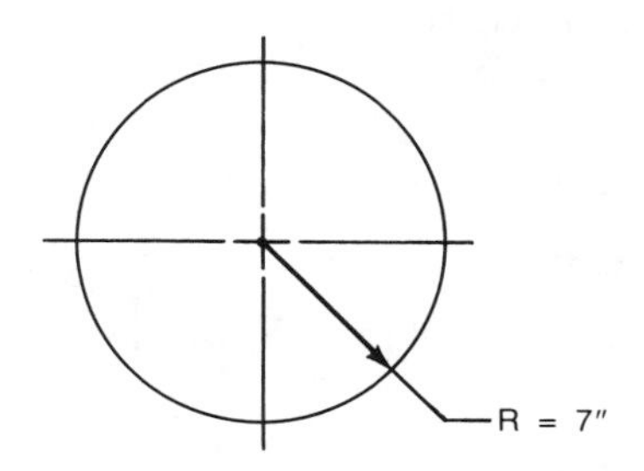

Solution

$A = \pi R^2$
$= 3.14 \times (7'' \times 7'')$
$= 3.14 \times 49$
$Area = 153.86$ sq in.

Example
Determine the area (A) of the circle using the diameter of the circle.

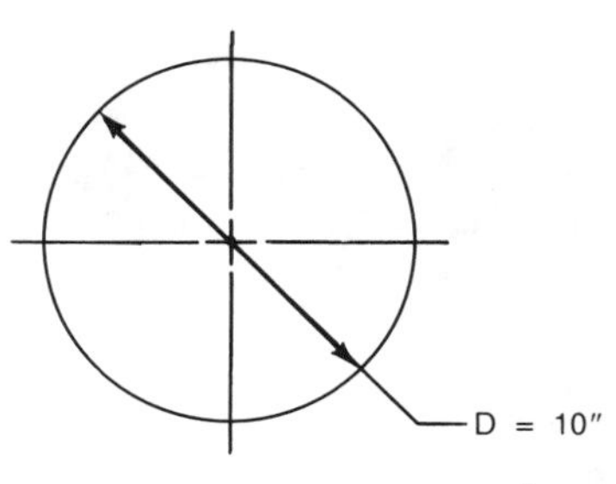

Solution

$A = .7854\, D^2$
$= .7854 \times (10'' \times 10'')$
$= .7854 \times 100$
$Area = 78.54$ sq in.

Volume Calculation

Volume is the amount of space that a three-dimensional figure or object occupies. Volume calculations are used to determine the amount of fill required for a building site or the amount of concrete required to form a footing or wall. Volume is expressed in cubic units, such as cubic feet (cu ft) or cubic yards (cu yd).

Rectangular Solids and Cubes. A rectangular solid is a six-sided solid object with a rectangular base. A cube is a solid object with six equal square faces. Examples of a rectangular solid or cube are a foundation wall, square pier footing, or square or rectangular column. The volume of a rectangular solid or cube is determined by multiplying its thickness, length, and height. The volume of a square column is determined by multiplying the width squared by the height.

Example
Determine the volume (V) of the rectangular solid.

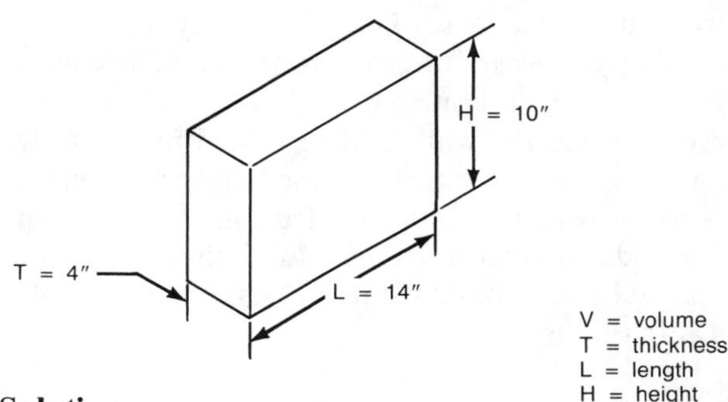

Solution
$V = T \times L \times H$
$= 4'' \times 14'' \times 10''$
$Volume = 560$ cu in.

When calculating large amounts of material, volume is commonly expressed in cubic feet or cubic yards. One cubic foot equals 1728 cubic inches and 1 cubic yard equals 27 cubic feet. When calculating volume for large amounts of material, thickness, length, and height are converted to decimal feet and then multiplied to obtain cubic feet. The result is divided by 27 to obtain cubic yards.

Example
Determine the volume (V) of the foundation wall.

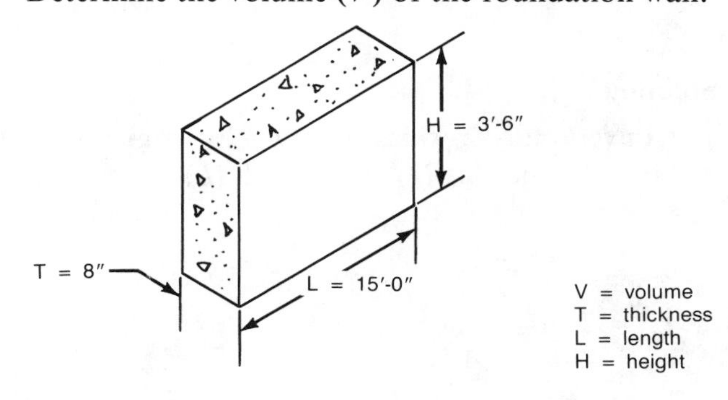

Solution

1. Convert the thickness (T), length (L), and height (H) to decimal foot equivalents.
   $T = 8'' = .67'$
   $L = 15'\text{-}0'' = 15.0'$
   $H = 3'\text{-}6'' = 3.5'$
2. Calculate the volume of the foundation wall.
   $V = T \times L \times H$
   $= .67' \times 15.0' \times 3.5'$
   $Volume = 35.18$ cu ft
3. Convert cubic feet to cubic yards.
   $Cu\ yd = cu\ ft \div 27$
   $= 35.18 \div 27$
   Cubic yards = 1.3

Frustums of Pyramids. A frustum is a pyramid cut off parallel to its base. The sides of a frustum, such as a tapered pier footing, are trapezoids. The volume of a frustum with one battered side, such as a battered foundation wall, is determined by multiplying the average thickness by the height and length. The approximate volume of a frustum with four battered sides is determined by adding the areas of the top and bottom and dividing by 2 and then multiplying by the height.

Example

Determine the volume (V) of the battered foundation wall.

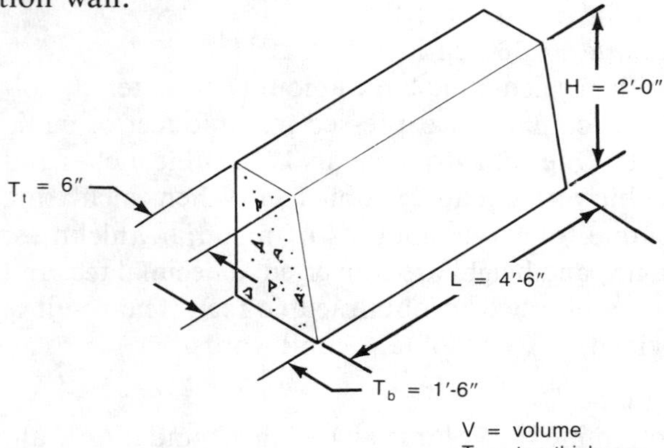

Solution

1. Convert the top thickness (T_t), bottom thickness (T_b), height (H), and length (L) to decimal foot equivalents.
   $T_t = 6'' = .5'$
   $T_b = 1'\text{-}6'' = 1.5'$
   $H = 2'\text{-}0'' = 2.0'$
   $L = 4'\text{-}6'' = 4.5'$
2. Calculate the volume of the battered foundation wall.
   $$V = \frac{T_t + T_b}{2} \times H \times L$$
   $$= \frac{.5' + 1.5'}{2} \times 2.0' \times 4.5'$$
   $= 1.0' \times 2.0' \times 4.5'$
   $Volume = 9$ cu ft
3. Convert cubic feet to cubic yards.
   $Cu\ yd = cu\ ft \div 27$
   $= 9 \div 27$
   Cubic yards = .33

Example

Determine the volume (V) of the tapered pier footing.

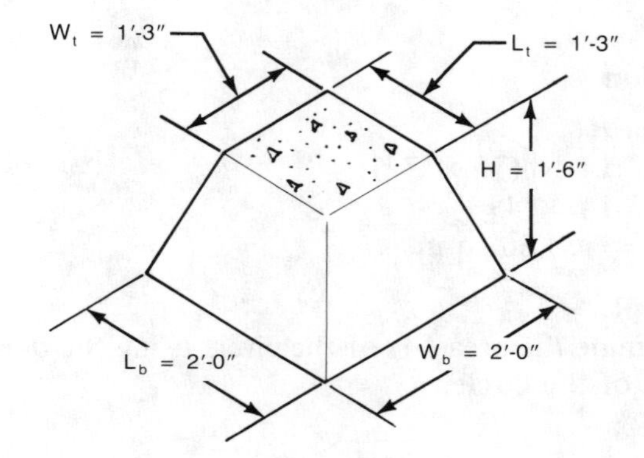

V = volume
A$_t$ = area of top
L$_t$ = length of top
W$_t$ = width of top
A$_b$ = area of bottom
L$_b$ = length of bottom
W$_b$ = width of bottom
H = height

Solution

1. Convert dimensions to decimal foot equivalents.
   $W_t = 1'\text{-}3'' = 1.25'$
   $L_t = 1'\text{-}3'' = 1.25'$
   $W_b = 2'\text{-}0'' = 2.0'$
   $L_b = 2'\text{-}0'' = 2.0'$
   $H = 1'\text{-}6'' = 1.5'$
2. Determine the areas of the top (A_t) and bottom (A_b).
   $A_t = L_t \times W_t$
   $= 1.25' \times 1.25'$
   Area of top = 1.56 sq ft
   $A_b = L_b \times W_b$
   $= 2.0' \times 2.0'$
   Area of bottom = 4.0 sq ft

3. Determine the volume of the tapered pier footing.

$$V = \frac{A_t + A_b}{2} \times H$$

$$= \frac{1.56 + 4.0}{2} \times 1.5'$$

$$= 2.78 \times 1.5'$$

Volume = 4.17 cu ft

4. Convert cubic feet to cubic yards.

Cu yd = *cu ft* ÷ 27

= 4.17 ÷ 27

Cubic yards = .15

Cylinders. A cylinder is a solid object with a circular cross-sectional area. Round columns and piers are cylinders. The volume of a cylinder is determined by multiplying the cross-sectional area by the height. The cross-sectional area is determined by using the diameter or radius.

Example

Determine the volume (V) of the round column.

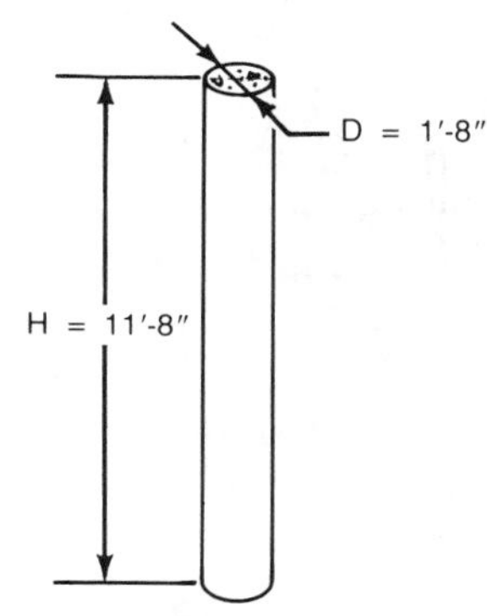

V = volume
D^2 = diameter^2
H = height

Solution

1. Convert the diameter (D) and height (H) to decimal foot equivalents.

D = 1'-8" = 1.67'

H = 11'-8" = 11.67'

2. Determine the volume of the round column.

$$V = .7854 \times D^2 \times H$$

$$= (.7854 \times 1.67^2) \times 11.67'$$

$$= (.7854 \times 2.79) \times 11.67'$$

$$= 2.19 \times 11.67'$$

Volume = 25.56 cu ft

3. Convert cubic feet to cubic yards.

Cu yd = *cu ft* ÷ 27

= 25.56 ÷ 27

Cubic yards = .95

Example

Determine the volume (V) of the circular pier footing.

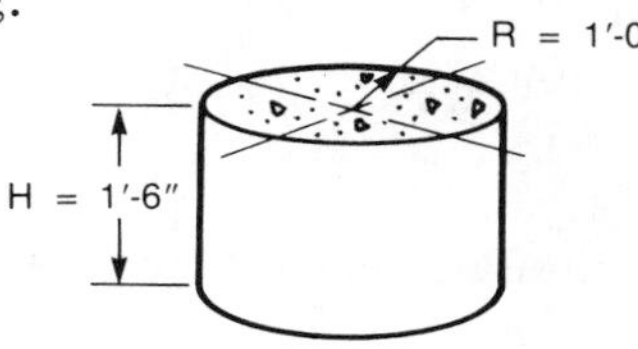

V = volume
R^2 = radius^2
H = height

Solution

1. Convert the radius (R) and height (H) to decimal foot equivalents.

R = 1'-0" = 1.0'

H = 1'-6" = 1.5'

2. Determine the volume of the circular pier footing.

$$V = \pi R^2 \times H$$

$$= (3.14 \times 1.0^2) \times 1.5'$$

$$= (3.14 \times 1.0) \times 1.5'$$

$$= 3.14 \times 1.5'$$

Volume = 4.71 cu ft

3. Convert cubic feet to cubic yards.

Cu yd = *cu ft* ÷ 27

= 4.71 ÷ 27

Cubic yards = .17

Curved Walls and Footings. Curved walls and footings are used in the construction of storage tanks and silos. When calculating the volume of a curved wall or footing, the circumference (distance around the outside edge) is multiplied by the thickness and height.

Example

Determine the volume (V) of the curved wall and footing. Include the keyway in the footing calculations.

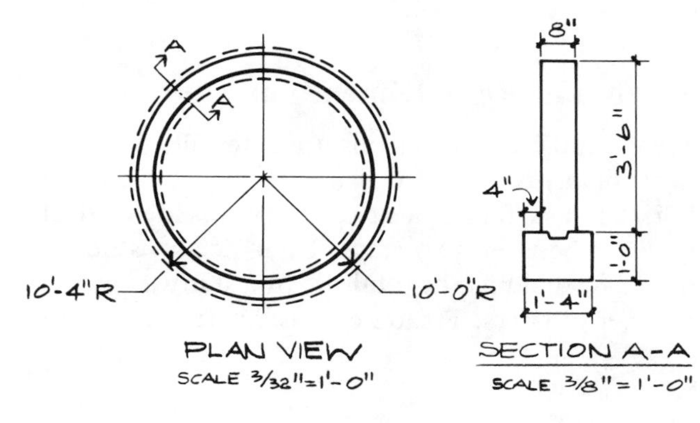

V = volume
R = radius
C = circumference
W = width
H = height

Solution

1. Convert the curved wall dimensions to decimal foot equivalents.
   $R = 10'\text{-}0'' = 10.0'$
   $H = 3'\text{-}6'' = 3.5'$
   $T = 8'' = .67'$
2. Determine the circumference (C) of the curved wall.
   $C = 2\pi R$
   $= 2 \times 3.14 \times 10.0'$
   Circumference $= 62.8'$
3. Determine the volume of the curved wall.
   $V = C \times T \times H$
   $= 62.8' \times .67' \times 3.5'$
   Volume = 147.27 cu ft
4. Convert cubic feet to cubic yards.
   Cu yd = *cu ft* ÷ 27
   = 147.27 ÷ 27
   Cubic yards = 5.45
5. Convert the footing dimensions to decimal foot equivalents.
   $R = 10'\text{-}4'' = 10.33'$
   $H = 1'\text{-}0'' = 1.0'$
   $W = 1'\text{-}4'' = 1.33'$
6. Determine the circumference of the footing.
   $C = 2\pi R$
   $= 2 \times 3.14 \times 10.33'$
   Circumference $= 64.87'$
7. Determine the volume of the footing.
   $V = C \times T \times H$
   $= 64.87' \times 1.33' \times 1.0'$
   Volume = 86.28 cu ft
8. Convert cubic feet to cubic yards.
   Cu yd = *cu ft* ÷ 27
   = 86.28 ÷ 27
   Cubic yards = 3.2

Tread and Riser Dimensions

Tread and riser dimensions are determined before the stairway forms are constructed. The tread is the horizontal surface of a step. The riser is the vertical member between two steps. Riser height is determined by dividing the total rise of a stairway by the number of risers. Tread depth is determined by dividing the total run of the stairway by the number of treads. The number of treads in a stairway is one less than the number of risers.

Example

Determine the number of treads and risers and the riser height and tread depth of a stairway. The total rise is 6′-7″ and the total run is 8′-6″. The riser height should be between 7″ and 7½″ and the tread depth should be 10″ minimum.

Solution

1. Convert the total rise to inches.
   Total rise =
   (*no. of feet* × 12) + *no. of inches*
   = (6′ × 12) + 7″
   Total rise = 79″
2. Determine the number of risers by dividing the total rise by the minimum desired riser height. Disregard the decimal remainder.
   No. of risers =
   total rise ÷ *minimum riser height*
   = 79 ÷ 7
   = 11.28
   No. of risers = 11
3. Determine the exact riser height by dividing the total rise by the number of risers.
   Riser height = *total rise* ÷ *no. of risers*
   = 79 ÷ 11
   Riser height = 7.18″
4. Convert the decimal inch value to a fractional equivalent.
   $7.18'' = 7\frac{3}{16}''$
5. Convert the total run to inches.
   Total run =
   (*no. of feet* × 12) + *no. of inches*
   = (8′ × 12) + 6″
   Total run = 102″
6. Determine the tread depth by dividing the total run by the number of treads.
   Tread depth = *total run* ÷ *no. of treads*
   = 102″ ÷ 10
   Tread depth = 10.2″
7. Convert the decimal inch value to a fractional equivalent.
   $10.2'' = 10\frac{3}{16}''$

Chapter 8–Section 1—Review Questions

Convert the decimal foot values to foot and inch equivalents.

_________ **1.** .47′

_________ **2.** .20′

_________ **3.** 2.04′

_________ **4.** .79′

_________ **5.** 1.56′

_________ **6.** .09′

_________ **7.** 11.35′

_________ **8.** .98′

_________ **9.** .63′

_________ **10.** 3.16′

Convert the inch values to decimal inch equivalents.

_________ **11.** $\frac{1}{2}''$

_________ **12.** $\frac{3}{16}''$

_________ **13.** $1\frac{5}{8}''$

_________ **14.** $4\frac{1}{4}''$

_________ **15.** $\frac{1}{16}''$

_________ **16.** $10\frac{7}{8}''$

_________ **17.** $\frac{3}{4}''$

_________ **18.** $\frac{7}{16}''$

_________ **19.** $3\frac{1}{8}''$

_________ **20.** $\frac{3}{8}''$

Convert the foot and inch values to decimal foot equivalents.

_________ **21.** 9″

_________ **22.** 3′-6″

_________ **23.** $1'\text{-}4\frac{3}{4}''$

_________ **24.** $1\frac{1}{2}''$

_________ **25.** $5'\text{-}9\frac{1}{4}''$

_________ **26.** 6′-0″

_________ **27.** 5′-8″

_________ **28.** $6'\text{-}2\frac{1}{2}''$

_________ **29.** $1'\text{-}0\frac{3}{4}''$

_________ **30.** $12'\text{-}6\frac{3}{8}''$

Determine the area of the geometric shapes. Express the answer in square feet.

_________ **1.** Height = 12″, Length = $4\frac{1}{2}''$

_________ **2.** Height = 1′-6″, Length = 4′-0″

_________ **3.** Height = 5′-2″, Length = 5′-2″

_________ **4.** Height = 5′-0″, Length = 34′-0″

_________ **5.** Height = 2′-6″, Length = 2′-6″

_________ **6.** Height = 1′-8″, Length = 3′-6″

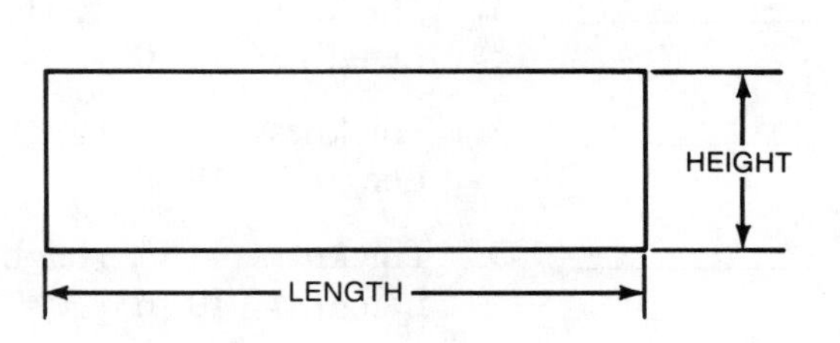

_______ **7.** Height = 4′-0″, Length = 11′-0″

_______ **8.** Height = 5′-6″, Length = 14′-0″

_______ **9.** Height = 8′-2″, Length = 10′-6″

_______ **10.** Height = 13′-4″, Length = 15′-10″

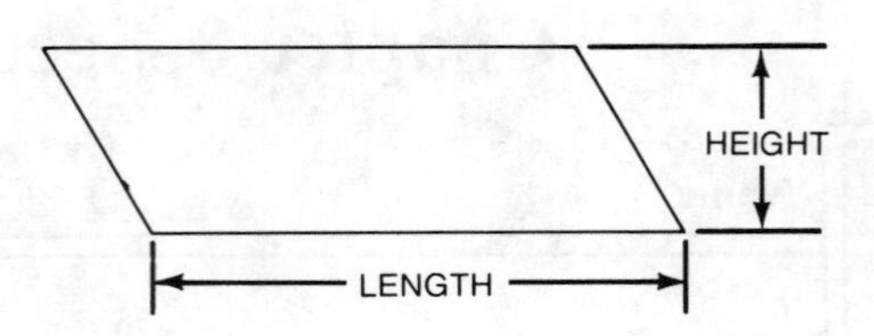

_______ **11.** Base = 5′-0″, Altitude = 6′-0″

_______ **12.** Base = 10′-0″, Altitude = 15′-6″

_______ **13.** Base = 9′-2″, Altitude = 10′-6″

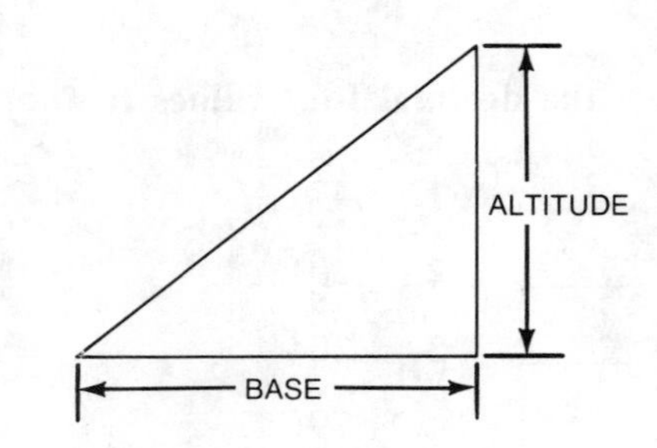

_______ **14.** Height = 4′-0″, Length of top = 8′-0″, Length of bottom = 10′-0″

_______ **15.** Height = 6′-0″, Length of top = 16′-4″, Length of bottom = 24′-8″

_______ **16.** Height = 4′-6″, Length of top = 13′-10″, Length of bottom = 21′-8″

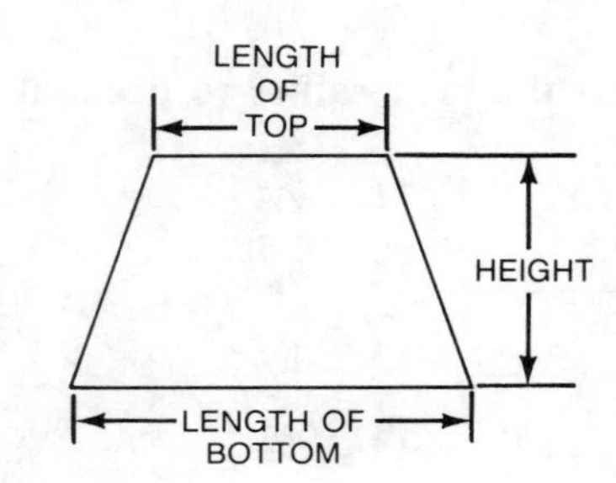

_______ **17.** Diameter = 2′-6″

_______ **18.** Radius = 5′-0″

_______ **19.** Diameter = 3′-4″

_______ **20.** Radius = 8′-6″

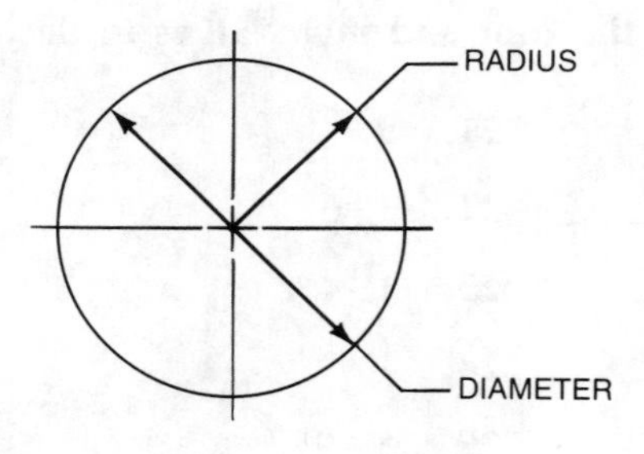

Determine the volume of the geometric solids. Express the answer in cubic yards.

_______ **1.** Thickness = 1′-0″, Height = 5′-0″, Length = 12′-0″

_______ **2.** Thickness = 7″, Height = 5′-10″, Length = 19′-0″

_______ **3.** Thickness = 8″, Height = 3′-6″, Length = 10′-6″

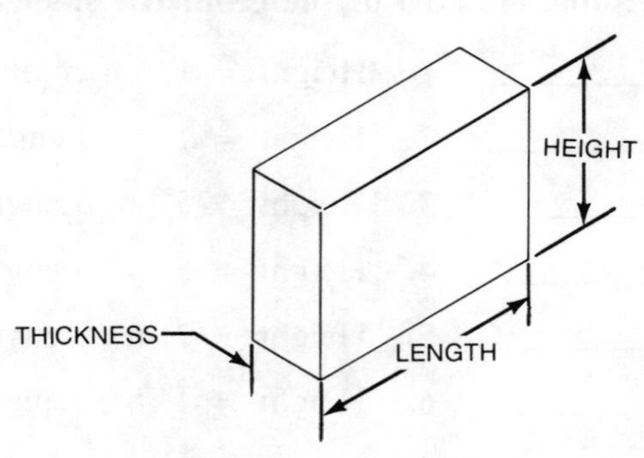

______ **4.** Width = 10″, Height = 12′-0″

______ **5.** Width = 1′-4″, Height = 9′-0″

______ **6.** Width = 2′-6″, Height = 15′-6″

______ **7.** Top thickness = 8″,
bottom thickness = 1′-4″,
height = 3′-0″, length = 10′-0″

______ **8.** Top thickness = 6″,
bottom thickness = 1′-0″,
height = 2′-0″, length = 15′-0″

______ **9.** Top thickness = 8″,
bottom thickness = 1′-8″,
height = 4′-0″, length = 18′-0″

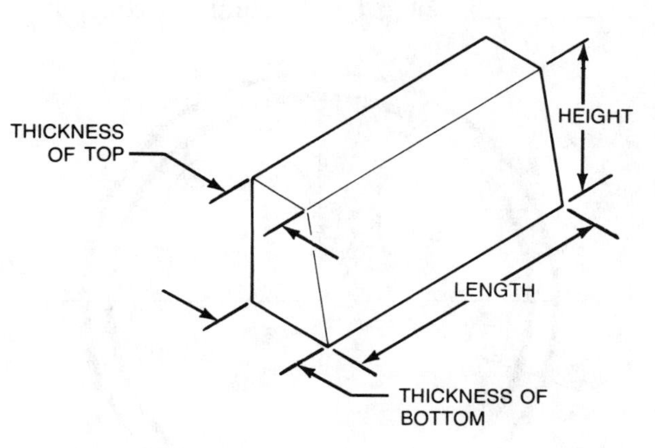

______ **10.** Area of top = 2.25 sq ft,
Area of bottom = 4.0 sq ft,
Height = 2′-0″

______ **11.** Area of top = 4.0 sq ft,
Area of bottom = 6.25 sq ft,
Height = 3′-6″

______ **12.** Area of top = 6.25 sq ft,
Area of bottom = 9.0 sq ft,
Height = 3′-0″

______ **13.** Area of top = 9.0 sq ft,
Area of bottom = 12.25 sq ft,
Height = 4′-0″

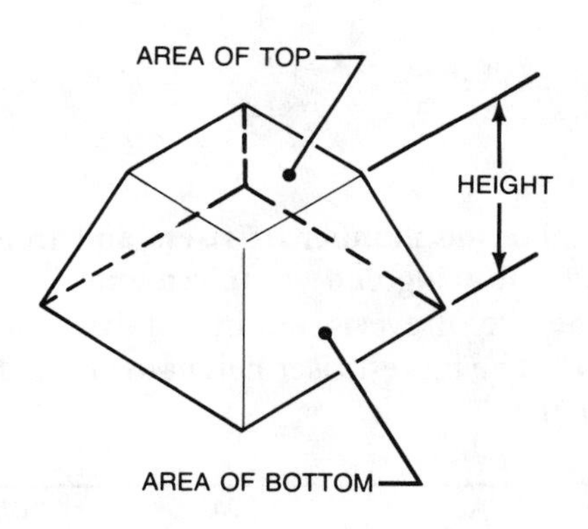

______ **14.** Diameter = 2′-0″, Height = 10′-0″

______ **15.** Radius = 1′-3″, Height = 8′-0″

______ **16.** Diameter = 1′-6″, Height = 10′-3″

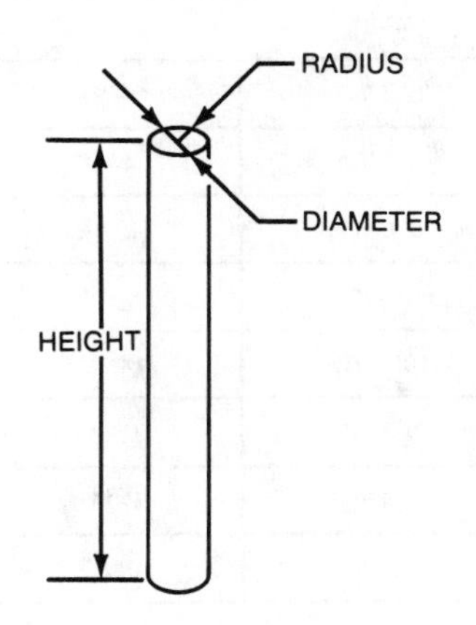

Determine the volume of the concrete required to form the secondary clarification tank walls and footing. Express the answers in cubic yards.

__________ **17.** Footing

__________ **18.** 8″ wall

__________ **19.** 12″ wall

__________ **20.** Total volume of concrete

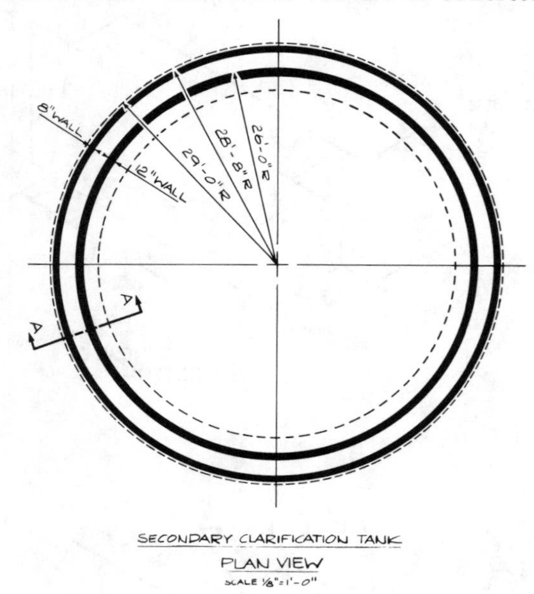

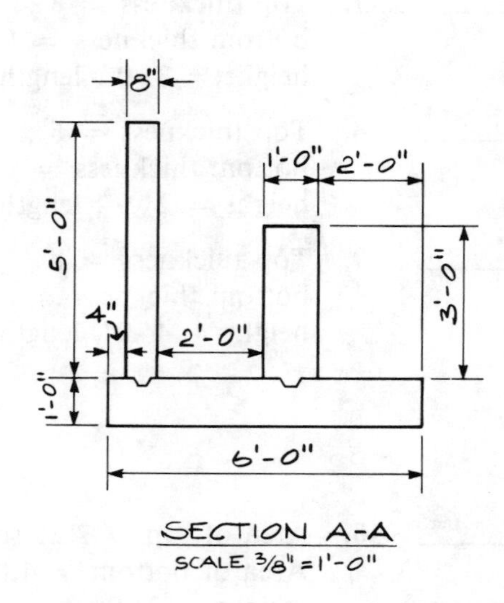

Symons Corporation

Determine the number of risers and treads and the riser height and tread depth for the stairways based on the following code requirements.

The rise of every step in a stairway shall not exceed 7½″. The tread depth shall not be less than 10″. The largest riser height or tread depth within a stairway shall not exceed the smallest by more than ¼″.

TOTAL RISE	TOTAL RUN	NUMBER OF RISERS	HEIGHT OF RISERS	NUMBER OF TREADS	DEPTH OF TREADS
3′-9″	4′-2″				
4′-8″	5′-11¾″				
6′-1¾″	8′-7½″				
4′-1⅞″	5′-0″				
8′-6⅜″	12′-8¾″				
5′-6″	6′-11″				
10′-3¼″	14′-11″				
7′-3¾″	9′-2¾″				
11′-4 9/16″	15′-11¼″				
7′-5¼″	10′-5⅛″				

Chapter 8–Section 2—Printreading Exercises

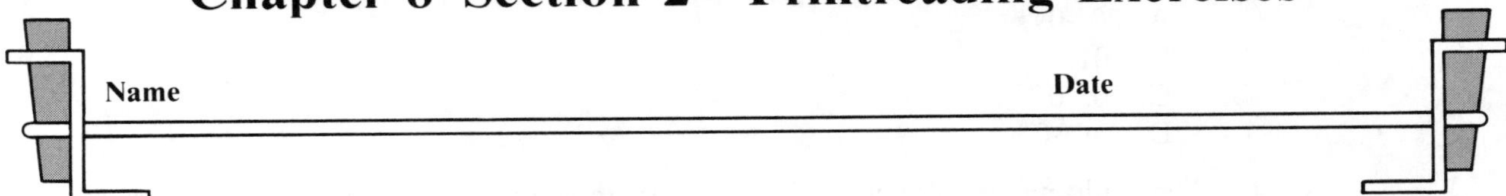

Plot Plan

A plot plan is the main source of information regarding preliminary site work. A plot plan shows the location of a building, finish grade levels and elevations, utilities, streets and roads, sidewalks, easements, and locations of trees, bushes, and shrubs. Bench marks are also identified and used as a reference for elevations.

Refer to the Plot Plan on page 233 to answer the following questions.

Completion

_______ **1.** The bench mark is located in the _______ corner of the building site.

_______ **2.** The finish floor elevation is _______′.

_______ **3.** The overall east to west dimension of the building site is _______.

_______ **4.** The difference in height between the bench mark and the grade at the northwest corner of the lot is _______′.

_______ **5.** The distance from the curb line to the front sidewalk is _______.

_______ **6.** The widest north to south dimension of the building is _______.

_______ **7.** The front setback of the building is _______.

_______ **8.** The gas line and water main are located below the surface of the _______.

_______ **9.** The driveway is _______ wide.

_______ **10.** The grade level at the northeast corner of the building is _______′.

Multiple Choice

_______ **1.** The driveway slopes _______′ from the north end to the southeast corner.

A. .4
B. .5
C. .6
D. .8

_______ **2.** The sidewalk is _______ wide.

A. 3′-0″
B. 4′-6″
C. 6′-0″
D. 7′-5″

_______ **3.** The west property line is _______ long.
A. 80′-0″
B. 101.8′
C. 103.0′
D. 140′-0″

_______ **4.** The building site slopes _______′ from the northeast to northwest corner.
A. 1.0
B. 2.0
C. 3.4
D. 4.0

_______ **5.** The grade level at the southwest corner of the building is _______′.
A. 101.8
B. 101.0
C. 102.2
D. 102.4

_______ **6.** The front property line is _______ from the sidewalk.
A. 2′-0″
B. 4′-0″
C. 6′-0″
D. 15′-0″

_______ **7.** The east property line is _______ from the side of the building.
A. 15′-0″
B. 18′-0″
C. 19′-6″
D. 20′-0″

_______ **8.** The driveway is sloped _______′ from the southwest to southeast corner.
A. .1
B. .2
C. 1.1
D. 1.4

_______ **9.** The lowest grade level of the building site is _______′.
A. 99.8
B. 100.0
C. 100.2
D. 102.8

_______ **10.** The building site is sloped _______′ from the southwest corner of the building to the southwest corner of the lot.
A. .2
B. .4
C. .6
D. .9

_______ **11.** The finish floor elevation is _______′ higher than the southwest corner of the lot.
A. 3.2
B. 4.3
C. 5.0
D. 6.2

Chapter 8–Section 2—Printreading Exercises

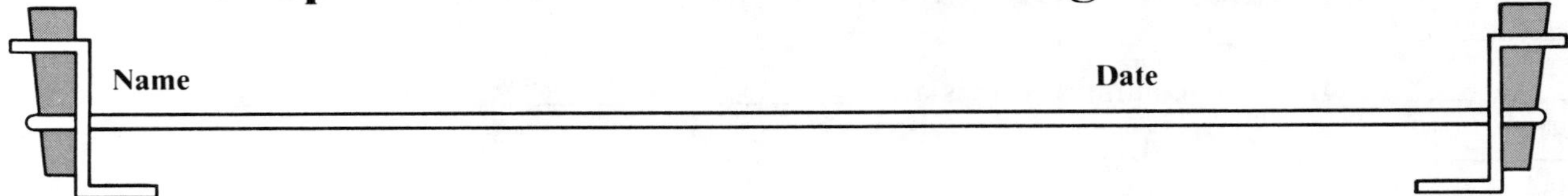

Full Basement Foundation Prints

A full basement foundation provides an area below the superstructure for living or storage space. The basement area is commonly below ground level. A set of prints for a full basement foundation includes a plan view and related section view drawings. A plan view provides elevations and overall dimensions of a building and locations and dimensions of structural components such as pier footings and stairways. The size and spacing of first floor joists and size and location of supporting beams and posts are also included in the plan view.

Section view drawings indicate the thickness and height of foundation footings and walls, type and spacing of reinforcement, and thickness of the basement floor slab. Information regarding specially formed features, such as pockets or shoulders, is also provided in the section view drawings.

Refer to the Full Basement Foundation print on pages 235 and 236 to answer the following questions.

Completion

______ **1.** The north foundation wall is ______ long.

______ **2.** The east garage wall is ______ from the east wall of the house.

______ **3.** The back stoop is ______ wide × ______ long.

______

______ **4.** The center of the basement window in the southeast corner of the south foundation wall is ______ from the southeast corner of the building.

______ **5.** The dimensions of the beam pocket are ______ × ______ × ______.

______

______

______ **6.** There are ______ treads and ______ risers in the stairway.

______

______ **7.** There are ______ No. 4 rebars running continuously in the foundation footings.

______ **8.** The east foundation wall is ______ long.

______ **9.** A(n) ______″ wide-flange, ______ pound beam is supported by the pipe columns.

______

______ **10.** ______ × ______ floor joists are supported by the wide-flange beam.

______

______ 11. The front stoop is ______ wide × ______ long.

______

______ 12. The column footings are ______ ″ × ______ ″ × ______ ″.

______

______

______ 13. The stair risers are ______ ″ high and the treads are ______ ″ deep.

______

______ 14. The concrete floor slab is ______ ″ thick.

______ 15. The floor joists are spaced ______ ″ OC.

______ 16. The garage opening is ______ wide.

______ 17. The wide-flange beam is ______ long.

______ 18. The north foundation wall is ______ ″ thick.

______ 19. The footing supporting the south foundation wall is ______ ″ wide.

______ 20. Three No. ______ rebars are spaced ______ ″ OC vertically in the foundation walls.

______

______ 21. The elevation at the garage door opening is ______.

______ 22. The footing beneath the garage door opening extends ______ ″ beyond the foundation wall.

______ 23. The elevation at the top of the front stoop wall is ______.

______ 24. The wide flange beam has a ______ ″ clearance at the end.

______ 25. The footing under the north foundation wall is ______ ″ wide.

Multiple Choice

______ 1. The pipe columns are ______ ″ in diameter.
A. 2
B. 4
C. 6
D. 8

______ 2. The stairway dimensions are ______.
A. 3′-0″ × 9′-0″
B. 3′-0″ × 12′-8″
C. 4′-0″ × 9′-0″
D. 5′-0″ × 12′-8″

______ 3. The front stoop wall is ______ ″ thick.
A. 4
B. 5
C. 6
D. 8

_________ **4.** The distance between the column centers is _________.
A. 9′-11″
B. 12′-0″
C. 12′-8″
D. 38′-0″

_________ **5.** The footing under the south foundation wall is _________″ high.
A. 6
B. 8
C. 10
D. 12

_________ **6.** The front and back stoop walls extend _________″ below the frost line.
A. 2
B. 4
C. 6
D. 10

_________ **7.** All foundation walls extend a minimum of _________″ above the grade line.
A. 4
B. 6
C. 8
D. 10

_________ **8.** The foundation footing extends _________″ beyond the west garage wall.
A. 4
B. 6
C. 8
D. 10

_________ **9.** Basement windows in the north foundation wall are _________ apart from each other.
A. 7′-8″
B. 9′-11″
C. 11′-8″
D. 12′-0″

_________ **10.** The top of the north foundation footing is _________ below the grade line.
A. 8″
B. 2′-6″
C. 7′-2″
D. 7′-10″

_________ **11.** The rebars in the north foundation footing are placed _________″ from the bottom of the footing.
A. ¾
B. 1
C. 1½
D. 2½

_________ **12.** The footing below the garage door opening is _________″ wide.
A. 10
B. 16
C. 20
D. 24

__________ **13.** The back stoop wall is __________ high.
A. 2′-6″
B. 3′-0″
C. 3′-6″
D. 4′-0″

__________ **14.** A concrete __________ is formed along the outside of the north section of the west foundation wall.
A. keyway
B. shoulder
C. pocket
D. core

__________ **15.** The distance from the southwest corner of the building to the garage door opening is __________.
A. 2′-0″
B. 2′-3⅝″
C. 16′-4¾″
D. 100′-8″

__________ **16.** The foundation walls under the brick veneer are __________″ thick.
A. 6
B. 8
C. 10
D. 12

__________ **17.** The foundation wall elevation at the garage opening is __________″ lower than the adjacent walls.
A. 6
B. 8
C. 10
D. 12

__________ **18.** The stairway has __________ risers.
A. 12
B. 13
C. 14
D. 15

Chapter 8-Section 2—Printreading Exercises

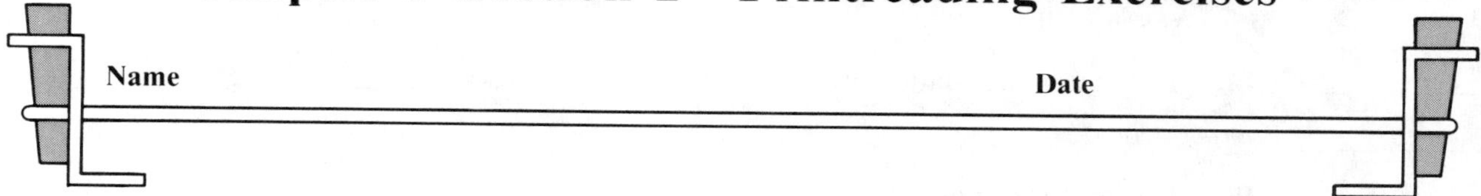

Crawl Space Foundation Prints

Crawl space foundations provide a crawl space beneath the floor joists for access to plumbing and other utilities. The prints for a crawl space foundation include a plan view and related section view drawings required to construct the foundation. The plan view provides overall dimensions of the foundation and locations and dimensions of foundation walls, footings, and pier footings. The size and direction of floor joists and posts and beams are also included in the plan view.

Section view drawings show the thickness and heights of foundation walls and footings and type and size of reinforcement. Section view drawings also provide information regarding the minimum clearance between the ground and floor joists.

Refer to the Crawl Space Foundation print on pages 237 to 240 to answer the following questions.

Completion

__________ **1.** The west foundation wall is __________ long.

__________ **2.** The minimum depth of excavation below the floor joists is __________.

__________ **3.** A(n) __________" I beam extends across the crawl space area.

__________ **4.** The beam pockets are __________" × __________" × __________".

____

____

__________ **5.** The foundation footing beneath the west wall is __________" × __________".

____

__________ **6.** The concrete piers in the crawl space area are __________" in diameter.

__________ **7.** The fireplace footing is __________ × __________ × __________.

____

____

__________ **8.** The east foundation wall is __________ long.

__________ **9.** The fireplace extends __________ outside the east foundation wall.

__________ **10.** The centers of the piers in the crawl space area are __________ from the north foundation wall.

__________ **11.** The west foundation wall is __________" thick × __________ high.

____

__________ **12.** The elevation of the garage door pockets is __________.

__________ **13.** The distance from the outside of the west garage wall to the east side of the crawl space wall is __________.

__________ **14.** The south foundation wall is __________" thick × __________ high.

____

__________ **15.** The pier footings are __________" × __________" × __________".

____

____

__________ **16.** The difference in elevation between the garage floor and the top of the foundation walls is __________.

__________ **17.** The west garage wall is __________ from the center of the west door pocket.

__________ **18.** There are __________ No. 4 rebars required in the foundation footings.

__________ **19.** The center of the pier footing in the garage is __________ from the north foundation wall.

__________ **20.** No. 4 rebars are placed __________" OC each way in the fireplace footing.

__________ **21.** The elevation at the top of the concrete pier is __________.

__________ **22.** A(n) __________" concrete slab is placed below the fireplace.

__________ **23.** The hearth extends __________" in front of the fireplace.

__________ **24.** The north foundation wall is __________" thick × 3′-0″ high.

__________ **25.** The sill plates are fastened to the top of the foundation wall with __________" × __________" anchor bolts.

____

Multiple Choice

__________ **1.** The north foundation wall is __________ long.

A. 24′-0″
B. 26′-0″
C. 50′-0″
D. 65′-0″

__________ **2.** A __________" concrete slab is placed in the garage.

A. 3
B. 4
C. 5
D. 6

__________ **3.** The center-to-center distance between the piers in the crawl space area is __________.

A. 8′-8″
B. 10′-6″
C. 13′-9″
D. 18′-8″

_______ **4.** The elevation of the door pocket in the north foundation wall is _______.
A. 95′-6″
B. 96′-6″
C. 99′-8″
D. 100′-0″

_______ **5.** The outside of the east foundation wall is _______ from the center foundation wall.
A. 8′-8″
B. 13′-9″
C. 24′-3″
D. 26′-0″

_______ **6.** The cavity between the walls under the fireplace is _______ wide.
A. 1′-4″
B. 1′-6″
C. 1′-10″
D. 2′-8″

_______ **7.** The foundation plan is drawn to a scale of _______″ = 1′-0″.
A. ⅛
B. ¼
C. ⅜
D. ½

_______ **8.** Vertical and horizontal rebars are placed _______″ OC in the foundation walls.
A. 12
B. 16
C. 18
D. 24

_______ **9.** Anchor bolts are placed _______″ OC.
A. 16
B. 24
C. 32
D. 48

_______ **10.** A _______ film vapor barrier is placed over the soil in the crawl space area.
A. polyurethane
B. polyethylene
C. rubberized
D. none of the above

_______ **11.** The concrete piers are _______ tall.
A. 4″
B. 1′-4″
C. 2′-0″
D. 2′-4″

_______ **12.** The center foundation wall is _______ high.
A. 8″
B. 3′-0″
C. 7′-4″
D. 9′-6″

_________ **13.** The divider between the two door pockets in the garage is _________" thick.
A. 9
B. 12
C. 15
D. 18

_________ **14.** The front entrance wall extends _________ in front of the south foundation wall.
A. 14′-4″
B. 24′-3″
C. 29′-6″
D. 30′-0″

_________ **15.** The south foundation wall extends _________" above the finished grade level.
A. 4
B. 6
C. 8
D. 10

_________ **16.** A 7" _________ I beam extends under the floor in the crawl space area.
A. wood
B. steel
C. concrete
D. none of the above

_________ **17.** The section views are drawn to a scale of _________" = 1′-0″.
A. ¼
B. ⅜
C. ½
D. ¾

_________ **18.** The I beam is supported by _________ piers.
A. steel
B. wood
C. concrete
D. masonry

Chapter 8–Section 2—Printreading Exercises

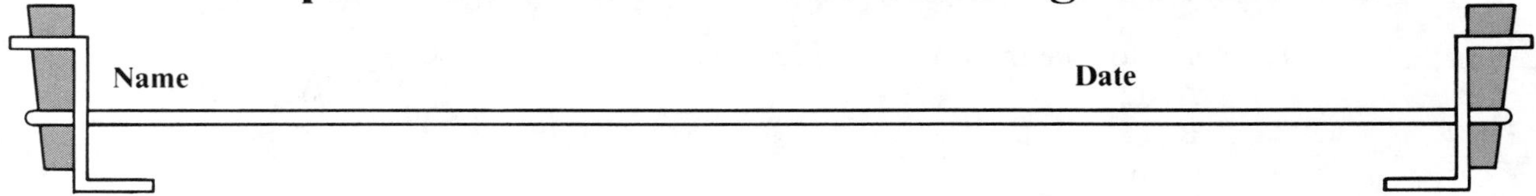

Slab-on-Grade Foundation Prints

Slab-on-grade foundations combine foundation walls with a concrete floor slab. The top of the floor slab is at the same elevation as the top of the foundation walls. The concrete floor slab receives its main support from the ground directly below. The plan view of a slab-on-grade foundation provides overall dimensions of the building and dimensions and locations of footings.

Section view drawings indicate the thickness and height of foundation footings, and thickness and elevation of the concrete floor slab. Section view drawings also include the type, size, and spacing of reinforcement in the footings and floor slab.

Refer to the Slab-on-Grade Foundation print on pages 241 and 242 to answer the following question.

Completion

______ **1.** The south foundation wall is ______ long.

______ **2.** The foundation walls extend ______″ above the finished grade line.

______ **3.** The concrete slab is ______″ thick.

______ **4.** The outside of the east garage wall is ______ from the center line of the footing under the west garage wall.

______ **5.** Two-inch ______ insulation is used around the perimeter of the slab.

______ **6.** The footings under the interior walls are ______″ wide at the lowest point.

______ **7.** The east foundation wall is ______ long.

______ **8.** The footings under the interior walls are ______″ high.

______ **9.** Anchor bolts along the exterior foundation walls are placed ______ OC.

______ **10.** The exterior foundation walls are ______ high.

______ **11.** The welded wire fabric in the concrete slab is designated as ______.

______ **12.** The difference in elevation between the garage floor and the house is ______″.

______ **13.** Two No. ______ rebars are used in the interior wall footings.

______ **14.** The exterior foundation walls are ______″ thick.

______ **15.** A ______ mil vapor barrier is placed under the concrete slab.

______ **16.** The fireplace foundation is ______ long × ______ wide.

______

Multiple Choice

_________ 1. Studs for the interior walls are placed _________" OC.
A. 12
B. 14
C. 16
D. 18

_________ 2. The outside of the north foundation wall is _________ from the fireplace foundation.
A. 9'-8"
B. 12'-2"
C. 12'-6"
D. 26'-7"

_________ 3. No. 4 rebars are placed _________" OC in the exterior foundation walls.
A. 16
B. 20
C. 24
D. 28

_________ 4. The porch slab is _________ wide.
A. 3'-0"
B. 3'-5"
C. 5'-0"
D. 6'-5"

_________ 5. The interior wall footings are detailed in Sections _________ and _________.
A. A, B
B. B, D
C. A, E
D. D, E

_________ 6. A(n) _________" corbel is used to support the front porch slab.
A. 1
B. 2
C. 4
D. 8

_________ 7. The finished grade line is _________ above the frost line.
A. 8"
B. 1'-4"
C. 2'-0"
D. 2'-4"

_________ 8. The footing below the south side of the porch slab is _________" wide.
A. 5
B. 8
C. 10
D. 20

Chapter 8–Section 2—Printreading Exercises

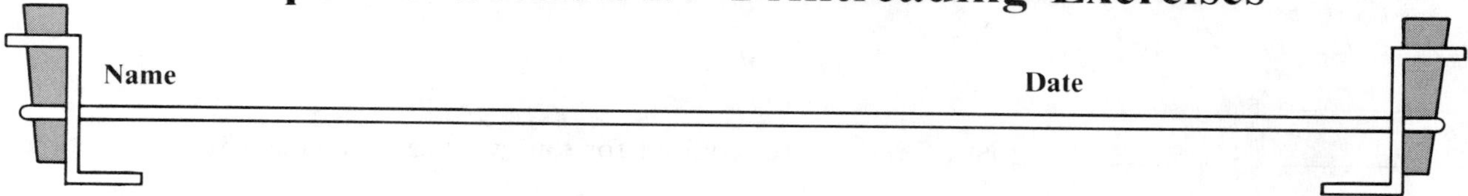

Heavy Construction Prints

A heavy construction print includes a plan view, schedules, and section view drawings. The plan view provides overall dimensions and locations of foundation walls and footings, columns, and pier footings. Specific dimensions for stairways, door and window openings, and areaways are also included in the plan view. Elevations for various floor levels, and tops of foundation footings and floor slabs are also indicated.

Footing schedules are commonly included in a heavy construction print. Footing schedules provide the dimensions of the footings and indicate the size and location of reinforcement.

Section view drawings indicate the thickness and heights of foundation walls and footings and the size and location of reinforcement. Stair details or other details are commonly used to show complex structural features.

Refer to the Heavy Construction print on pages 243 to 246 to answer the following questions.

Completion

_____ **1.** The overall length of the building from east to west is _____.

_____ **2.** The elevation at the top of the footing in the southwest corner of the building is _____.

_____ **3.** The center of the first row of the column footings is _____ from the outside of the north foundation wall.

_____ **4.** The radius of the curved southwest wall is _____.

_____ **5.** The two openings in the areaway in the east wall are _____ × _____.

_____

_____ **6.** The elevation at the top of the F8 footings is _____.

_____ **7.** The center-to-center distance between F8 footings is _____ from east to west.

_____ **8.** The elevation at the top of the first floor slab is _____.

_____ **9.** The elevation at the top of the F4 footing along the interior stairway is _____.

_____ **10.** The elevation at the top of the basement slab along the curved southwest wall is _____.

_____ **11.** _____″ of gravel is required beneath the floor slab.

_____ **12.** A(n) _____″ clearance must be maintained between the bottom footing rebars and soil.

_____ **13.** No. _____ rebars are required in the F6 footing.

_______ 14. The dimensions of an F10 footing are _______ × _______ × _______.
____
____

_______ 15. _______ No. 8 rebars are required for long way reinforcement for F8 footings.

_______ 16. Typical stepped footings require No. 5 rebars spaced _______" vertically.

_______ 17. A(n) _______° angle is the maximum slope for the bottom of a stepped footing.

_______ 18. The north and south foundation footings are _______ wide × _______ deep.
____

_______ 19. No. 4 rebars are placed horizontally at _______" intervals along the interior of the foundation wall in Section 8.

_______ 20. The top of the floor slab in Section 1 is _______ above the top of the foundation footing.

_______ 21. Exterior foundation walls are _______" thick.

_______ 22. The total run of the exterior stairway is _______.

_______ 23. No. _______ nosing rebars are embedded in the stairway.

_______ 24. The total rise of the exterior stairway is _______.

_______ 25. The concrete landing at the foot of the exterior stairway is _______ wide × _______ long.
____

_______ 26. The elevation of the basement slab adjacent to the exterior stairway is _______.

_______ 27. The elevation at the top of the wall footings next to the exterior stairway is _______.

_______ 28. The areaway along the east wall is _______ wide.

_______ 29. The exterior foundation footing along the east wall is _______ wide × _______ deep.
____

_______ 30. The exterior wall of the areaway along the east foundation wall is _______" thick.

_______ 31. The dimensions of the sump pit are _______ × _______ × _______.
____
____

_______ 32. The footing beneath the southwest curved wall is _______ wide × _______ deep.
____

_______ 33. Section 7 is drawn at a scale of _______" = 1'-0".

_______ 34. The design of the foundation footings is based on a soil bearing capacity of _______ PSF.

_______ 35. Footing F_______ is the smallest foundation footing shown on the basement and foundation plan.

Multiple Choice

__________ **1.** The basement slab along the areaway is __________" thick.
A. 4
B. 5
C. 6
D. 7

__________ **2.** The wall adjacent to the sump pit is __________" thick.
A. 8
B. 10
C. 12
D. 16

__________ **3.** The foundation footing beneath the southwest curved wall projects __________.
A. 8¾"
B. 1'-0"
C. 1'-4½"
D. 1'-8"

__________ **4.** The vertical rebars placed along the inside of the southwest curved wall are spaced at __________" intervals.
A. 4
B. 5
C. 8
D. 12

__________ **5.** The foundation footing beneath the interior wall projects __________".
A. 8
B. 10
C. 12
D. 16

__________ **6.** F6 footings are __________ wide.
A. 5'-0"
B. 8'-6"
C. 10'-0"
D. 13'-6"

__________ **7.** The step height for the stepped foundation walls is __________.
A. 1'-3"
B. 2'-0"
C. 4'-6"
D. 5'-0"

__________ **8.** The first floor slab along the interior wall is __________" thick.
A. 5½
B. 7
C. 8½
D. 10

_________ **9.** The landing for the exterior stairway is _________" below the basement floor slab.
A. 6
B. 8
C. 10
D. 12

_________ **10.** The outside foundation wall along the exterior stairway is _________ thick.
A. 6"
B. 8"
C. 11"
D. 1'-0"

_________ **11.** The wall footing beneath the north end of the exterior stairway projects _________ to the north.
A. 6"
B. 1'-0"
C. 2'-8"
D. 4'-8"

_________ **12.** The elevation at the top of the exterior foundation wall along the areaway is _________.
A. −12'-0"
B. −0'-3"
C. +0'-0"
D. +0'-3"

_________ **13.** The basement slab is reinforced with No. _________ rebars spaced at 18" intervals.
A. 4
B. 5
C. 6
D. 8

_________ **14.** The footing for the elevator shaft (area southeast of the interior stairway) is _________ wide.
A. 2'-6"
B. 6'-11"
C. 8'-9"
D. 11'-9"

_________ **15.** The sump pit is on the _________ side of the elevator shaft.
A. north
B. south
C. east
D. west

SECTION 3

ESTIMATING FORM MATERIALS AND CONCRETE

Estimating form materials and concrete is a rough calculation of the amount of form materials and volume of concrete required for a specific construction project. Professional estimators commonly estimate the form materials and concrete for heavy construction projects. On small construction projects, such as the construction of foundation footings and walls for a residence, the estimating is performed by the contractor or job supervisor.

ESTIMATING FORM MATERIALS

Form materials are estimated separately for each section of the concrete work. When estimating, dimensions of the form materials are rounded to the next highest foot increment before calculations are performed. For example, the dimension of a wall section measuring 5′-4½″ is rounded to 6′-0″.

When estimating plywood form components, the total surface area of the forms is determined by multiplying the length of the forms by the height. When estimating dimensional lumber such as planks, studs, walers, braces, and stakes, the total length of the lumber is calculated.

Waste occurs when form components are cut from standard sizes of form materials. Estimators add from 5% to 15% of the total amount of form materials to compensate for waste. Form materials for the foundation footings and wall forms can be reused for framing materials in the structure or for future formwork. Underestimation of form materials results in a delay in form construction.

Sheathing

Sheathing is the form material in direct contact with the concrete. Plywood or 2″ thick planks are used as sheathing for foundation and pier footing forms. Foundation walls are sheathed with plywood reinforced with studs and/or walers, or 2″ thick planks reinforced with cleats and strongbacks.

The following examples describe a fundamental approach to estimating form materials and are based on the Foundation Plan and Section A-A on page 247.

Foundation Footing Forms. Foundation footing forms consist of outer and inner form walls. Foundation footing forms are constructed with 2″ thick planks or plywood reinforced with stakes and braces. When using 2″ thick planks as the sheathing, the total length of the form walls must be determined. The lengths of the outer form walls are calculated by adding the length of the foundation wall and the footing projections at the ends of the foundation walls. The lengths of the inner form walls are calculated by adding the adjacent wall thicknesses to the footing projections and subtracting the sum from the length of the foundation wall.

Example
Determine the total length of 2 × 10 planks required to sheath the foundation footing forms.

Solution

1. Calculate the lengths of outer footing form walls A and B.

   Lengths of outer footing form walls A and B = lengths of foundation walls A and B + (footing projections)

   = 48′-0″ + (5″ + 5″)
   = 48′-0″ + 10″
   = 48′-10″

   Lengths of outer footing form walls are rounded to 49′-0″.

2. Calculate the lengths of outer footing form walls C and D.

   Lengths of outer footing form walls C and D = lengths of foundation walls C and D + (footing projections)

   = 26′-0″ + (5″ + 5″)
   = 26′-0″ + 10″
   = 26′-10″

   Lengths of outer footing form walls C and D are rounded to 27′-0″.

3. Calculate the lengths of inner footing forms A and B.

   Lengths of inner footing form walls A and B = lengths of foundation walls A and B − [(adjacent wall thicknesses) + (footing projections)]

= 48′-0″ − [(10″ + 10″) + (5″ + 5″)]
= 48′-0″ − [20″ + 10″]
= 48′-0″ − 30″
= 48′-0″ − 2′-6″
= 45′-6″

Lengths of inner footing form walls A and B are rounded to 46′-0″.

4. Calculate the lengths of inner footing form walls C and D.

Lengths of inner footing form walls C and D = lengths of foundation walls C and D − [(adjacent wall thicknesses) + (footing projections)]

= 26′-0″ − [(10″ + 10″) + (5″ + 5″)]
= 26′-0″ − [20″ + 10″]
= 26′-0″ − 30″
= 26′-0″ − 2′-6″
= 23′-6″

Lengths of inner footing form walls C and D are rounded to 24′-0″.

5. Calculate the total length of 2 × 10 planks required.

Total length of 2 × 10 planks = sum of lengths of individual footing form planks

= 49′ + 49′ + 27′ + 27′ + 46′ + 46′ + 24′ + 24′

Total length of 2 × 10 planks = 292′-0″

When using plywood as sheathing for foundation footing forms, the surface area of individual footing forms is calculated by multiplying the foundation footing form lengths by the height of the forms. The individual surface areas are added together to determine the total surface area. The total surface area is divided by the area of a plywood panel to determine the number of panels required to sheath the foundation footing forms.

Example
Determine the number of 4 × 8 plywood panels required to sheath the foundation footing forms.

Solution

1. Determine the lengths of the individual footing forms. (See previous example for individual footing form lengths.)

Footing Form	Length
Outer footing form wall A	49′-0″
Outer footing form wall B	49′-0″
Outer footing form wall C	27′-0″
Outer footing form wall D	27′-0″
Inner footing form wall A	46′-0″
Inner footing form wall B	46′-0″
Inner footing form wall C	24′-0″
Inner footing form wall D	24′-0″

2. Calculate the surface area of the individual footing forms. Round the footing height to the next highest foot.

Footing Form	Length	Height	Surface Area (sq ft)
Outer footing form wall A	49′-0″	1′-0″	49
Outer footing form wall B	49′-0″	1′-0″	49
Outer footing form wall C	27′-0″	1′-0″	27
Outer footing form wall D	27′-0″	1′-0″	27
Inner footing form wall A	46′-0″	1′-0″	46
Inner footing form wall B	46′-0″	1′-0″	46
Inner footing form wall C	24′-0″	1′-0″	24
Inner footing form wall D	24′-0″	1′-0″	24

3. Calculate the total surface area of the foundation footing forms.

Total surface area = sum of individual surface areas

= 49 + 49 + 27 + 27 + 46 + 46 + 24 + 24

Total surface area = 292 sq ft

4. Calculate the number of 4 × 8 plywood panels required to sheath the foundation footing forms.

No. of plywood panels = total surface area ÷ area of panel

= 292 ÷ (4 × 8)
= 292 ÷ 32
= 9.125

The number of 4 × 8 plywood panels is rounded to 10.

Pier Footing Forms. Pier footings support steel columns, wood posts, and masonry or concrete piers. Pier footing design determines the type of forms to be constructed. Rectangular and tapered pier footings require forms around the perimeter of the pier footing. Stepped pier footings require a base form and a step form for each additional step. Circular pier footings are formed using the required length of tubular fiber form.

The surface area of a rectangular or tapered pier footing form is determined by multiplying the combined length of all sides of the pier footing form by the height of the form. If multiple pier footings with the same dimensions are constructed, the individual surface area is multiplied by the number of piers.

Example
Determine the number of 4 × 8 plywood panels required to sheath the pier footings.

Solution

1. Calculate the combined length of the four sides of one pier footing form.

   Combined length = sum of all side lengths

   = 2′-0″ + 2′-0″ + 2′-0″ + 2′-0″

   Combined length = 8′-0″

2. Calculate the surface area of one pier footing form. Round the pier footing height to the next highest foot increment.

   Surface area of one pier footing form = combined length × height

   = 8′-0″ × 1′-0″

   Surface area of one pier footing form = 8 sq ft

3. Calculate the total surface area of two pier footing forms.

   Total surface area = surface area of one pier footing form × no. of pier footings

   = 8 sq ft × 2

   Total surface area of two pier footing forms = 16 sq ft

4. Calculate the number of 4 × 8 plywood panels required.

   No. of 4 × 8 plywood panels = total surface area ÷ area of panel

   = 16 sq ft ÷ (4 × 8)
   = 16 sq ft ÷ 32
   = .5

   The number of plywood panels required to sheath the pier footing forms is rounded to 1.

Foundation Wall Forms. Foundation wall forms consist of the inner and outer form walls. The foundation wall forms are constructed with plywood and reinforced with studs, walers, and/or strongbacks. The surface areas of the outer form walls are determined by multiplying the length of the form wall by the height. The surface areas of the inner form walls are determined by subtracting the adjacent wall thicknesses from the outer form wall lengths and multiplying by the height.

Example
Determine the number of 4 × 8 plywood panels required to sheath the foundation walls.

Solution

1. Calculate the lengths of the outer form walls.

   Lengths of outer form walls A and B = 48′-0″
   Lengths of outer form walls C and D = 26′-0″

2. Calculate the lengths of the inner form walls A and B.

   Lengths of inner form walls A and B = lengths of outer form walls A and B − (adjacent wall thicknesses)

   = 48′-0″ − (10″ + 10″)
   = 48′-0″ − 20″
   = 48′-0″ − 1′-8″
   = 46′-4″

   The lengths of inner form walls A and B are rounded to 47′-0″.

3. Calculate the lengths of inner form walls C and D.

   Lengths of inner form walls C and D = lengths of outer form walls C and D − (adjacent wall thicknesses)

   = 26′-0″ − (10″ + 10″)
   = 26′-0″ − 20″
   = 26′-0″ − 1′-8″
   = 24′-4″

   The lengths of inner form walls C and D are rounded to 25′-0″.

4. Determine the surface areas of the individual form walls.

Form Wall	Length	Height	Surface Area (sq ft)
Outer form wall A	48′-0″	8′-0″	384
Outer form wall B	48′-0′	8′-0″	384
Outer form wall C	26′-0″	8′-0″	208
Outer form wall D	26′-0″	8′-0″	208
Inner form wall A	47′-0″	8′-0″	376
Inner form wall B	47′-0″	8′-0″	376
Inner form wall C	25′-0″	8′-0″	200
Inner form wall D	25′-0″	8′-0″	200

5. Calculate the total surface area of the foundation wall forms.

   Total surface area =
   sum of individual surface areas

   = 384 + 384 + 208 + 208 + 376 + 376 + 200 + 200

   Total surface area = 2336 sq ft

6. Calculate the number of 4 × 8 plywood panels required.

   No. of 4 × 8 plywood panels =
   total surface area ÷ area of panel

   = 2336 ÷ (4 × 8)
   = 2336 ÷ 32

   The total number of 4 × 8 plywood panels required to sheath the form walls is 73.

Low form walls are often sheathed with planks instead of plywood. When form walls are sheathed with planks, divide the total surface area by the width of the planks (in feet) to determine the total length of planks required.

Example
Determine the total length of 2 × 10 planks required to sheath the foundation wall forms.

Solution
Total length of 2 × 10 planks =
total surface area ÷ width of planks (ft)

= 2336 ÷ (10″ ÷ 12)
= 2336 ÷ .83′
= 2814.5

The total length of 2 × 10 planks required to sheath the foundation wall forms is rounded to 2815′-0″.

Stiffeners and Supports

Stiffeners and supports such as base plates, studs, walers, braces, and stakes are used to reinforce wall form sheathing. Base plates secure the bottom edge of the sheathing and provide a nailing surface for studs. Studs stiffen and directly reinforce wall form sheathing. Walers reinforce the studs and align the wall forms. Braces secure and align the tops of form walls. Stakes secure foundation and pier footing forms and the lower ends of braces.

Base Plates. Base plates are nailed to the foundation footing and secure the outer form wall for most forming operations. Base plates secure the inner form wall when forming a round structure such as a storage tank. The total length of base plate material required is determined by calculating the perimeter (distance around the outside) of the foundation walls.

Example
Determine the total length of base plate material required to form the foundation wall.

Solution

1. Calculate the perimeter of the foundation walls.

   Perimeter of foundation walls =
   length of foundation wall A + B + C + D

   = 48′-0″ + 48′-0″ + 26′-0″ + 26′-0″

   Perimeter of foundation walls = 148′-0″

   The total length of base plate material required is 148′-0″.

Studs. The number of studs required to reinforce a form wall is determined by the stud spacing. Stud spacing is based on the stud size and other type of stiffeners to be used. The number of studs for outer and inner form walls is determined by dividing the length of the form wall by the recommended spacing, and adding one stud. For example, a 20′-0″ form wall with studs spaced 2′-0″ OC requires 11 studs [(20′-0″ ÷ 2′-0″) + 1 = 11]. The length of odd-length walls is divided by the recommended spacing and the answer is rounded to the next highest whole number. One additional stud is then added. For example, a 23′-0″ form wall with studs spaced 2′-0″ OC requires 13 studs [(23′-0″ ÷ 2′-0″) + 1 = 13]. The total length of stud material is determined by multiplying the total number of studs by the form wall height.

Example
Determine the total length of stud material required to reinforce the outer and inner form walls if the studs are spaced 2′-0″ OC.

Solution

1. Calculate the number of studs required for each form wall.

Outer Form Wall	Length	OC Spacing	Number of Studs
A	48′-0″	2′-0″	25
B	48′-0″	2′-0″	25
C	26′-0″	2′-0″	14
D	26′-0″	2′-0″	14

2. Calculate the total number of studs required for the outer form wall.

   Total number of studs for outer form wall = sum of individual form wall studs

   = 25 + 25 + 14 + 14

   Total number of studs for outer form wall = 78 studs

3. Calculate the number of studs required for the inner form wall.

Inner Form Wall	Length	OC Spacing	Number of Studs
A	47′-0″	2′-0″	25
B	47′-0″	2′-0″	25
C	25′-0″	2′-0″	14
D	25′-0″	2′-0″	14

4. Calculate the total number of studs required for the inner form wall.

   Total number of studs for inner form wall = sum of individual form wall studs

   = 25 + 25 + 14 + 14

   Total number of studs for inner form wall = 78 studs

5. Calculate the total number of studs for the outer and inner form walls.

   Total number of studs for outer and inner form walls = total number of studs for outer form wall + total number of studs for inner form wall

   = 78 + 78

   Total number of studs for outer and inner form walls = 156 studs

6. Calculate the total length of stud material required.

   Length of stud material = total number of studs × form wall height

   = 156 × 8′-0″

   Total length of stud material = 1248′-0″

Walers. Single or double walers are secured against the studs for reinforcement. When studs are not used, the walers are secured against the sheathing. The spacing of walers determines the number of rows of walers required. The length of waler material for a single row of walers for the outer form wall is determined by calculating the perimeter of the foundation walls and multiplying by the number of rows of walers required. The length of waler material for a double waler system is twice the length of the waler material for a single row of walers.

The length of waler material required for a single row of walers for the inner form wall is determined by calculating the perimeter of the inner form wall and multiplying by the number of rows of walers required. The length of waler material required for a double waler system is twice the length of waler material for a single row of walers. The total length of waler material for the outer and inner form walls is determined by adding the lengths of waler material for the form walls.

Example

Determine the total length of waler material required for a double waler system. The double walers are secured to the the outer and inner form walls and are spaced 12″ from the top and bottom of the sheathing with the intervening rows spaced 24″ OC (four rows of double walers).

Solution

1. Calculate the perimeter of the foundation wall.

   Perimeter of foundation wall = length of foundation walls A + B + C + D

   = 48′-0″ + 48′-0″ + 26′-0″ + 26′-0″

   Perimeter of foundation wall = 148′-0″

2. Calculate the length of waler material required for a single row of walers for the outer form wall.

   Length of walers = perimeter of foundation wall × no. of rows

   = 148′-0″ × 4

   Length of waler material for a single row of walers for the outer form wall = 592′-0″

3. Calculate the length of waler material for a double waler system for the outer form wall.

   Length of walers = 2 × length of single waler material

   = 2 × 592′-0″

   Length of waler material for a double waler system = 1184′-0″

4. Calculate the perimeter of the inner form wall.

   Perimeter of inner form wall = lengths of inner form walls A + B + C + D

   = 47′-0″ + 47′-0″ + 25′-0″ + 25′-0″

   Perimeter of inner form wall = 144′-0″

5. Calculate the length of waler material required for a single row of walers for the inner form wall.

   Length of walers = perimeter of inner form wall × no. of rows

   = 144′-0″ × 4

   Length of waler material for a single row of walers for the inner form wall = 576′-0″

6. Calculate the length of waler material for a double waler system for the inner form wall.

   Length of walers = 2 × length of single waler material

   = 2 × 576′-0″

   Length of waler material for a double waler system for the inner form wall = 1152′-0″

7. Calculate the total length of waler material required.

   Total length of waler material = length of waler material for outer form wall + length of waler material for inner form wall

   = 1184′-0″ + 1152′-0″

   Total length of waler material required = 2336′-0″

Braces. Braces support form walls and are secured at the lower ends by stakes. The number of braces required is based on the spacing of the braces. The number of braces is determined by dividing the length of a form wall by the recommended spacing and adding one brace. The number of braces for individual form walls are added to obtain the total number of braces required.

The length of brace required is based on the angle that it forms with the form wall. Braces are commonly attached at approximately a 45° angle. The length of braces attached at a 45° angle is determined by multiplying the height from the base of the footing to the brace attachment point by 1.41. The total length of brace material required is determined by multiplying the number of braces by the individual brace length.

Example

Determine the total length of brace material required for the foundation wall. The braces are spaced 6′-0″ OC and are attached at a 45° angle. The distance from the base of the footing to the brace attachment point is 8′-0″.

Solution

1. Calculate the number of braces required for each form wall.

Outer Form Wall	Length	OC Spacing	Number of Braces
A	48′-0″	6′-0″	9
B	48′-0″	6′-0″	9
C	26′-0″	6′-0″	6
D	26′-0″	6′-0″	6

2. Calculate the number of braces required for the foundation wall.

   Total number of braces = sum of individual braces

   = 9 + 9 + 6 + 6

   Number of braces for the foundation wall = 30 braces

3. Calculate the length of an individual brace.

   Brace length = height × 1.41

   = 8′-0″ × 1.41

   Height of an individual brace = 11.28′

   Round 11.28′ to 12′-0″

4. Calculate the total length of brace material required.

   Total length of brace material = no. of braces × individual brace length

   = 30 × 12′-0″

   = 360′-0″

   Total length of brace material is rounded to 360′-0″.

Stakes. Stakes secure foundation footing and pier forms and the lower ends of braces. Foundation footing forms require stakes along the outer surface of both form walls. The number of stakes for individual outer form walls is determined by dividing the length of the outer form wall by the recommended spacing and adding one stake. The number of stakes required for the outer form wall of an entire foundation footing is determined by adding the number of stakes for the individual form walls.

The number of stakes required for individual inner form walls is determined by dividing the length of the inner form wall by the recommended spacing and adding one stake. The number of stakes required for the inner form wall of an entire foundation is determined by adding the number of stakes for the individual form walls. The total length of stake material required for the outer and inner form walls is determined by adding the number of stakes for the outer and inner form walls and multiplying by the individual stake length.

Example
Determine the total length of stake material required for the foundation footing forms. The stakes are 2′-0″ long and are spaced 2′-0″ OC.

Solution

1. Calculate the number of stakes required for the outer form wall of each foundation footing.

Form Wall	Length	OC Spacing	Number of Stakes
A	49′-0″	2′-0″	26
B	49′-0″	2′-0″	26
C	27′-0″	2′-0″	15
D	27′-0″	2′-0″	15

2. Calculate the total number of stakes required for the outer form wall of the entire foundation footing.

   Total number of stakes = *sum of individual footing stakes*

   = 26 + 26 + 15 + 15

   Total number of stakes for the outer form wall = 82 stakes

3. Calculate the number of stakes required for the inner form wall of each foundation footing.

Form Wall	Length	OC Spacing	Number of Stakes
A	46′-0″	2′-0″	24
B	46′-0″	2′-0″	24
C	24′-0″	2′-0″	13
D	24′-0″	2′-0″	13

4. Calculate the number of stakes required for the inner form wall of the entire foundation footing.

   Number of stakes = *sum of individual footing stakes*

   = 24 + 24 + 13 + 13

   Number of stakes for the inner form wall = 74 stakes

5. Calculate the total number of stakes for the inner and outer form walls.

   Total number of stakes for outer and inner form walls = *sum of footing stakes for outer and inner form walls*

   = 82 + 74

   Total number of stakes for outer and inner form walls = 156 stakes

6. Calculate the total length of stake material required for the foundation footing forms.

   Total length of stake material = *no. of stakes* × *individual stake length*

   = 156 × 2′-0″

   Total length of stake material for the foundation footing forms = 312′-0″

Pier footing forms require stakes only along the perimeter of the pier box. Small pier footing forms require four corner stakes. Large pier footing forms require corner stakes and intermediate stakes. The total length of stake material required is determined by multiplying the number of stakes by the individual stake length.

Example
Determine the total length of stake material required for the pier footing forms. The stakes are 2′-0″ long and are placed at the four corners of the pier box.

Solution

1. Calculate the length of stake material required for one pier box.

   Length of stake material = *no. of stakes* × *individual stake length*

   = 4 × 2′-0″

   Length of stake material for one pier box = 8′-0″

2. Calculate the total length of stake material required.

   Total length of stake material = *no. of pier boxes* × *length of stake material for one pier box*

   = 2 × 8′-0″

   Total length of stake material for two pier boxes = 16′-0″

Stakes secure the lower ends of braces to the ground. The total length of stake material required to secure the braces is determined by multiplying the number of braces by the individual stake length.

Example
Determine the total length of stake material required for the form wall. The stakes are 2′-0″ long and placed at the lower end of every brace.

Solution

1. Calculate the total length of stake material.

   Total length of stake material = no. of braces × stake length

   = 30 × 2′-0″

   Total length of stake material = 60′-0″

ESTIMATING CONCRETE

Concrete is estimated by volume and is expressed in cubic yards. The volume of each section of a structure is calculated separately and added to other sections to obtain the total volume of concrete required. The volume of a horizontal structural member such as a floor slab is determined by multiplying the thickness, width, and height. The volume of a vertical member such as a foundation wall is determined by multiplying thickness, length, and height. Since dimensions for structural members are expressed in feet, the volume of concrete required is initially expressed in cubic feet. The volume is divided by 27 to obtain the volume of concrete in cubic yards. (One cubic yard equals 27 cubic feet.)

The following examples are based on the Foundation Plan and Section A-A on page 247.

Foundation Footings

Foundation footings support foundation walls. Since foundation footings commonly project beyond foundation walls, the lengths of the footing projections must be added to the length of the foundation walls to obtain a total foundation footing length. The volume is then determined by multiplying the height, width, and length.

Example
Determine the volume of concrete required for the foundation footings.

Solution

1. Calculate the lengths of foundation footings A and B.

   Lengths of footings A and B = lengths of foundation footings A and B + (footing projections)

   = 48′-0″ + (5″ + 5″)

   = 48′-0″ + 10″

   Lengths of foundation footings A and B = 48′-10″

2. Calculate the lengths of foundation footings C and D. To avoid calculating the corners of the footings twice, subtract the sum of the wall thicknesses and footing projections of adjacent walls A and B from the length of foundation walls C and D.

   Lengths of footings C and D = lengths of foundation walls C and D − [(wall A and B thicknesses) + (footing projections)]

   = 26′-0″ − [(10″ + 10″) + (5″ + 5″)]

   = 26′-0″ − [20″ + 10″]

   = 26′-0″ − 30″

   = 26′-0″ − 2′-6″

   Lengths of foundation footings C and D = 23′-6″

3. Calculate the volume of concrete required for foundation footings A and B.

   Volume of footings A and B = height × width × length

   = 10″ × 1′-8″ × 48′-10″

   = .83′ × 1.67′ × 48.83′

   Volume of footings A and B = 67.68 cu ft

4. Calculate the volume of concrete required for foundation footings C and D.

   Volume of footings C and D = height × width × length

   = 10″ × 1′-8″ × 23′-6″

   = .83′ × 1.67′ × 23.5′

   Volume of footings C and D = 32.57 cu ft

5. Calculate the total volume of concrete required for the foundation footings in cubic feet.

   Total volume (cu ft) = sum of individual volumes

   = 67.68 + 67.68 + 32.57 + 32.57

   Total volume (cu ft) = 200.5 cu ft

6. Calculate the total volume of the concrete required for the foundation footings in cubic yards.

Total volume (cu yd) = *total volume (cu ft)* ÷ 27

= 200.5 ÷ 27

Total volume (cu yd) = 7.43 cu yd

Pier Footings

Pier footings support steel columns, wood posts, and concrete or masonry piers. The volume of a pier footing is determined by multiplying the height, width, and length.

Example
Determine the volume of concrete required for the pier footings.

Solution

1. Calculate the total volume of concrete required for the pier footings in cubic feet.

Total volume (cu ft) = *no. of pier footings* × *(height* × *width* × *length)*

= 2 × (10″ × 2′-0″ × 2′-0″)

= 2 × (.83′ × 2.0′ × 2.0′)

= 2 × 3.32

Total volume (cu ft) = 6.64 cu ft

2. Calculate the volume of concrete required for the pier footings in cubic yards.

Volume of pier footings (cu yd) = *total volume (cu ft)* ÷ 27

= 6.64 ÷ 27

Volume of pier footings (cu yd) = .25 cu yd

Foundation Walls

Foundation walls support the superstructure. Foundation walls for crawl space foundations are lower than foundation walls for full basement foundations. The volume of foundation walls is determined by multiplying the thickness, length, and height.

Example
Determine the volume of concrete required for the foundation walls.

Solution

1. Calculate the volume of concrete required for foundation walls A and B.

Volume of foundation walls A and B = *thickness* × *length* × *height*

= 10″ × 48′-0″ × 8′-0″

= .83′ × 48.0′ × 8.0′

Volume of foundation walls A and B = 318.72 cu ft

2. Calculate the lengths of foundation walls C and D.

Lengths of foundation walls C and D = *lengths of foundation walls C and D* − *(thicknesses of foundation walls A and B)*

= 26′-0″ − (10″ + 10″)

= 26′-0″ − 20″

= 26′-0″ − 1′-8″

Lengths of foundation walls C and D = 24′-4″

3. Calculate the volume of concrete required for foundation walls C and D.

Volume of foundation walls C and D = *thickness* × *length* × *height*

= 10″ × 24′-4″ × 8′-0″

= .83′ × 24.33′ × 8.0′

Volume of foundation walls C and D = 161.55 cu ft

4. Calculate the total volume of the concrete required for the foundation walls in cubic feet.

Total volume (cu ft) = *sum of individual volumes*

= 318.72 + 318.72 + 161.55 + 161.55

Total volume (cu ft) = 960.54 cu ft

5. Calculate the total volume of concrete required for the foundation walls in cubic yards.

Total volume (cu yd) = *total volume (cu ft)* ÷ 27

= 960.54 ÷ 27

Total volume (cu yd) = 35.58 cu yd

Complete Foundation

The complete foundation consists of foundation and pier footings, and the foundation walls. The volume of concrete required for the complete foundation is determined by adding the volumes of all individual sections of the foundation.

Example

Determine the volume of concrete required for the complete foundation.

Solution

1. Calculate the total volume of concrete for the complete foundation.

Total volume = sum of individual volumes

= foundation footings + pier footings + foundation walls

= 7.43 + .25 + 35.58

= 43.26 cu yd

The total volume of concrete is rounded to 44 cubic yards.

Chapter 8–Section 3—Review Questions

Estimate the total amount of form material required to construct the foundation forms for the full basement foundation on pages 235 and 236. Use the encircled boldface letters on the foundation plan to identify the foundation footings, walls, and pier footings on the quantity take-offs.

Foundation wall G extends from the outside of the west foundation wall to the outside of the foundation wall between the garage and the basement. When estimating, round the height of foundation wall G to 4′-0″. Foundation wall A extends 39′-4″ from the east foundation wall. Foundation wall J extends 20′-8″ from the west foundation wall. The perimeters of walls E and K are determined by adding the lengths of the sides not adjacent to the foundation walls.

Foundation Footing Forms. Foundation footing forms are constructed with 2 × 8 and 2 × 10 planks. The footing forms are secured with 2 × 4 × 2′-0″ long stakes spaced 2′-0″ OC.

Foundation Wall Forms. Foundation wall forms are sheathed with $\frac{3}{4}$″ × 4 × 8 Plyform® Class 1 panels with the face grain running vertically. The Plyform® panels are reinforced with 2 × 4 studs spaced 2′-0″ OC and 2 × 4 base plates. The stud length is equal to the height of the finished concrete wall. Four rows of 2 × 4 double walers stiffen the inner and outer form walls of foundation walls F, D, B, and A. The remaining foundation walls are stiffened with two rows of 2 × 4 double walers along the inner and outer form walls. Braces are attached to the outer form walls. Braces for foundation walls F, D, B, and A are 2 × 4 × 12′-0″ long and are spaced 8′-0″ OC. Braces for the remaining foundation walls are 2 × 4 × 6′-0″ long and spaced 8′-0″. All braces are secured at the lower ends with 2 × 4 × 1′-6″ long stakes.

Pier Footing Forms. Pier footing forms are constructed with 2 × 12 planks reinforced with four 2 × 4 × 2′-0″ long stakes. Additional bracing is not required.

QUANTITY TAKE-OFF FOUNDATION FOOTINGS—OUTER FORM WALL				
Wall	Planks		Stakes	
	Thickness × Width	Length	Quantity	Total Length
A				
B				
C				
D				
E	FOOTINGS NOT REQUIRED			
F				
G				
H				
I				
J				
K	FOOTINGS NOT REQUIRED			

QUANTITY TAKE-OFF FOUNDATION FOOTINGS—INNER FORM WALLS				
Wall	Planks		Stakes	
	Thickness × Width	Length	Quantity	Total Length
A				
B				
C	INNER FORM WALL NOT REQUIRED			
D				
E	FOOTINGS NOT REQUIRED			
F				
G				
H				
I				
J				
K	FOOTINGS NOT REQUIRED			

FOUNDATION FOOTINGS

_______ 1. Length of 2 × 4 stakes (in feet)

_______ 2. Length of 2 × 8 planks (in feet)

_______ 3. Length of 2 × 10 planks (in feet)

_______ 4. Length of 2 × 12 planks (in feet)

QUANTITY TAKE-OFF FOUNDATION WALLS—OUTER FORM WALL											
Wall	Length	Height	Area (sq ft)	Studs		Base Plates	Walers	Braces		Stakes	
				Quantity	Total Length	Total Length	Total Length	Quantity	Total Length	Quantity	Total Length
A											
B											
C					FORM WALL NOT REQUIRED						
D											
E											
F											
G											
H											
I											
J											
K											

QUANTITY TAKE-OFF FOUNDATION WALLS—INNER FORM WALL											
Wall	Length	Height	Area (sq ft)	Studs		Base Plates	Walers	Braces		Stakes	
				Quantity	Total Length	Total Length	Total Length	Quantity	Total Length	Quantity	Total Length
A											
B											
C			FORM WALL NOT REQ'D			NOT					
D						REQ'D					
E						FOR			NOT REQ'D		
F						INNER			FOR INNER		
G						FORM			FORM WALL		
H						WALL					
I											
J											
K											

FOUNDATION WALLS

_______ **5.** Total surface area of outer and inner form walls (in square feet)

_______ **6.** Number of ¾″ × 4 × 8 Plyform® Class 1 panels

_______ **7.** Length of 2 × 4 studs (in feet)

_______ **8.** Length of 2 × 4 base plates (in feet)

_______ **9.** Length of 2 × 4 double walers (in feet)

_______ **10.** Length of 2 × 4 braces (in feet)

_______ **11.** Length of 2 × 4 stakes (in feet)

_______ **12.** Total length of 2 × 4 material (in feet)

Estimate the total volume of concrete required to construct the foundation footings, walls, and pier footings for the full basement foundation on pages 235 and 236. Use the encircled boldface letters on the foundation plan to identify the foundation footings, walls, and pier footings on the quantity take-offs. Decimal equivalents and volumes are rounded to two places after the decimal point.

When determining the volume of concrete required for the foundation, first calculate the volumes of foundation walls and footings G, D, A, and J. When calculating the volume of foundation wall G, use the height shown in Section 6. Next, calculate the volumes of foundation walls and footings F, B, H, and I to eliminate calculating the volume of concrete for the corners twice. Finally, calculate the volumes of stoops E and K, and pier footings C.

QUANTITY TAKE-OFF FOUNDATION FOOTINGS							
Footing	Height		Width		Length		Volume
	Inches	Decimal Foot	Inches	Decimal Foot	Ft and In.	Decimal Foot	Cu Ft
A							
B							
C							
D							
E				FOOTINGS NOT REQUIRED			
F							
G							
H							
I							
J							
K				FOOTINGS NOT REQUIRED			

QUANTITY TAKE-OFF FOUNDATION WALLS							
Footing	Thickness		Length		Height		Volume
	Inches	Decimal Foot	Ft and In.	Decimal Foot	Ft and In.	Decimal Foot	Cu Ft
A							
B							
C				WALLS NOT REQUIRED			
D							
E							
F							
G							
H							
I							
J							
K							

_________ **13.** Total volume of foundation footings (in cubic feet)

_________ **14.** Total volume of foundation footings (in cubic feet)

_________ **15.** Total volume of foundation walls (in cubic feet)

_________ **16.** Total volume of foundation walls (in cubic yards)

_________ **17.** Total volume of foundation (in cubic yards)

_________ **18.** Total volume of concrete to be ordered (in cubic yards)

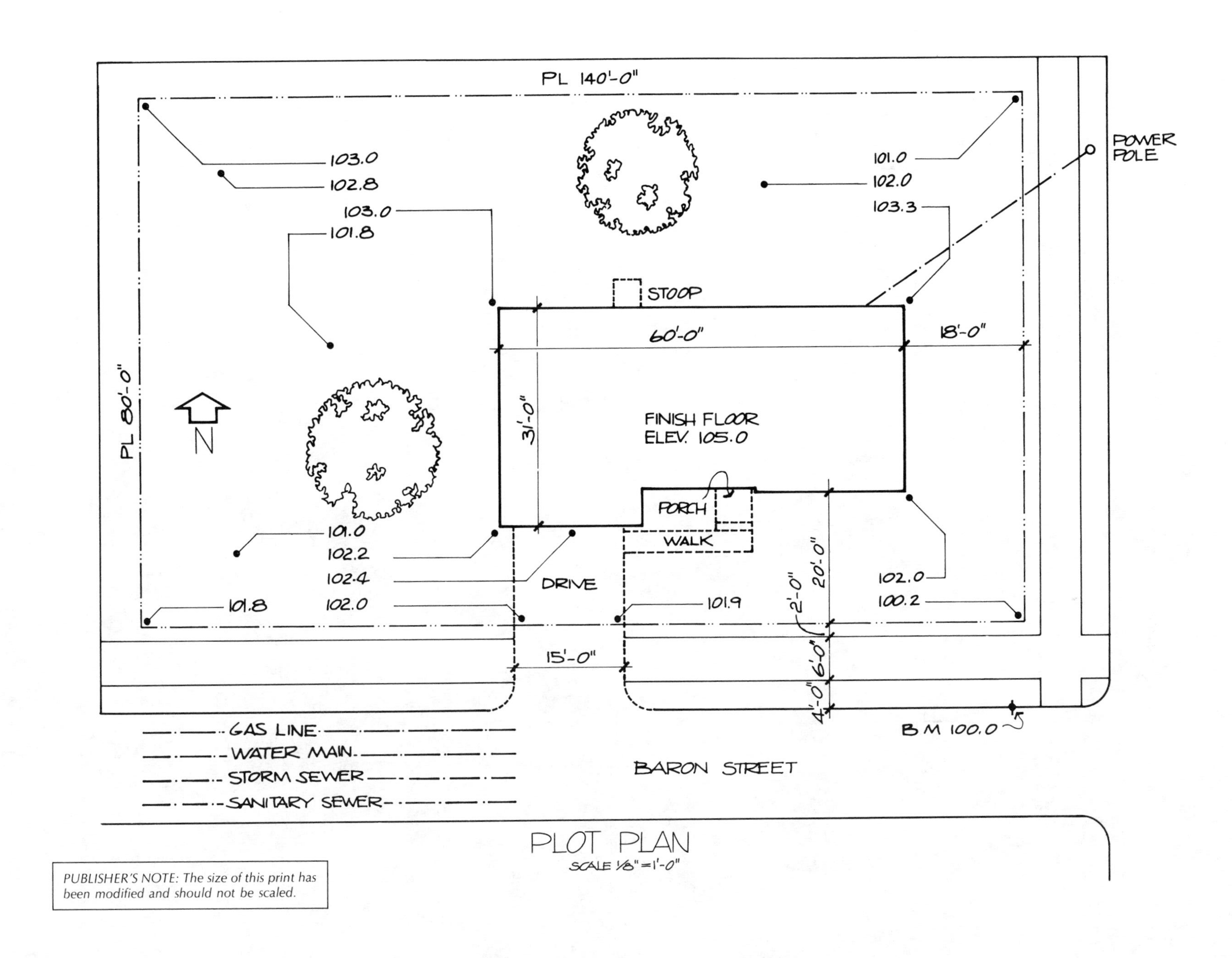

PUBLISHER'S NOTE: The size of this print has been modified and should not be scaled.

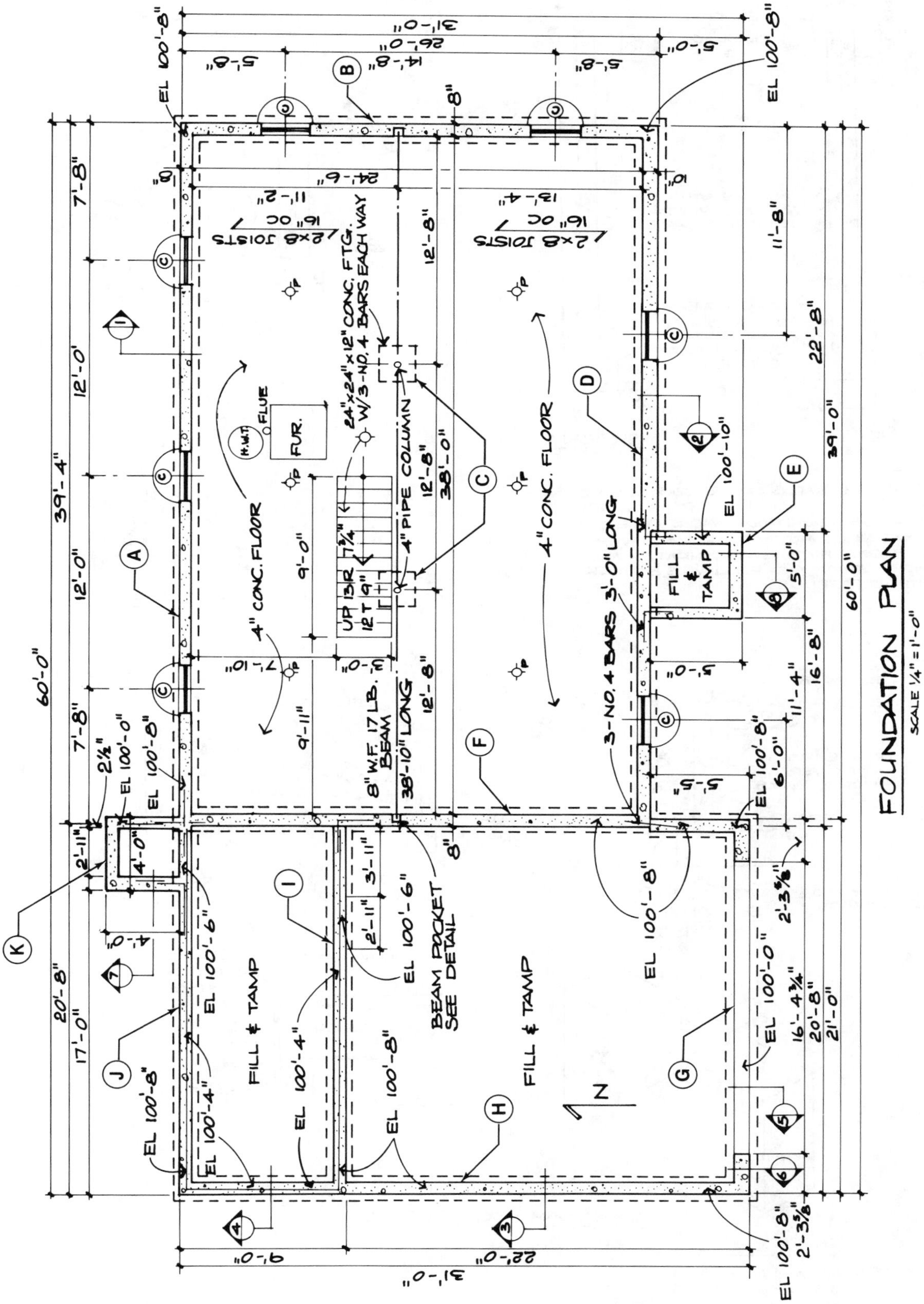

The Garlinghouse Company

FOUNDATION PLAN

SCALE 1/4" = 1'-0"

PUBLISHER'S NOTE: *The size of this print has been modified and should not be scaled.*

Full Basement Foundation

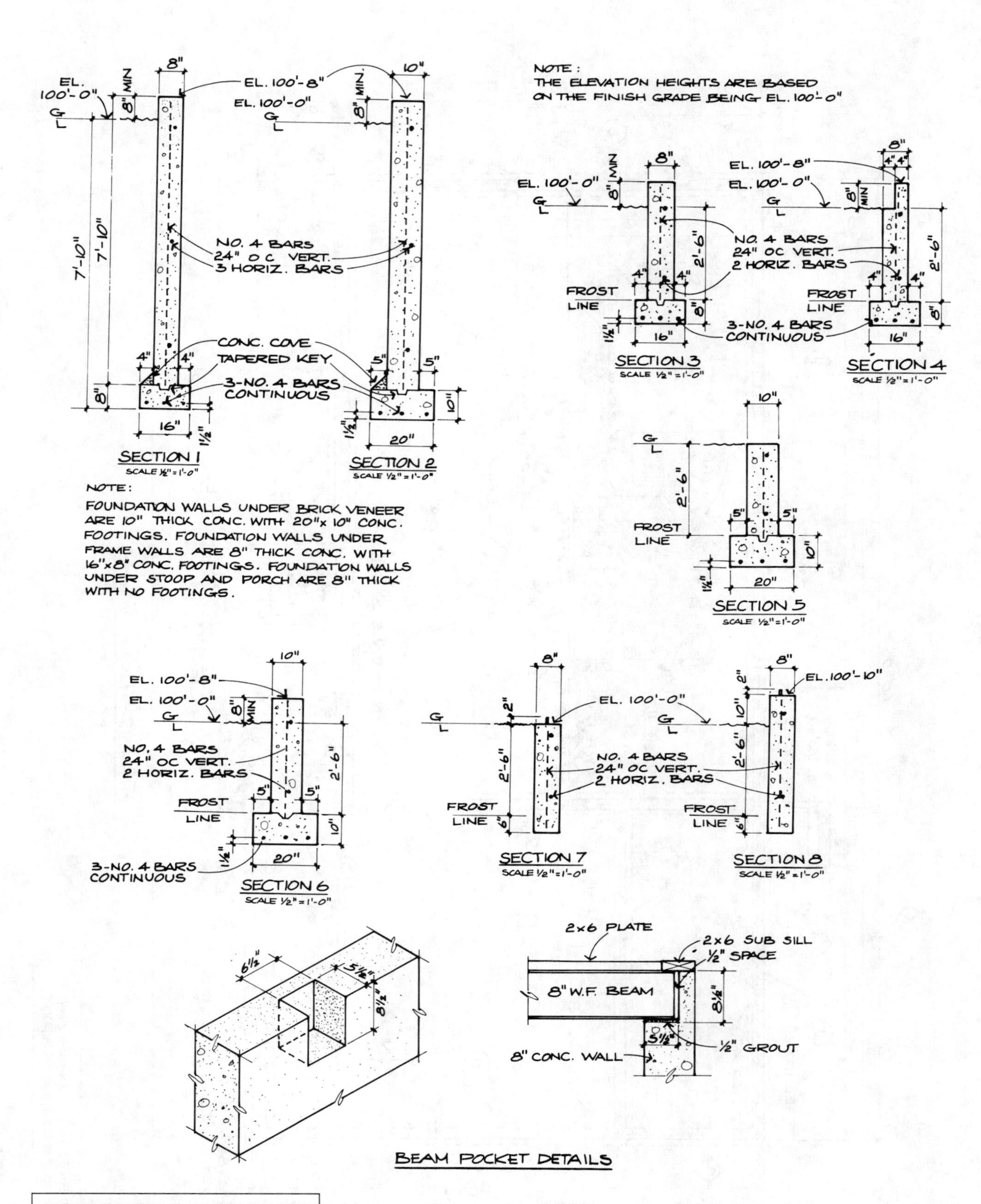

PUBLISHER'S NOTE: The size of this print has been modified and should not be scaled.

The Garlinghouse Company

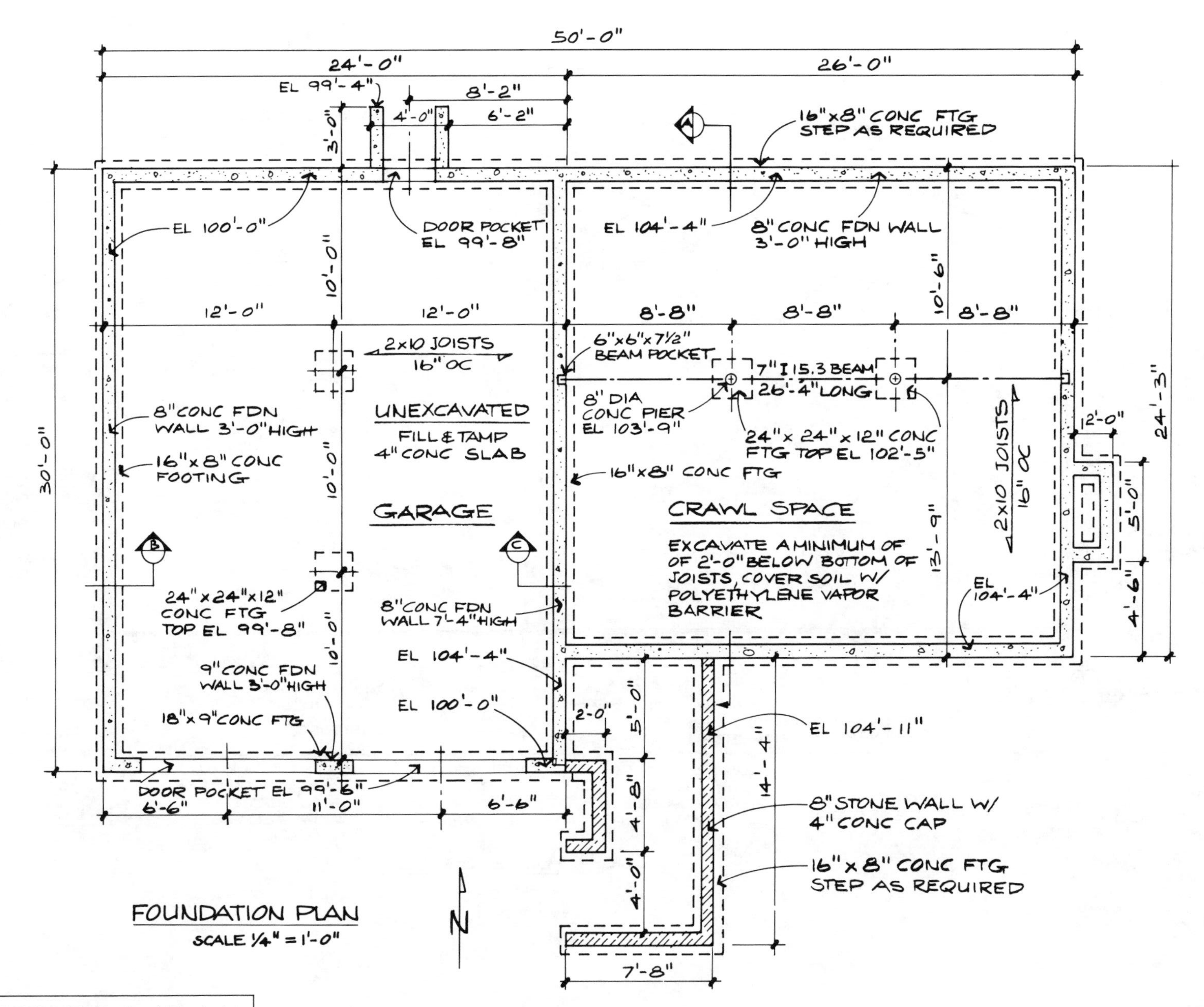

PUBLISHER'S NOTE: The size of this print has been modified and should not be scaled.

The Garlinghouse Company

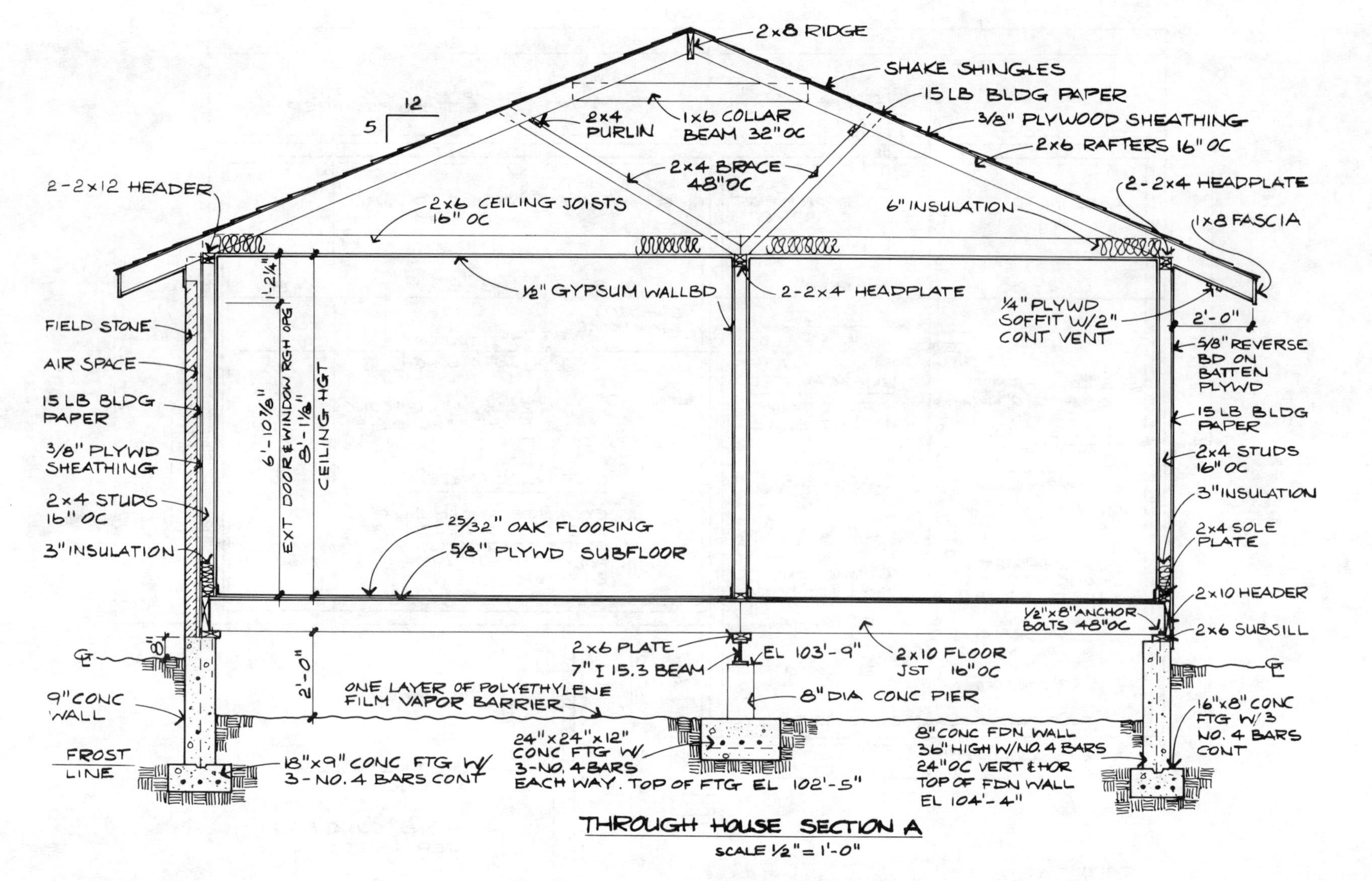

PUBLISHER'S NOTE: The size of this print has been modified and should not be scaled.

The Garlinghouse Company

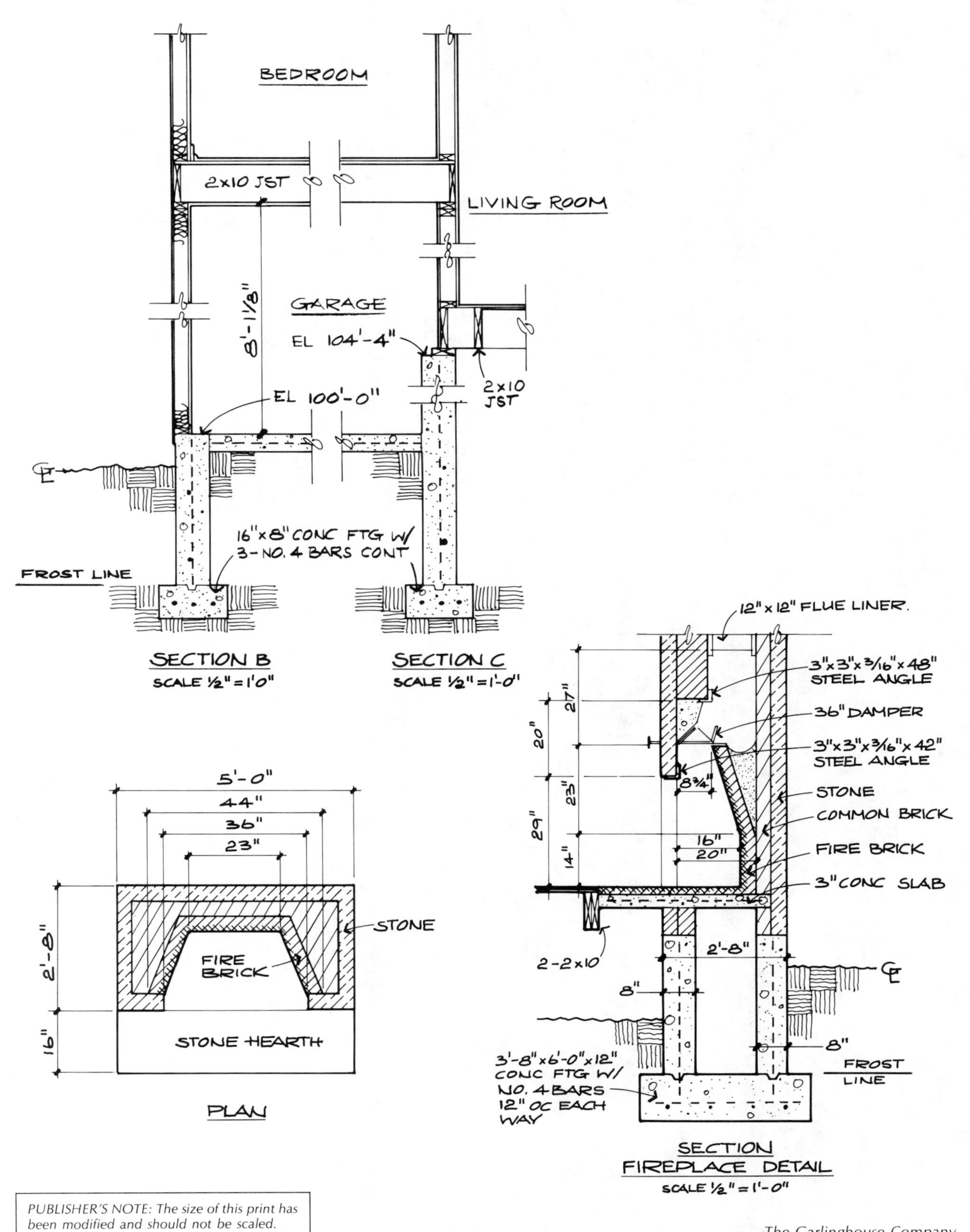

PUBLISHER'S NOTE: The size of this print has been modified and should not be scaled.

The Garlinghouse Company

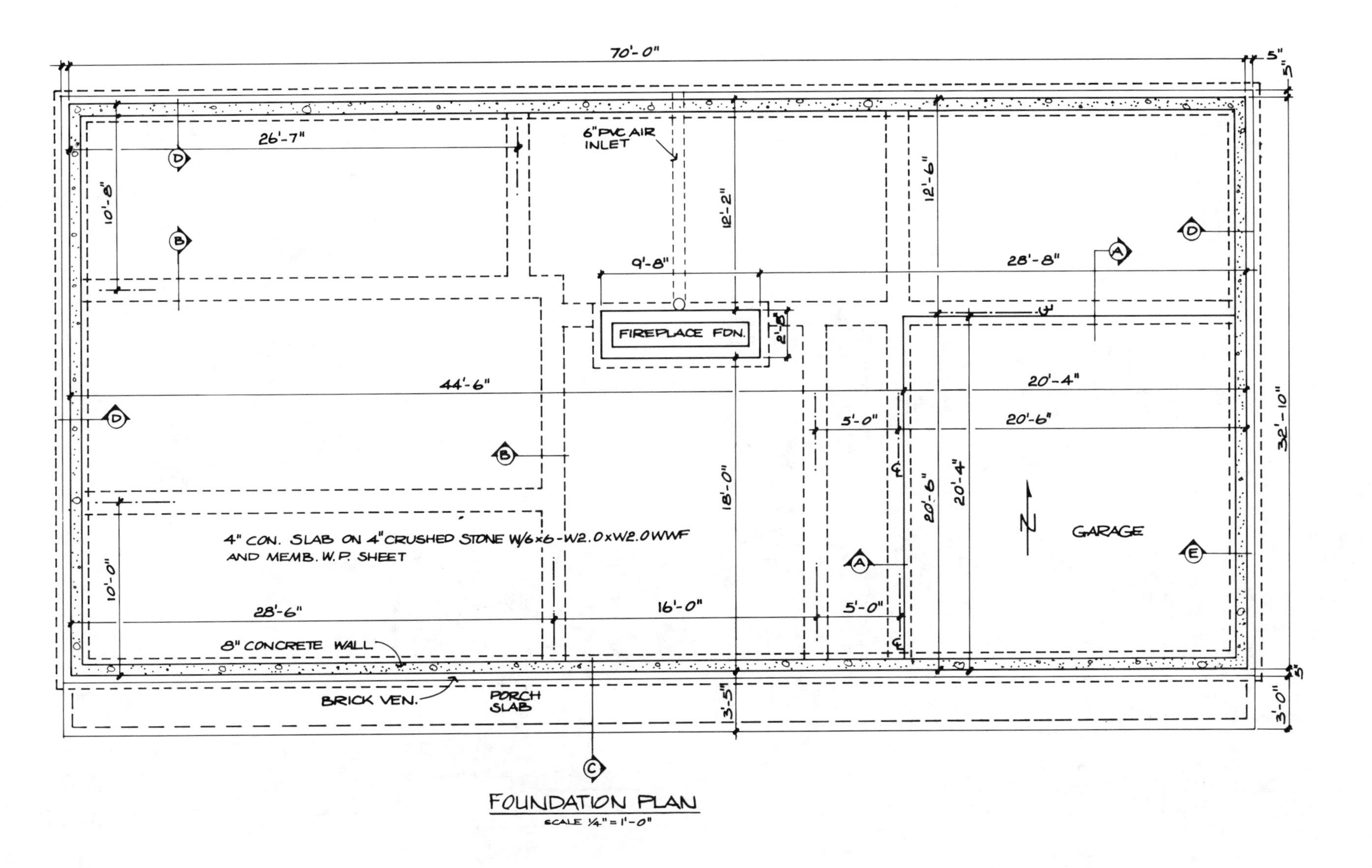

FOUNDATION PLAN
SCALE 1/4" = 1'-0"

PUBLISHER'S NOTE: The size of this print has been modified and should not be scaled.

The Garlinghouse Company

Slab-on-grade Foundation

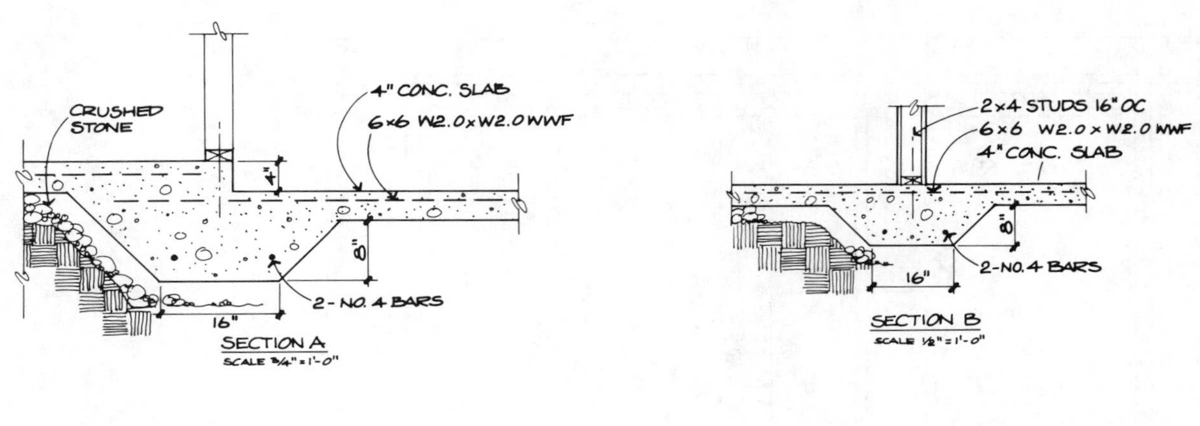

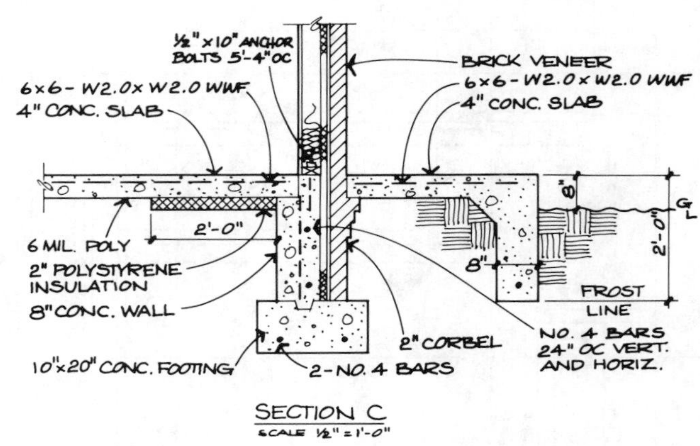

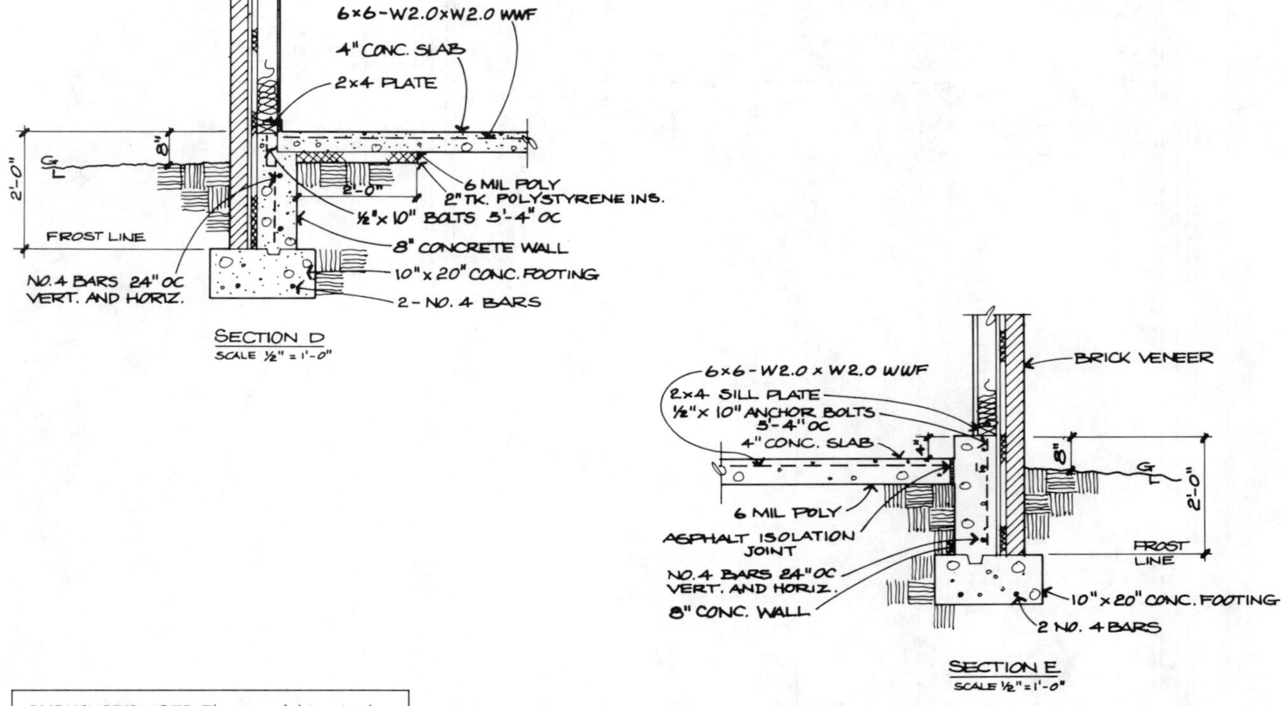

PUBLISHER'S NOTE: The size of this print has been modified and should not be scaled.

The Garlinghouse Company

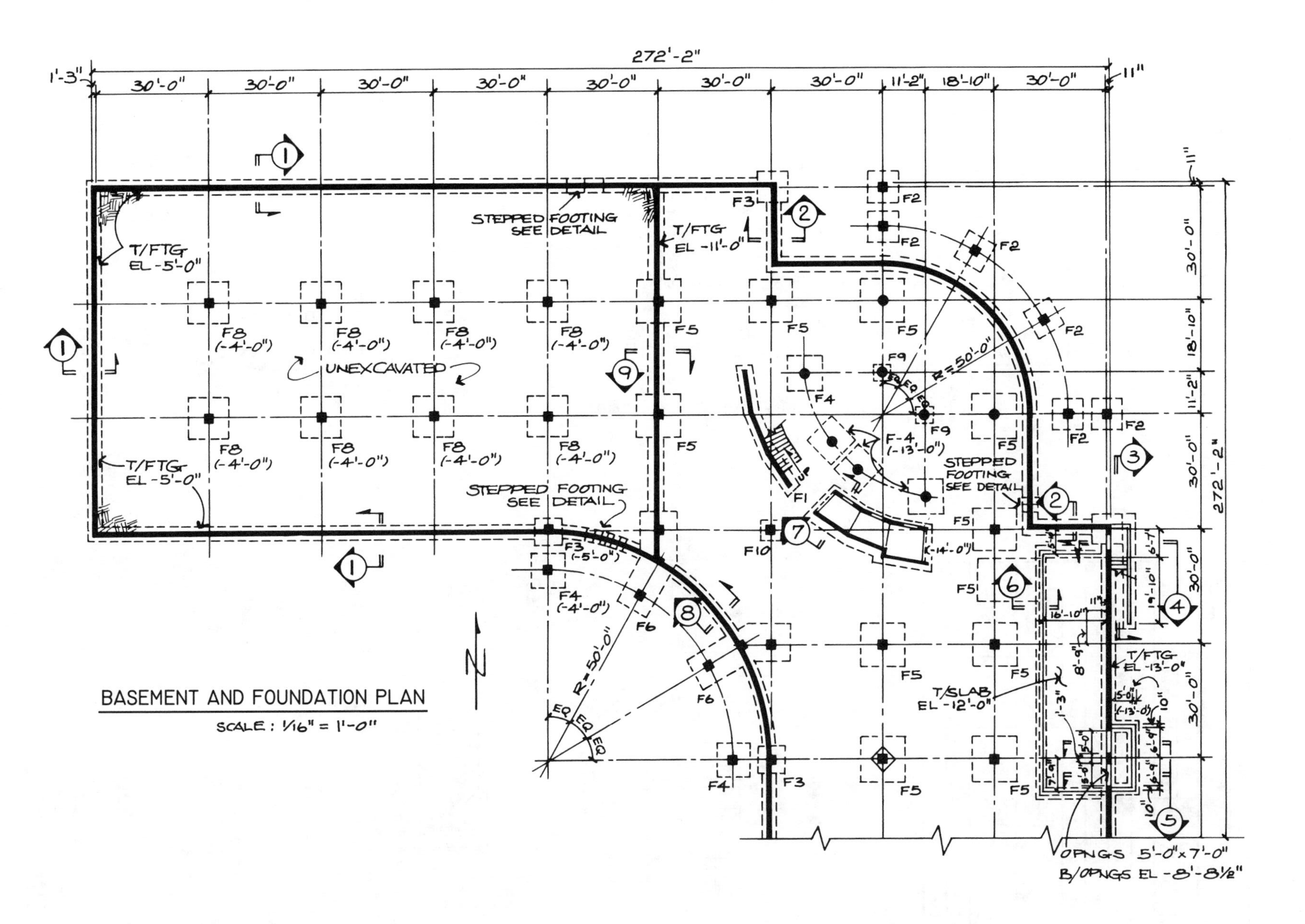

Chris P. Stefanos Associates, Inc.

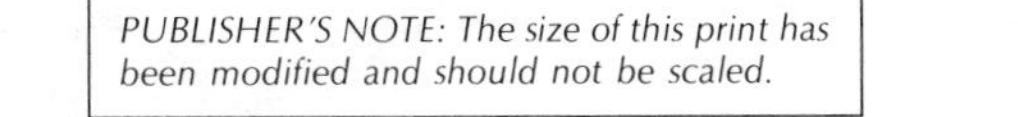

Heavy Construction Foundation

SOIL DATA 5000 PSF BEARING	FOOTING SCHEDULE			LC = 4000 PSI FY = 80,000 PSI
MARK	SIZE W×L×D	REINFORCEMENT LONG WAY	REINFORCEMENT SHORT WAY	REMARKS
F1	5′-0″×36′-0″×14″	8-# 4	34-# 6	
F2	7′-3″×7′-3″×16″	7-# 7	7-# 7	
F3	8′-3″×8′-3″×20″	8-# 7	8-# 7	
F4	9′-9″×9′-9″×26″	9-# 8	9-# 8	
F5	12′-0″×12′-0″×30″	11-# 9	11-# 9	
F6	SEE DETAIL	11-# 9	30-# 9	
F7	15′-3″×15′-3″×40″	13-# 9	18-# 9	
F8	11′-0″×11′-0″×28″	11-# 8	11-# 8	
F9	4′-6″×4′6″×16″	5-# 5	5-# 5	
F10	5′-6″×5′-6″×18″	7-# 5	7-# 5	

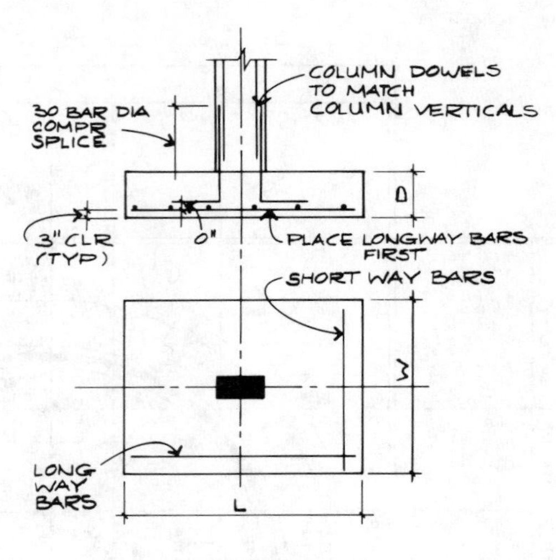

TYPICAL FOOTING DETAIL

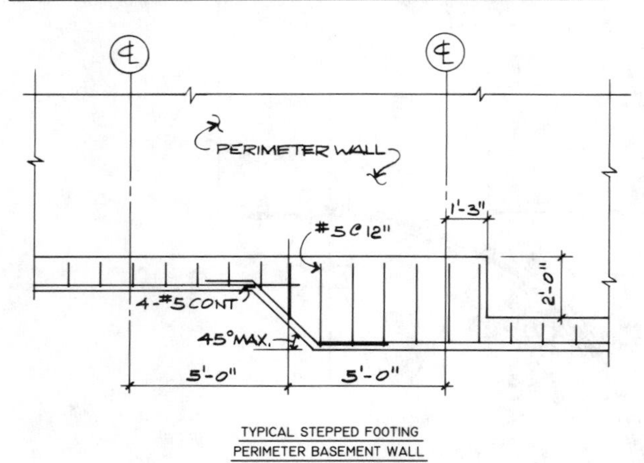

TYPICAL STEPPED FOOTING
PERIMETER BASEMENT WALL
SCALE: 3/8" = 1'-0"

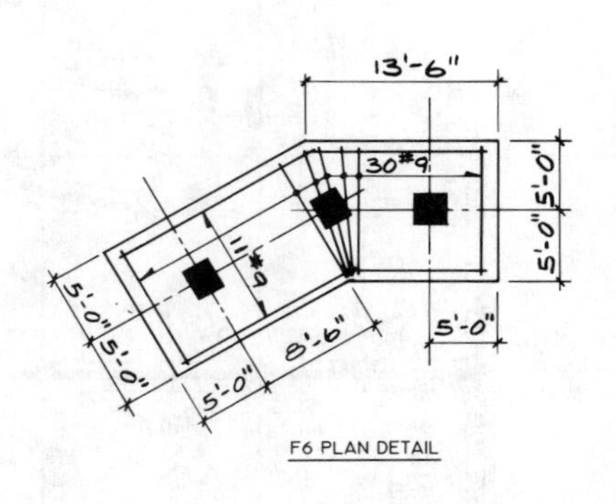

F6 PLAN DETAIL

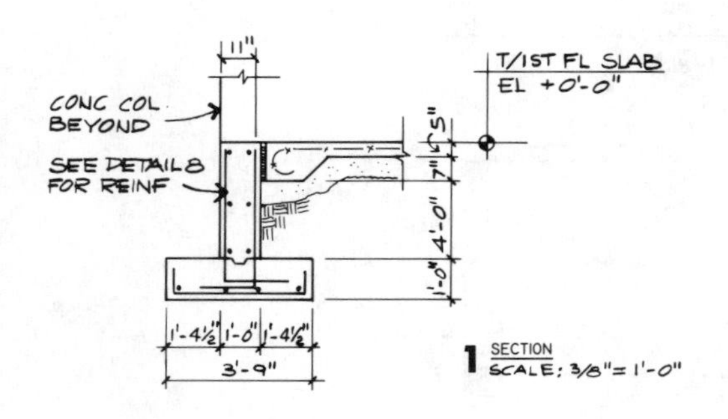

1 SECTION
SCALE: 3/8" = 1'-0"

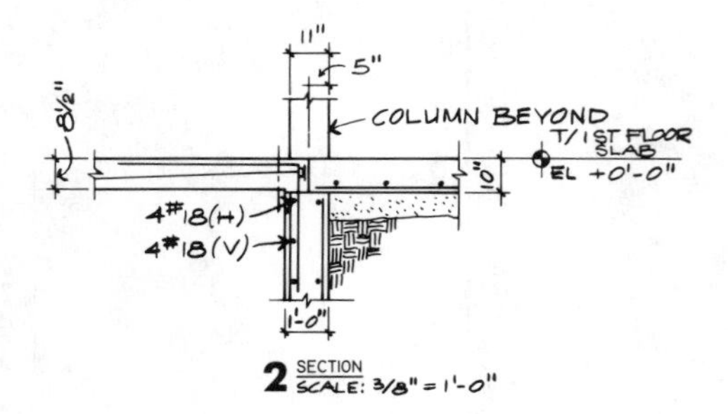

2 SECTION
SCALE: 3/8" = 1'-0"

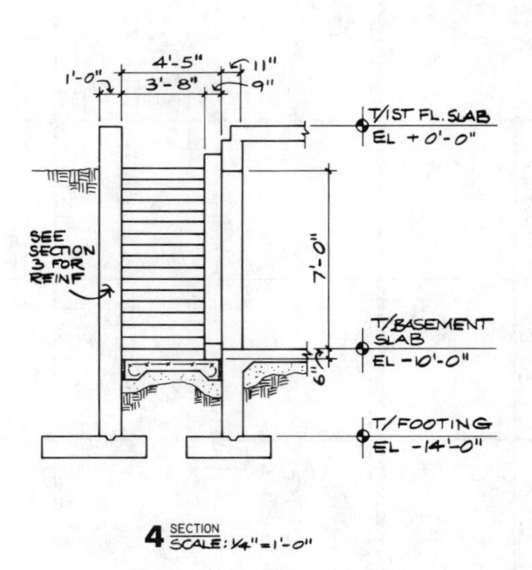

4 SECTION
SCALE: 1/4" = 1'-0"

PUBLISHER'S NOTE: The size of this print has been modified and should not be scaled.

Chris P. Stefanos Associates, Inc.

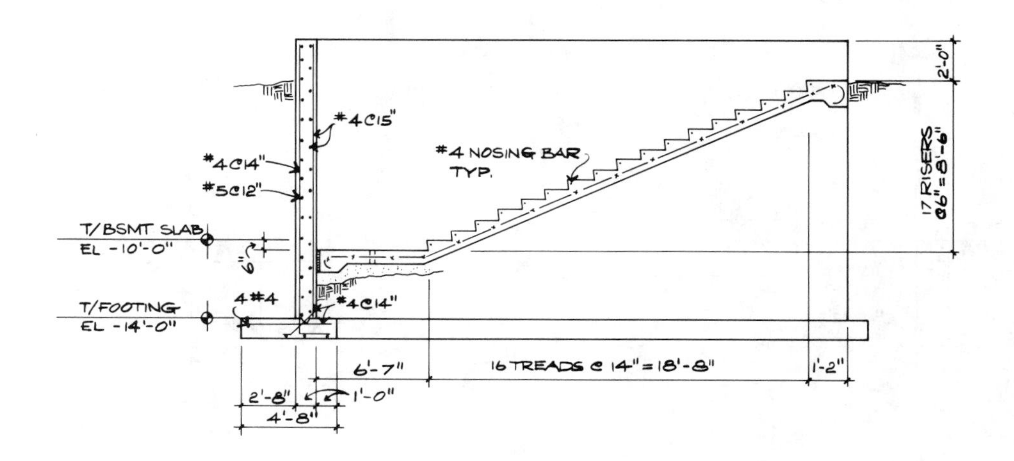

3 SECTION SCALE 1/4" = 1'-0"

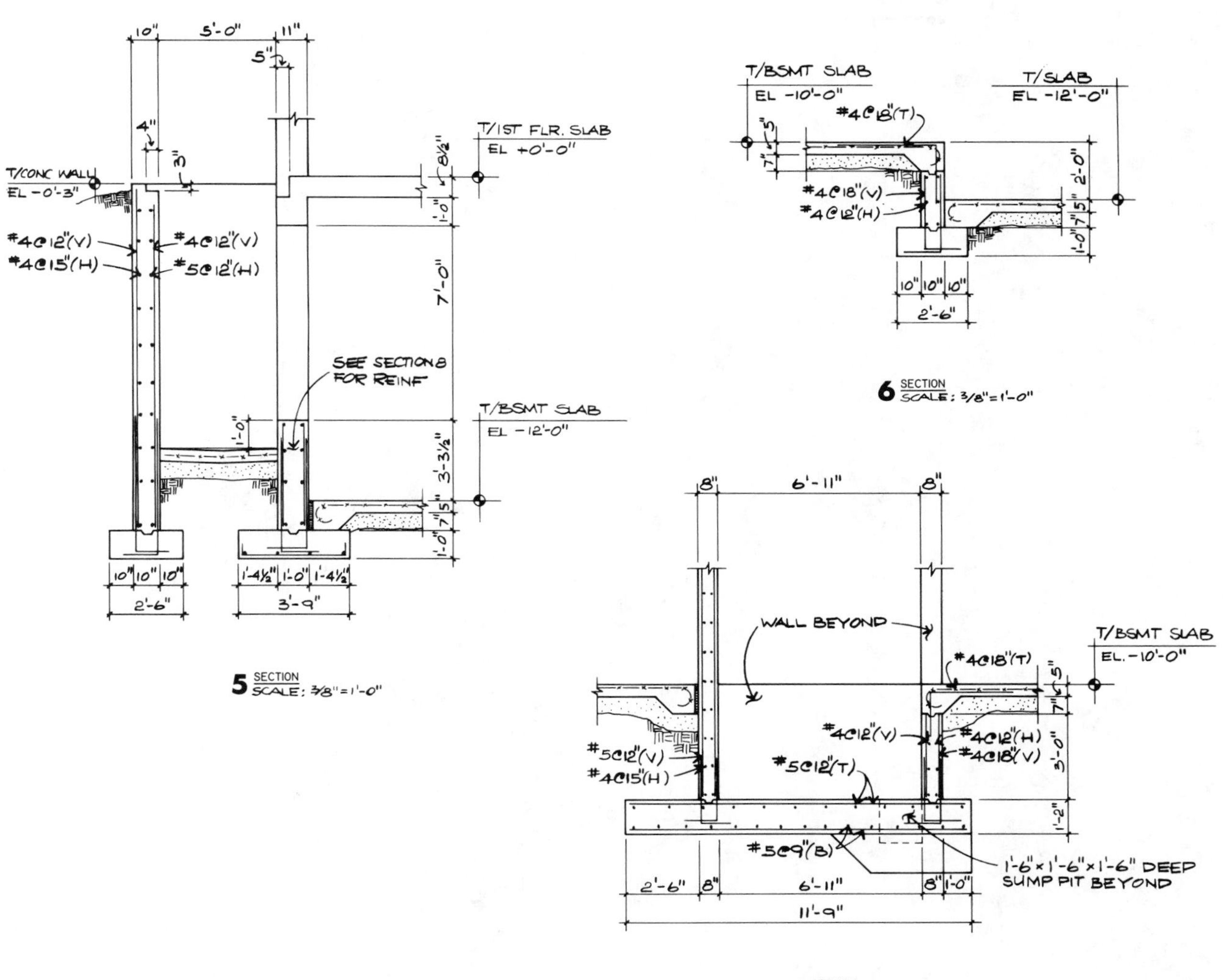

5 SECTION SCALE: 3/8" = 1'-0"

6 SECTION SCALE: 3/8" = 1'-0"

7 SECTION SCALE: 3/8" = 1'-0"

PUBLISHER'S NOTE: The size of this print has been modified and should not be scaled.

Chris P. Stefanos Associates, Inc.

Heavy Construction Foundation

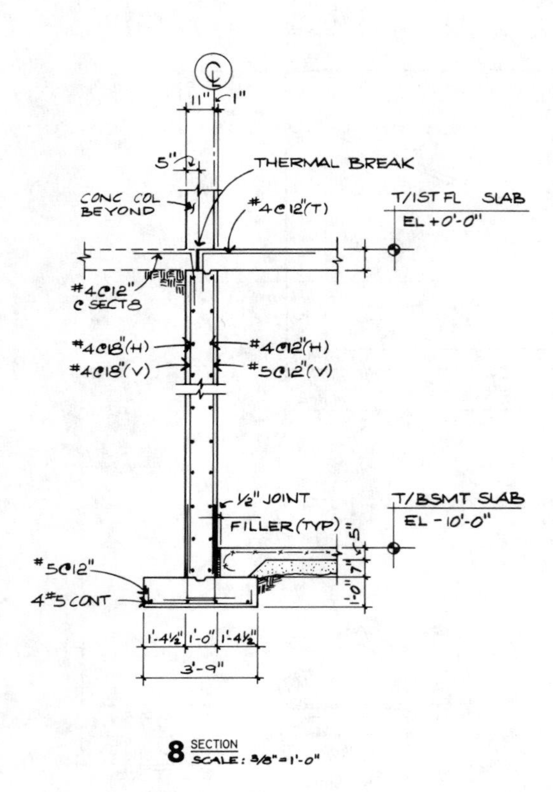

8 SECTION
SCALE: 3/8" = 1'-0"

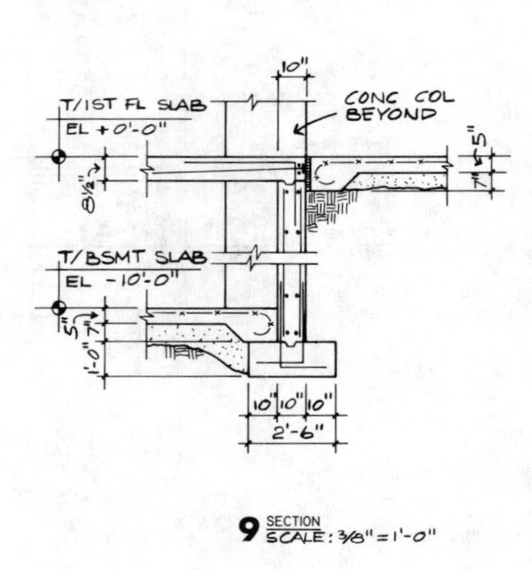

9 SECTION
SCALE: 3/8" = 1'-0"

PUBLISHER'S NOTE: The size of this print has been modified and should not be scaled.

Chris P. Stefanos Associates, Inc.

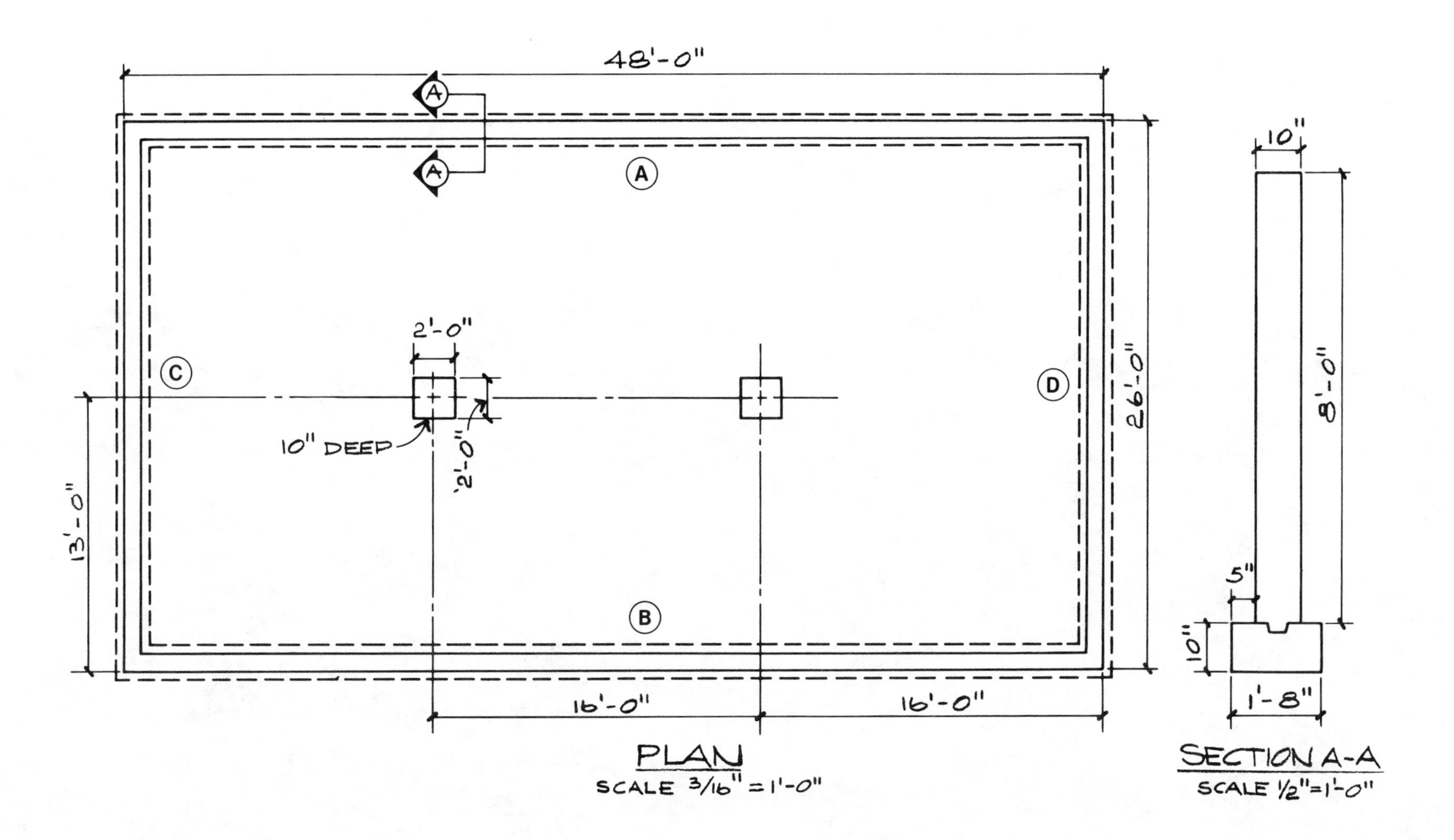

PUBLISHER'S NOTE: The size of this print has been modified and should not be scaled.

Appendix A: Tables and Formulas

DECIMAL EQUIVALENTS OF COMMON FRACTIONS	
Fraction (inches)	Decimal Equivalent
1/16	.0625
1/8	.125
3/16	.1875
1/4	.25
5/16	.3125
3/8	.375
7/16	.4375
1/2	.5
9/16	.5625
5/8	.625
11/16	.6875
3/4	.75
13/16	.8125
7/8	.875
15/16	.9375
1	1.0

CONVERSION TABLE—DECIMAL FEET TO INCHES												
	Inches											
8th of an Inch	0	1	2	3	4	5	6	7	8	9	10	11
0	.00	.08	.17	.25	.33	.42	.50	.58	.67	.75	.83	.92
1	.01	.09	.18	.26	.34	.43	.51	.59	.68	.76	.84	.93
2	.02	.10	.19	.27	.35	.44	.52	.60	.69	.77	.85	.94
3	.03	.11	.20	.28	.36	.45	.53	.61	.70	.78	.86	.95
4	.04	.13	.21	.29	.38	.46	.54	.63	.71	.79	.88	.96
5	.05	.14	.22	.30	.39	.47	.55	.64	.72	.80	.89	.97
6	.06	.15	.23	.31	.40	.48	.56	.65	.73	.81	.90	.98
7	.07	.16	.24	.32	.41	.49	.57	.66	.74	.82	.91	.99

AREA

Square, Rectangle, or Parallelogram

$Area = Length \times Height^*$

$A = L \times H$

**Width* may be substituted for *Height*

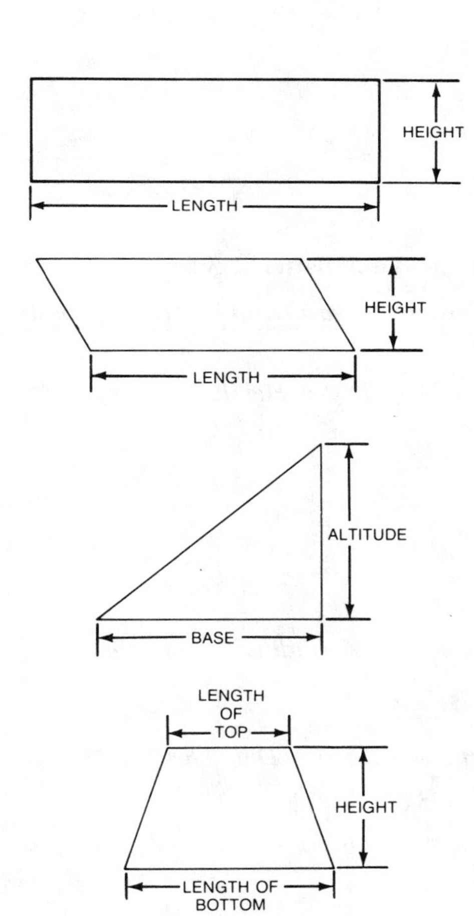

Triangle

$$Area = \frac{base \times altitude}{2}$$

$$A = \frac{ba}{2}$$

Trapezoid

$$Area = \frac{Height\ (Length\ of\ top + Length\ of\ bottom)}{2}$$

$$A = \frac{H\ (L_t + L_b)}{2}$$

Circle

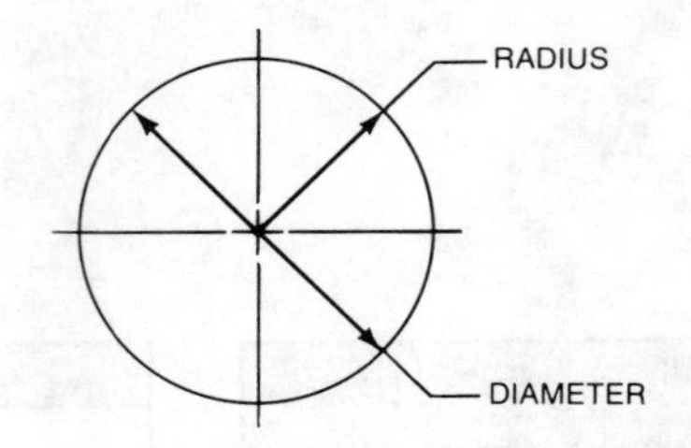

$Area = \pi \times Radius^2$

$A = \pi R^2$

$Area = .7854 \times Diameter^2$

$A = .7854 D^2$

VOLUME

Rectangular Solid

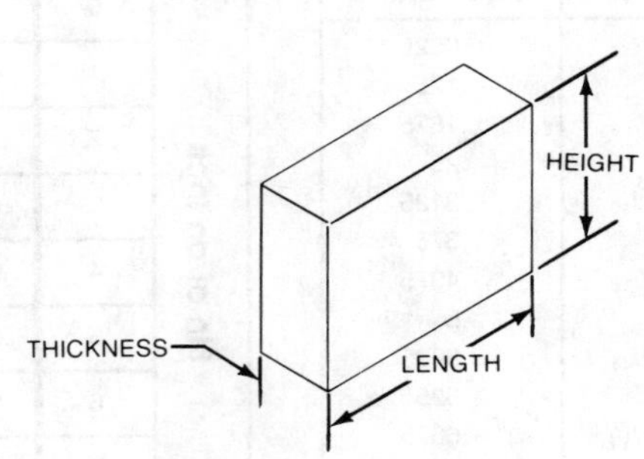

$Volume = Thickness \times Length \times Height$

$V = T \times L \times H$

Frustum—One Battered Side

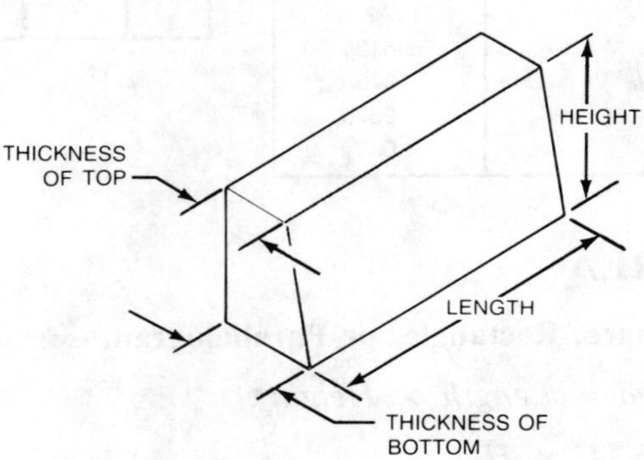

$$Volume = \frac{Top\ Thickness + Bottom\ Thickness}{2} \times Length \times Height$$

$$V = \frac{T_t + T_b}{2} \times L \times H$$

Frustum—Four Battered Sides

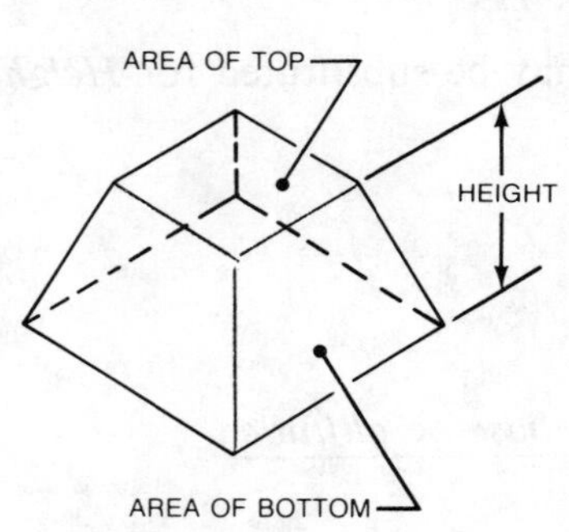

$$Volume = \frac{Area\ of\ top + Area\ of\ bottom}{2} \times Height$$

$$V = \frac{A_t + A_b}{2} \times Height$$

Cylinder

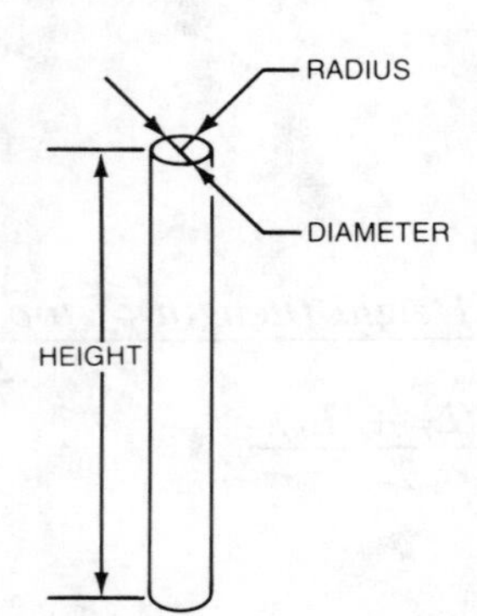

$Volume = \pi \times Radius^2 \times Height$

$V = \pi R^2 \times H$

$Volume = .7854 \times Diameter^2 \times Height$

$V = .7854 D^2 \times H$

Appendix B: Construction Materials

FORM MATERIALS

STANDARD LUMBER SIZES				
Type	Thickness		Width	
	Nominal Size	Actual Size	Nominal Size	Actual Size
COMMON BOARDS	1"	3/4"	2" 4" 6" 8" 10" 12"	1 1/2" 3 1/2" 5 1/2" 7 1/4" 9 1/4" 11 1/4"
DIMENSION	2"	1 1/2"	2" 4" 6" 8" 10" 12"	1 1/2" 3 1/2" 5 1/2" 7 1/4" 9 1/4" 11 1/4"
TIMBERS	4" 6" 8"	3 1/2" 5 1/2" 7 1/2"	4" 6" 8" 10"	3 1/2" 5 1/2" 7 1/2" 9 1/2"
	6"	5 1/2"	6" 8" 10"	5 1/2" 7 1/2" 9 1/2"
	8"	7 1/2"	8" 10"	7 1/2" 9 1/2"

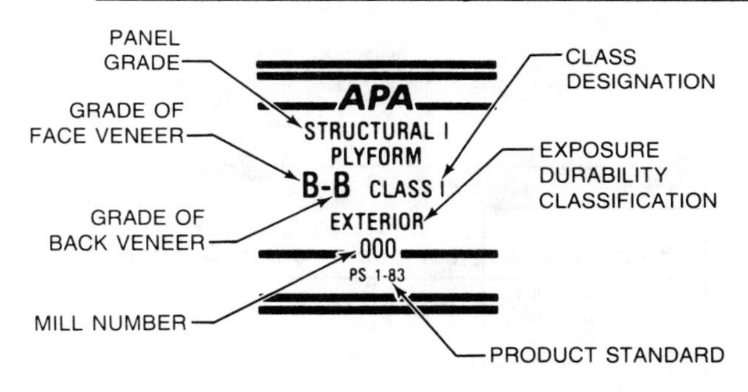

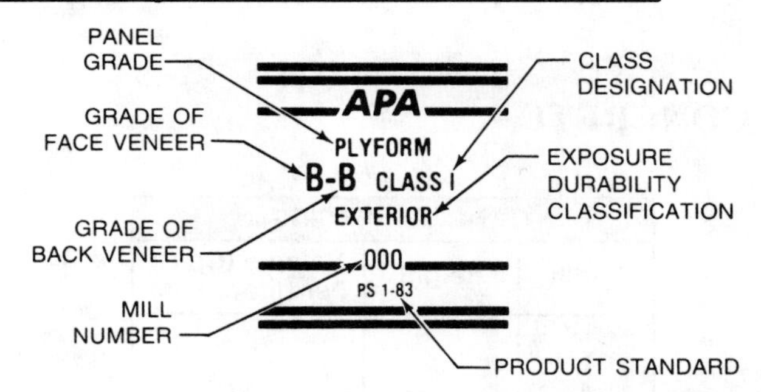

REINFORCEMENT

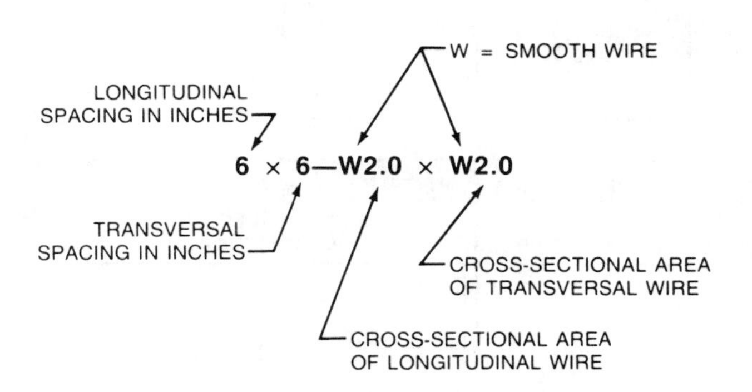

COMMON STOCK SIZES OF WELDED WIRE FABRIC				
Style Designation		Steel Area sq. in. per ft.		Weight Approx. lbs. per 100 sq. ft.
New Designation (by W-Number)	Old Designation (by Steel Wire Gauge)	Longit.	Trans.	
ROLLS				
6 x 6—W1.4 x W1.4	6 x 6—10 x 10	.028	.028	21
6 x 6—W2.0 x W2.0	6 x 6—8 x 8*	.040	.040	29
6 x 6—W2.9 x W2.9	6 x 6—6 x 6	.058	.058	42
SHEETS				
6 x 6—W2.9 x W2.9	6 x 6—6 x 6	.058	.058	42
6 x 6—W4.0 x W4.0	6 x 6—4 x 4	.080	.080	58
6 x 6—W5.5 x W5.5	6 x 6—2 x 2**	.110	.110	80

*Exact W-number size for 8 gauge is W2.1.
**Exact W-number size for 2 gauge is W5.4.

Wire Reinforcement Institute

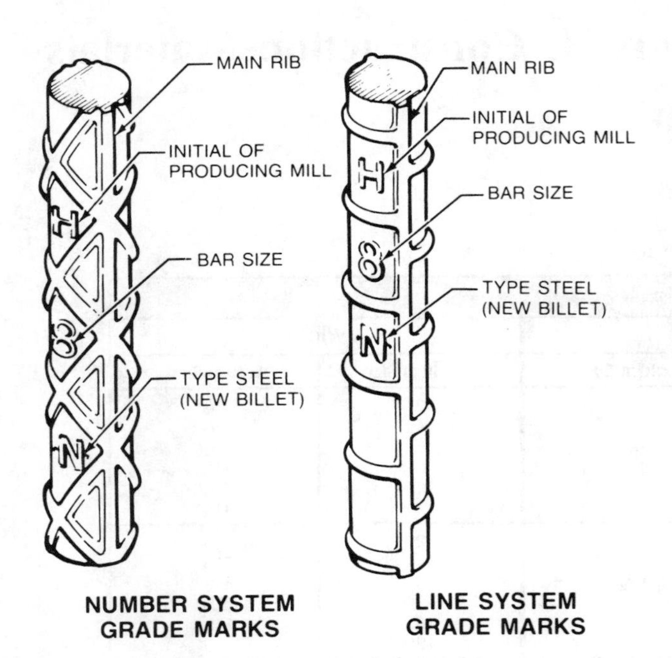

STANDARD REBAR SIZES						
Bar Size Designation	Weight Per Foot		Diameter		Cross-Sectional Area Squared	
	LB	KG	IN.	CM	IN.	CM
#3	0.376	0.171	0.375	0.953	0.11	0.71
#4	0.668	0.303	0.500	1.270	0.20	1.29
#5	1.043	0.473	0.625	1.588	0.31	2.00
#6	1.502	0.681	0.750	1.905	0.44	2.84
#7	2.044	0.927	0.875	2.223	0.60	3.87
#8	2.670	1.211	1.000	2.540	0.79	5.10
#9	3.400	1.542	1.128	2.865	1.00	6.45
#10	4.303	1.952	1.270	3.226	1.27	8.19
#11	5.313	2.410	1.410	3.581	1.56	10.07
#14	7.650	3.470	1.693	4.300	2.25	14.52
#18	13.600	6.169	2.257	5.733	4.00	25.81

American Society for Testing and Materials

CONCRETE

CONCRETE FOOTINGS		
Width	Height	Volume (Cu Ft) Per Linear Foot
1′-0″	6″	.50
1′-2″		.59
1′-4″		.67
1′-6″		.75
1′-0″	8″	.67
1′-2″		.78
1′-4″		.89
1′-6″		1.00
1′-0″	10″	.83
1′-2″		.97
1′-4″		1.11
1′-6″		1.25
1′-8″		1.39
1′-10″		1.53
2′-0″		1.67
1′-2″	12″	1.17
1′-4″		1.33
1′-6″		1.50
1′-8″		1.67
1′-10″		1.83
2′-0″		2.00

CONCRETE SLABS	
Slab Thickness (in.)	Area (sq ft) Coverage Per Cubic Yard of Concrete
1	324
2	162
3	108
4	81
5	65
6	54
7	46
8	40
9	36
10	32
11	29.5
12	27

CONCRETE WALLS	
Wall Thickness (in.)	Volume (Cu Ft) per 100 Sq Ft Wall
4	33.3
6	50.0
8	66.7
10	83.3
12	100.0

Appendix C: ACI Recommended Practices

Reproduced from Recommended Practice for Concrete Formwork (ACI 347-78)

3.1—Safety precautions

In addition to the very real moral and legal responsibility to maintain safe conditions for workers and the public, safe construction is in the final analysis more economical than any short-term cost savings from cutting corners on safety provisions. Attention to safety is particularly significant in formwork construction as these structures support the concrete during its plastic state and as it is developing in strength, until the concrete becomes structurally self-sufficient. Following the design criteria contained in this standard is essential to assuring safe performance of the forms. All structural members and their connection should be carefully planned so that a sound determination of loads thereon may be accurately made and stresses calculated.

In addition to the adequacy of the formwork, special structures such as multistory buildings require consideration of the behavior of newly completed beams and slabs which are used to support formwork and other construction loads. It must be kept in mind that the flexural, shear and bond strength of freshly cast slabs or beams are less than corresponding properties of an aged slab.

Many formwork failures can be attributed to some human error or omission rather than basic inadequacy in design. Careful supervision and continuous inspection of formwork erection can prevent many accidents.

Construction procedures must be planned in advance to insure the unqualified safety of personnel engaged in formwork and concrete placement and the integrity of the finished structure. Some of the safety provisions which should be considered are:

(a) Erection of safety signs and barricades to keep unauthorized personnel clear of areas in which erection or stripping is under way

(b) Providing experienced form watchers during concrete placement to assure early recognition of possible form displacement or failure. A supply of extra shores or other material and equipment that might be needed in an emergency by form watchers should be readily available. Means for adequate illumination of the formwork should be provided on those concrete placements which will, or may, extend into darkness hours

(c) Including lifting points in the design and detailing of all forms which will be crane-handled. This is especially important in flying forms or climbing forms. In the case of wall formwork consideration should be given to an independent scaffold bolted to the previous lift.

(d) Incorporation of scaffolds, working platforms, guardrails, etc., into formwork design and all formwork drawings

(e) A program of field safety inspections of formwork

3.1.1 *Construction deficiencies*—Some common construction deficiencies leading to form failures are:

(a) Inadequate diagonal bracing of shores

(b) Inadequate lateral and diagonal bracing and poor splicing of "double-tier shores" or "multiple-story shores"

(c) Failure to control rate of placing concrete vertically without regard to drop in temperature

(d) Failure to regulate properly the rate and sequence of placing concrete horizontally to avoid unbalanced loadings on the formwork

(e) Unstable soil under mudsills, sometimes caused by washwater from the forms

(f) Failure to inspect formwork during and after concrete placement to detect abnormal deflections or other signs of imminent failure which could be corrected

(g) Insufficient nailing, bolting, or fastening

(h) Failure to provide adequate support for lateral pressure on formwork

(i) Shoring not plumb and thus inducing lateral loading as well as reducing vertical load capacity

(j) Locking devices on metal shoring not locked, inoperative, or missing

(k) Vibration from adjacent moving loads or load carriers

(l) Inadequately tightened or secured form ties or wedges

(m) Form damage in excavation by reason of embankment failure

(n) Loosening of reshores under floors below

(o) Premature removal of supports, especially under cantilevered sections

(p) Inadequate bearing under mudsills; in no case should mudsills or spread footings rest on frozen ground

(q) Connection of shores to joists, stringers, or wales which are inadequate to resist uplifts or torsion at joints

(r) Failure to comply with manufacturer's recommendations for standard components

3.1.2 *Design deficiencies*—Some common design deficiencies leading to failure are:

(a) Lack of proper field inspection by qualified persons to see that form design has been properly interpreted by form builders

(b) Lack of allowance in design for such loadings as winding, power buggies, placing equipment, and temporary material storage

(c) Inadequate reshoring

(d) Improperly stressed reshoring

(e) Improperly positioning of shores from floor to floor which creates bending in slabs not designed for such stresses

(f) Inadequate provisions to prevent rotation of beam forms where slabs frame into them on only one side

(g) Inadequate anchorage against uplift due to battered form faces

(h) Insufficient allowance for eccentric loading due to placement sequences

(i) Failure to investigate bearing stresses in members in contact with shores or struts

3.2—Construction practices and workmanship

3.2.1—Bulkheads for control joints or construction joints should preferably be made by splitting the bulkhead so that each portion may be positioned and removed separately without applying undue pressure on the reinforcing rods which cause spalling or cracking of the concrete. When required on the engineer/architect's drawing, beveled inserts at control joints must be left undisturbed when forms are stripped, and removed only after the concrete has been sufficiently cured and dried out. Wood strips inserted for architectural treatment should be kerfed to permit swelling with pressure on the concrete.

3.2.2—Sloped surfaces steeper than 1.5 horizontal to 1 vertical should be provided with a top form to hold the shape of the concrete during placement, unless it can be demonstrated that the top forms can be omitted.

3.2.3—Loading of new slabs should be avoided in the first few days after placement. Loads such as aggregate, timber, boards, reinforcing steel, or support devices must not be thrown on new construction, nor be allowed to pile up in quantity.

3.2.4—Building materials must not be thrown or piled on the formwork in such manner as to damage or overload it.

3.2.5—Studs, walers, or shores should be properly spliced.

3.2.6—Joints or splices in sheathing, plywood panels, and bracing should be staggered.

3.2.7—Shores should be properly seated and anchored.

3.2.8—Install and properly tighten all form ties or clamps as specified.

3.2.9—Use specified size and capacity of form ties or clamps.

3.4—Shoring and centering

3.4.1 *Shoring*—Shoring must be supported on satisfactory foundations such as spread footings, mudsills, or piling.

Shoring resting on intermediate slabs or other construction already in place need not be located directly above shores or reshores below unless thickness of slab and the location of its reinforcement are inadequate to take the reversal of stresses and punching shear. Where the latter conditions are questionable the shoring location should be approved by the engineer/architect.

All members must be straight and true without twists or bends. Special attention should be given to beam and slab, or one-way and two-way joist construction to prevent local overloads when a heavily loaded shore rests on the thin slab.

Multitier shoring assemblies supporting forms for high stories must be set plumb and the separate parts of each shore located in a straight line over each other, with two-way horizontal bracing at each splice in the shore unless the entire assembly is designed as a structural framework or truss. Particular care must also be taken to transfer the horizontal loads to the ground or to completed construction of adequate strength.

Where a slab load is supported on one side of the beam only, edge beam forms should be carefully planned to prevent tipping of the beam due to unequal loading.

Shores or vertical posts must be erected so that they cannot tilt, and must have firm bearing. Inclined shores must be braced securely against slipping or sliding. The bearing ends of shores should be cut square and have a tight fit at splices. Splices must be secure against bending and buckling. Connections of shore heads to other frames should be adequate to prevent the shores from falling out when reversed bending causes upward deflection of the forms.

3.4.2 *Centering*—Centering is the highly specialized temporary support system used in the construction of arches, shells, space structures, or any continuous structure where the entire temporary support is lowered (struck or decentered) as a unit to avoid introducing injurious stress in any part of the structure. The lowering of the centering is generally accomplished by the use of sand boxes, jacks, or wedges beneath the supporting members.

3.5—Adjustment of formwork

3.5.1 *Before concreting*

3.5.1.1 Telltale devices should be installed on supported forms and elsewhere as required to facilitate detection and measurement of formwork movements during concreting.

3.5.1.2 Wedges used for final alignment before concrete placement should be secured in position after the final check.

3.5.1.3 Formwork must be anchored to the shores below so that upward or horizontal movement of any part of the formwork system will be prevented during concrete placement.

3.5.1.4 To ensure that lines and grades of finished concrete work will be within the required tolerances, the forms must be constructed to the elevation shown on the formwork drawings. If camber is required in the hardened concrete to resist deflection it should be so stipulated on the structural drawings. Additional elevation or camber should be provided to allow for closure of form joints, settlements of mudsills, shrinkage of lumber, dead load deflections, and elastic shortening of form members.

Where camber requirements may become cumulative, such as in cases where beams frame into other beams or girders at right angles, and at midspan of the latter, the engineer/architect should specify exactly the manner in which this condition is to be handled.

3.5.1.5 Positive means of adjustment (wedges or jacks) should be provided to permit realignment or adjustment of shores if excessive settlement occurs.

3.5.1.6 Runways for moving equipment should be provided with struts or legs as required and should be supported directly on the formwork or structural member. They should not bear on nor be supported by the reinforcing steel unless special bar supports are provided. The formwork must be suitable for the support of such runways without significant deflections, vibrations, or lateral movements.

3.5.2 *During and after concreting*—During and after concreting, but before initial set of the concrete, the elevations, camber, and plumbness of formwork systems should be checked, using tell-tale devices. Appropriate adjustments should be promptly made where necessary. If, during construction, any weakness develops and the formwork shows any undue settlement or distortion, the work should be stopped, the affected construction removed if permanently damaged, and the formwork strengthened.

Formwork must be continuously watched so that any corrective measures found necessary may be promptly taken. Form watchers must always work under safe conditions and should establish in advance a method of communication with placing crews in case of emergency.

Appendix D: OSHA Concrete and Shoring Regulations

Reproduced from the United States Occupational Safety and Health Administration provisions, U.S. Department of Labor.

Subpart Q—Concrete, Concrete Forms, and Shoring
§ 1926.700 General provisions.

(a) *General.* All equipment and materials used in concrete construction and masonry work shall meet the applicable requirements for design, construction, inspection, testing, maintenance and operations as prescribed in ANSI A10.9—1970, Safety Requirements for Concrete Construction and Masonry Work.

(b) *Reinforcing steel.* (1) Employees working more than 6 feet above any adjacent working surfaces, placing and tying reinforcing steel in walls, piers, columns, etc., shall be provided with a safety belt, or equivalent device.

(2) Employees shall not be permitted to work above vertically protruding reinforcing steel unless it has been protected to eliminate the hazard of impalement.

(3) *Guying:* Reinforcing steel for walls, piers, columns, and similar vertical structures shall be guyed and supported to prevent collapse.

(4) *Wire mesh rolls:* Wire mesh rolls shall be secured at each end to prevent dangerous recoiling action.

(c) *Bulk concrete handling.* Bulk storage bins, containers, or silos shall have conical or tapered bottoms with mechanical or pneumatic means of starting the flow of material.

(d) *Concrete placement*—(1) *Concrete mixers.* Concrete mixers equipped with 1-yard or larger loading skips shall be equipped with a mechanical device to clear the skip of material.

(2) *Guardrails.* Mixers of 1-yard capacity or greater shall be equipped with protective guardrails installed on each side of the skip.

(3) *Bull floats.* Handles on bull floats, used where they may contact energized electrical conductors, shall be constructed of nonconductive material, or insulated with a nonconductive sheath whose electrical and mechanical characteristics provide the equivalent protection of a handle constructed of nonconductive material.

(4) *Powered concrete trowels.* Powered and rotating-type concrete troweling machines that are manually guided shall be equipped with a control switch that will automatically shut off the power whenever the operator removes his hands from the equipment handles.

(5) *Concrete buggies.* Handles of buggies shall not extend beyond the wheels on either side of the buggy. Installation of knuckle guards on buggy handles is recommended.

(6) *Pumpcrete systems.* Pumpcrete or similar systems using discharge pipes shall be provided with pipe supports designed for 100 percent overload. Compressed air hose in such systems shall be provided with positive fail-safe joint connectors to prevent separation of sections when pressurized.

(7) *Concrete buckets.* (i) Concrete buckets equipped with hydraulic or pneumatically operated gates shall have positive safety latches or similar safety devices installed to prevent aggregate and loose material from accumulating on the top and sides of the bucket.

(ii) Riding of concrete buckets for any purpose shall be prohibited, and vibrator crews shall be kept out from under concrete buckets suspended from cranes or cableways.

(8) When discharging on a slope, the wheels of ready-mix trucks shall be blocked and the brakes set to prevent movement.

(9) Nozzlemen applying a cement, sand, and water mixture through a pneumatic hose shall be required to wear protective head and face equipment.

(e) *Vertical shoring—(1) General requirements.* (i) When temporary storage of reinforcing rods, material, or equipment on top of formwork becomes necessary, these areas shall be strengthened to meet the intended loads.

(ii) The sills for shoring shall be sound, rigid, and capable of carrying the maximum intended load.

(iii) All shoring equipment shall be inspected prior to erection to determine that it is as specified in the shoring layout. Any equipment found to be damaged shall not be used for shoring.

(iv) Erected shoring equipment shall be inspected immediately prior to, during, and immediately after the placement of concrete. Any shoring equipment that is found to be damaged or weakened shall be immediately reinforced or reshored.

(v) Reshoring shall be provided when necessary to safely support slabs and beams after stripping, or where such members are subjected to superimposed loads due to construction work done.

(2) *Tubular welded frame shoring.* (i) Metal tubular frames used for shoring shall not be loaded beyond the safe working load recommended by the manufacturer.

(ii) All locking devices on frames and braces shall be in good working order; coupling pins shall align the frame or panel legs; pivoted cross braces shall have their center pivot in place; and all components shall be in a condition similar to that of original manufacture.

(iii) When checking the erected shoring frames with the shoring layout, the spacing between towers and cross brace spacing shall not exceed that shown on the layout, and all locking devices shall be in the closed position.

(iv) Devices for attaching the external lateral stability bracing shall be securely fastened to the legs of the shoring frames.

(v) All base plates, shore heads, extension devices, or adjustments screws shall be in firm contact with the footing sill and the form.

§ 1926.701 Forms and shoring.

(a) *General provisions.* (1) Formwork and shoring shall be designed, erected, supported, braced, and maintained so that it will safely support all vertical and lateral loads that may be imposed upon it during placement of concrete.

(2) Drawings or plans showing the jack layout, formwork, shoring, working decks, and scaffolding, shall be available at the jobsite.

(3) Stripped forms and shoring shall be removed and stockpiled promptly after stripping, in all areas in which persons are required to work or pass. Protruding nails, wire ties, and other form accessories not necessary to subsequent work shall be pulled, cut, or other means taken to eliminate the hazard.

(4) Imposition of any construction loads on the partially completed structure shall not be permitted unless such loading has been considered in the design and approved by the engineer-architect.

(b) *Vertical slipforms.* (1) The steel rods or pipe on which the jacks climb or by which the forms are lifted shall be specifically designed for purpose. Such rods shall be adequately braced where not encased in concrete.

(2) Jacks and vertical supports shall be positioned in such a manner that the vertical loads are distributed equally and do not exceed the capacity of the jacks.

(3) The jacks or other lifting devices shall be provided with mechanical dogs or other automatic holding devices to provide protection in case of failure of the power supply or the lifting mechanism.

(4) Lifting shall proceed steadily and uniformly and shall not exceed the predetermined safe rate of lift.

(5) Lateral and diagonal bracing of the forms shall be provided to prevent excessive distortion of the structure during the jacking operation.

(6) During jacking operations, the form structure shall be maintained in line and plumb.

(7) All vertical lift forms shall be provided with scaffolding or work platforms completely encircling the area of placement.

(c) *Tube and coupler shoring.* (1) Couplers (clamps) shall not be used if they are deformed, broken, or have defective or missing threads on bolts, or other defects.

(2) The material used for the couplers (clamps) shall be of a structural type such as drop-forged steel, malleable iron, or structural grade aluminum. Gray cast iron shall not be used.

(3) When checking the erected shoring towers with the shoring layout, the spacing between posts shall not exceed that shown on the layout, and all interlocking of tubular members and tightness of couplers shall be checked.

(4) All base plates, shore heads, extension devices, or adjustment screws shall be in firm contact with the footing sill and the form material and shall be snug against the posts.

(d) *Single post shores.* (1) For stability, single post shores shall be horizontally braced in both the longitudinal and transverse directions, and diagonal bracing shall also be installed. Such bracing shall be installed as the shores are being erected.

(2) All base plates or shore heads of single post shores shall be in firm contact with the footing sill and the form materials.

(3) Whenever single post shores are used in more than one tier, the layout shall be designed and inspected by a structural engineer.

(4) When formwork is at an angle, or sloping, or when the surface shored is sloping, the shoring shall be designed for such loading.

(5) Adjustment of single post shores to raise formwork shall not be made after concrete is in place.

(6) Fabricated single post shores shall not be used if heavily rusted, bent, dent, rewelded, or having broken weldments or other defects. If they contain timber, they shall not be used if timber is split, cut, has sections removed, is rotted, or otherwise structurally damaged.

(7) All timber and adjusting devices to be used for adjustable timber single post shores shall be inspected before erection.

(8) Timber shall not be used if it is split, cut, has sections removed, is rotted, or is otherwise structurally damaged.

(9) Adjusting devices shall not be used if heavily rusted, bent, dent, rewelded, or having broken weldments or other defects.

(10) All nails used to secure bracing or adjustable timber single post shores shall be driven home and the point of the nail bent over if possible.

§ 1926.702 Definitions applicable to this subpart.

(a) "Bull float"—A tool used to spread out and smooth the concrete.

(b) "Formwork" or "falsework"—The total system of support for freshly placed concrete, including the mold or sheathing which contacts the concrete as well as all supporting members, hardware, and necessary bracing.

(c) "Guy"—A line that steadies a high piece of structure by pulling against an off-center load.

(d) "Shore"—A supporting member that resists a compressive force imposed by a load.

(e) "Vertical slipforms"—Forms which are jacked vertical and continuously during placing of the concrete.

Appendix E: Leveling Instruments

Preliminary site work must be done accurately to ensure that the building is in the proper location and at the appropriate elevation. Building lines are established and the location of columns, piers, and other structural components is determined from the building lines. The *builder's level, transit-level,* and *laser transit-level* are commonly used to lay out and establish building lines and other reference points on the job site. Each of the leveling instruments rotates horizontally and can be used for horizontal measurements. In addition, the transit-level rotates vertically to plumb and align vertical surfaces.

THE BUILDER'S LEVEL

A builder's level is used to establish and verify grades and elevations and set up reference points over long distances. It is used extensively for grading operations and general foundation layout.

The main parts of a builder's level are the *telescope, spirit level,* and *leveling screws.* The telescope is used to sight objects and contains *lenses,* an *eyepiece,* a *detachable sun shade,* and a *focusing knob.* The lenses magnify the object being sighted. The eyepiece is rotated to bring the cross hairs into focus. The detachable sun shade protects the objective lens from damage and reduces glare. The focusing knob is used to adjust the telescope until the object being viewed appears sharp and clear. A spirit level is a sensitive device used to indicate the levelness of the telescope. The spirit level is located above or below the telescope. If the spirit level indicates that the telescope is out-of-level, the leveling screws are used to level the telescope in all directions. See Figure E-1.

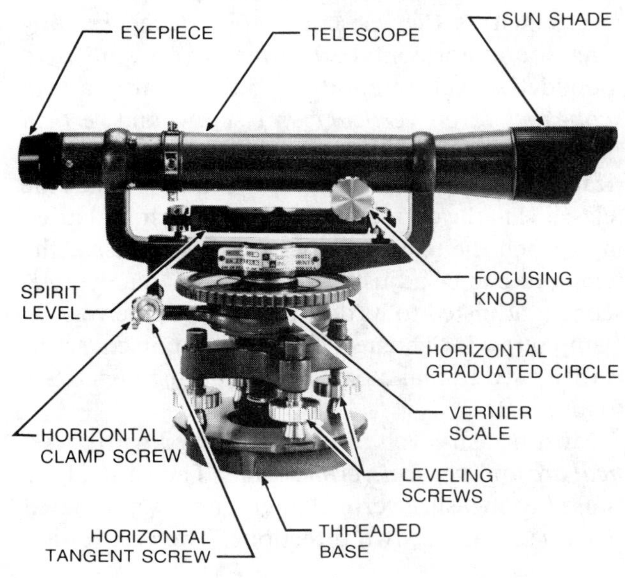

David White Instruments

Figure E-1. A builder's level is used to establish grade levels and elevations on a job site. The telescope can only be moved horizontally.

A *horizontal clamp screw* holds the builder's level in a fixed horizontal position. A *horizontal tangent screw* is used to make slight adjustments to the telescope in a horizontal direction after the horizontal clamp screw has been tightened.

Good quality builder's levels and transit-levels have an adjoining *horizontal circle* and a *horizontal vernier scale* to measure horizontal angles. The horizontal circle is moved manually, but does not move as the telescope is rotated. The horizontal circle is divided into four *quadrants,* each quadrant indicating 0° to 90°. The horizontal vernier scale is attached to the leveling instrument's frame and moves along the inside of the horizontal circle as the telescope is turned. The horizontal vernier scale is commonly graduated in 15 minute increments, but 5 minute increments are used for more precise readings. See Figure E-2.

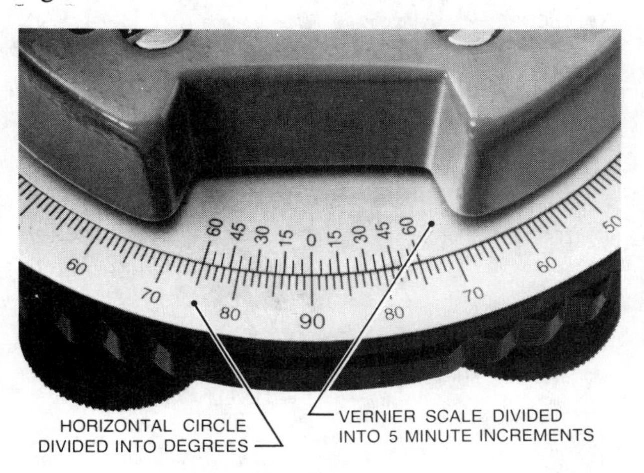

Figure E-2. The horizontal circle and vernier scale are used to measure horizontal angles.

THE TRANSIT-LEVEL

A transit-level can perform all of the functions of a builder's level. In addition, the telescope of the transit-level can be tilted vertically to plumb and align vertical surfaces. Many of the parts of a transit-level, such as the telescope, spirit level, leveling screws, and horizontal adjustments, are similar to a builder's level. In addition to these parts, a *telescope lock lever, vertical clamp screw,* and *vertical tangent screw* are used to make and maintain vertical adjustments. The telescope lock lever holds the telescope in the correct position for horizontal leveling. When the telescope lock lever is released, the telescope can be adjusted vertically. When the telescope is adjusted to its desired position, the vertical clamp screw is tightened. The vertical tangent screw is then used to make fine vertical adjustments. See Figure E-3.

Most transit-levels are also equipped with a *vertical arc* and *vertical vernier scale.* The vertical arc is used to measure vertical angles and is graduated from 0° to 45° in two directions. The vertical arc moves as the telescope is adjusted vertically. The vertical vernier scale is attached to the transit-level's frame and allows the transit-level to be adjusted in 5 minute increments. Other models of transit-levels have a fixed pointer that indicates the angle to which the telescope has been set. The fixed pointer gives the measurement to the nearest whole degree. See Figure E-4.

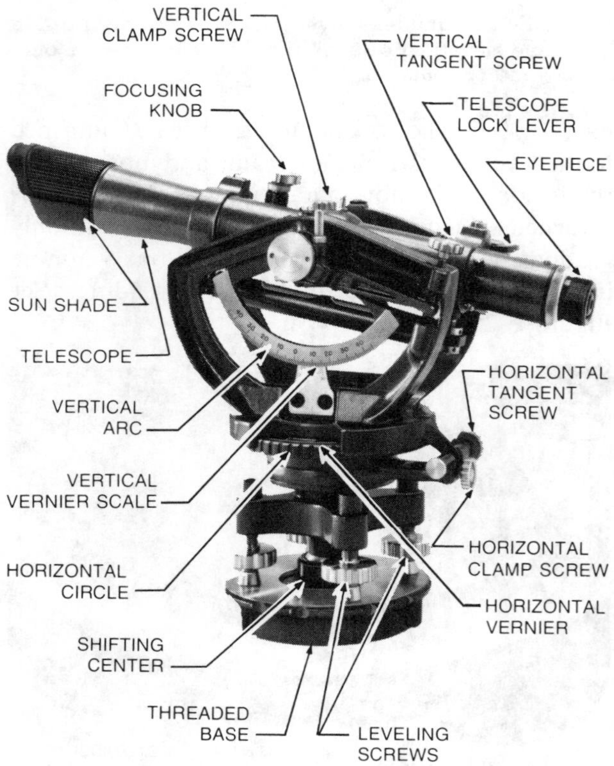

VERTICAL CLAMP SCREW
VERTICAL TANGENT SCREW
FOCUSING KNOB
TELESCOPE LOCK LEVER
EYEPIECE
SUN SHADE
TELESCOPE
HORIZONTAL TANGENT SCREW
VERTICAL ARC
VERTICAL VERNIER SCALE
HORIZONTAL CLAMP SCREW
HORIZONTAL CIRCLE
HORIZONTAL VERNIER
SHIFTING CENTER
THREADED BASE
LEVELING SCREWS

David White Instruments

Figure E-3. A transit-level is used to make vertical and horizontal angular measurements.

Figure E-4. A fixed pointer may be used to indicate an angular measurement on a vertical arc. Measurements are made to the nearest degree.

Tripods

A builder's level or transit-level must be mounted on a tripod. See Figure E-5. Threaded or cup assemblies are commonly used to fasten the leveling

Figure E-5. A leveling instrument is mounted on a tripod. The leveling head must be kept as level as possible.

instrument to the tripod head. A leveling instrument with a threaded base screws into a threaded tripod head. If the tripod head has a cup assembly, a threaded mounting stud at the base of the leveling instrument is screwed into the cup assembly.

When setting up the tripod, position the legs about 3′ apart and keep the tripod head as level as possible. Push the legs firmly into the ground and tighten the wing nuts. On sloping ground, force the uphill leg of the tripod into the slope. A triangular wood base frame helps prevent movement when placing a tripod over concrete or other smooth, hard surfaces. See Figure E-6.

Using the Builder's Level and Transit-Level

The builder's level and transit-level must be leveled in all positions over the base to ensure accurate measurements. When leveling a transit-level, the telescope lock lever must be in the closed position. The leveling screws are adjusted so that the telescope is level when it rotates on top of the base. See Figure E-7. Firm contact must be maintained between the leveling screws and the base.

When adjusting the leveling screws, turn the leveling screws equal amounts at the same time and in opposite directions. The direction in which the left thumb moves when turning the screws is the direction that the bubble moves. The leveling screws should not be overtightened. Repeated overtightening will damage the leveling instrument. See Figure E-8.

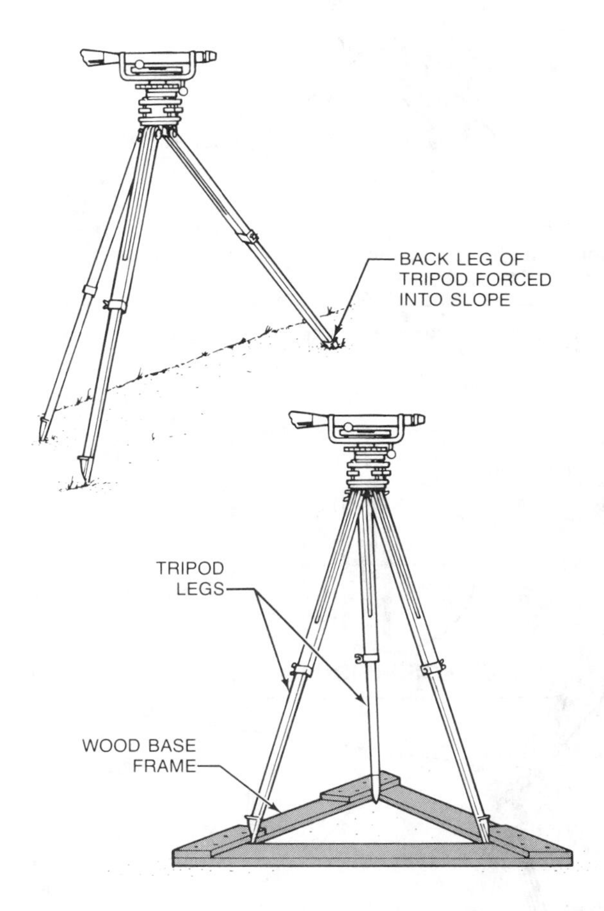

Figure E-6. The back leg of a tripod should be forced into the ground when setting up a leveling instrument on a slope. A triangular wood base frame can be fabricated when setting up a leveling instrument on a hard, smooth surface.

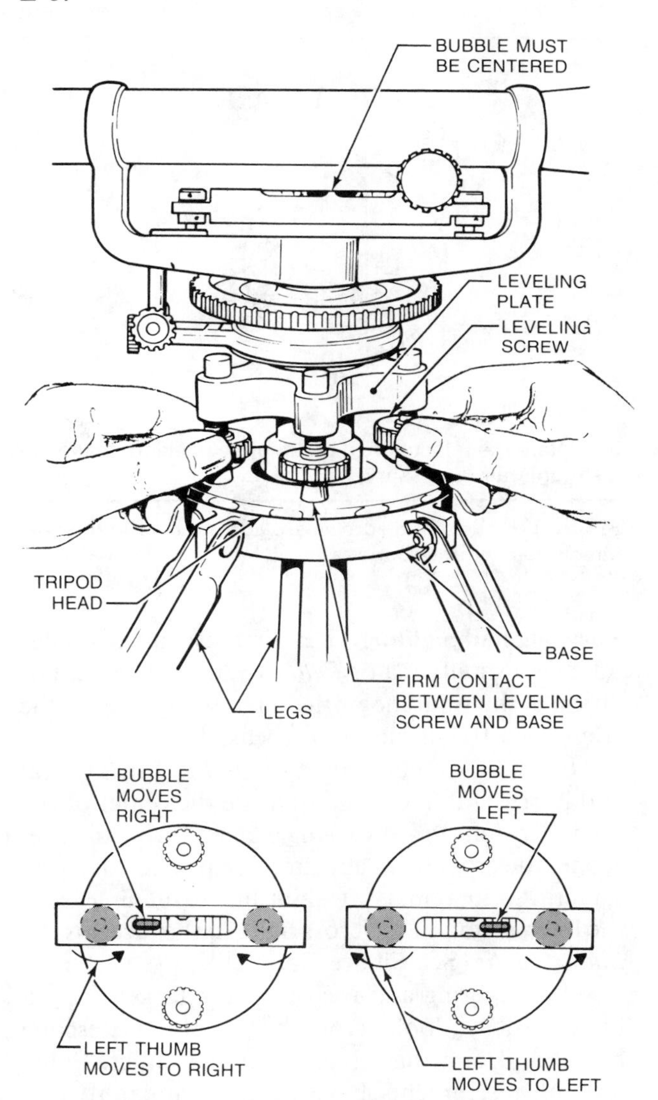

Figure E-7. Leveling screws must be adjusted to level the instrument. The spirit level bubble moves in the direction the left thumb moves.

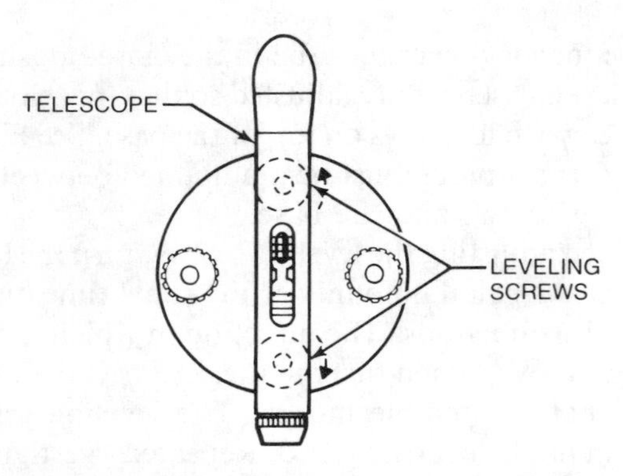

1. Position the telescope directly over a pair of leveling screws. Turn the two screws at the same time and in opposite directions until the spirit level bubble is centered.

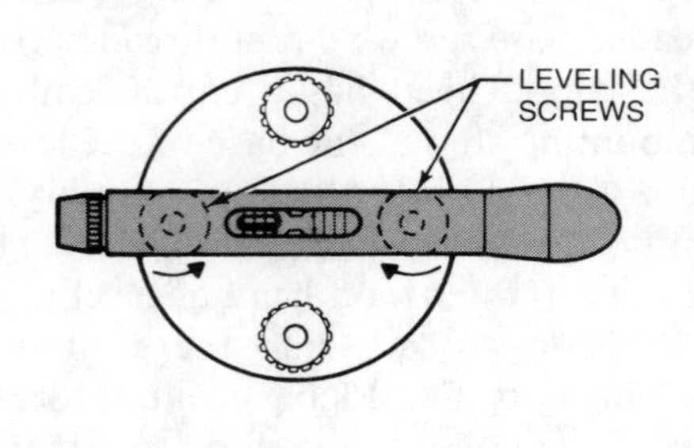

2. Rotate the telescope 90° and level the telescope.

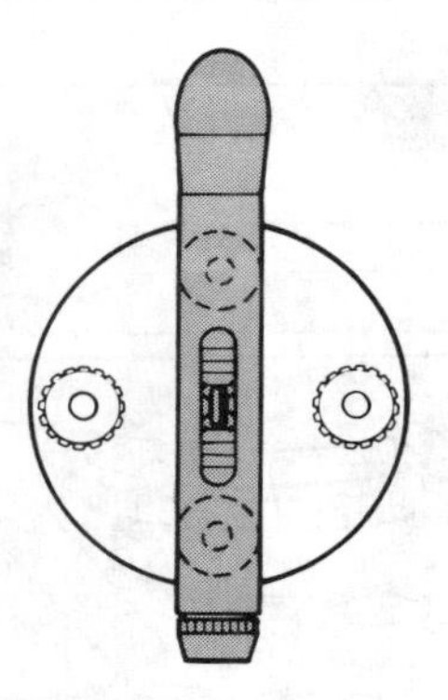

3. Rotate the telescope to the original position. Make adjustments as necessary.

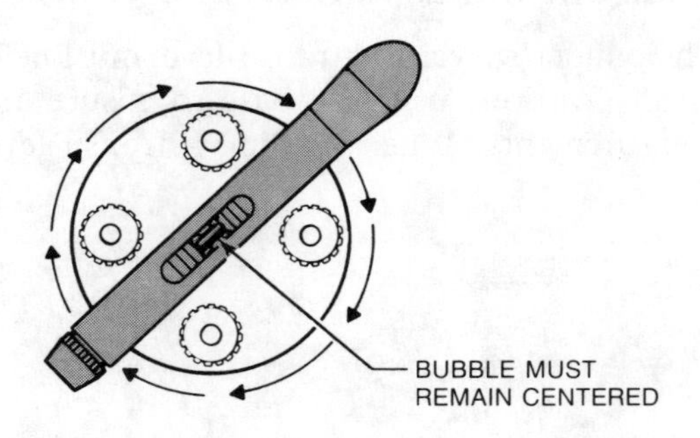

4. Rotate the telescope over all four leveling screws making sure the spirit level bubble remains centered in all positions.

Figure E-8. Leveling screws are adjusted while the telescope is positioned over them. The telescope is rotated 360° to check that the telescope is level.

Focusing and Sighting. Leveling instruments focus on a very small area or *field of vision*. The field of vision is the total magnified area seen through the telescope after it has been focused.

The telescope contains *cross hairs* (fine vertical and horizontal lines) that indicate the center of the field of vision. When viewing a target through a telescope, the cross hairs appear to be projected on the target. An imaginary straight line extending from the intersection of the cross hairs to the target is the *line of sight*. See Figure E-9.

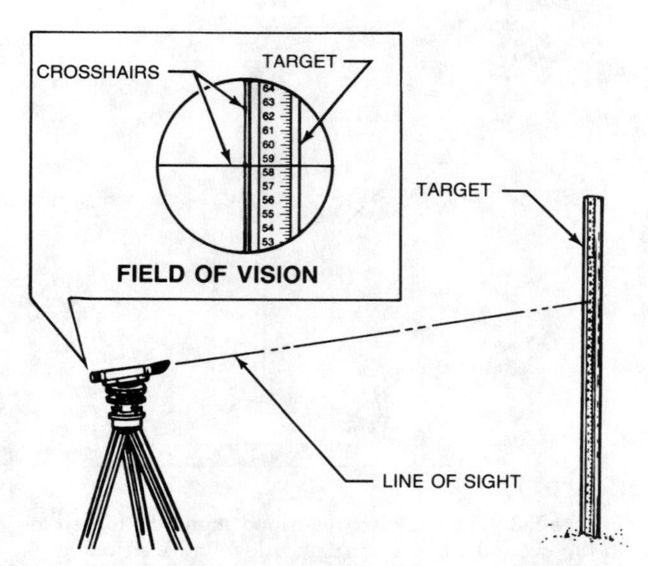

Figure E-9. Leveling instruments focus on a small field of vision. The line of sight extends from the intersection of the crosshairs to the target.

When sighting and focusing on a target using a builder's level, look across the top of the telescope barrel and aim the telescope at the target. Some leveling instruments have devices similar to gun sights to assist in aiming the telescope. Look through the telescope and turn the focusing knob until the target is sharp and clear. Move the telescope until the cross hairs are aligned with the target. Tighten the horizontal clamp screw and make final adjustments to the field of vision using the horizontal tangent screw.

When adjusting the telescope of a transit-level vertically, release the telescope lock lever that secures the telescope. Sight and focus the telescope on the target. Move the telescope until the cross hairs align with the target, and tighten the vertical clamp screw. Make final adjustments to the field of vision by turning the vertical tangent screw.

Leveling Rods

Leveling rods are made of wood, plastic, or aluminum. They commonly consist of two or three sections and are adjustable from 8 ′ to 14 ′ in length. When using a builder's level or transit-level, one person sights through the telescope while a second person holds the leveling rod at the desired location. See Figure E-10.

Large numbers and graduation marks on the leveling rod facilitate reading measurements on the rod. The foot measurements are red and the graduation marks and other figures are usually black. Movable metal targets may be used when sighting the telescope from a long distance. The center of the metal target aligns with the desired position on the leveling rod. See Figure E-11.

Figure E-10. A leveling rod is held vertically when being sighted. The hands should be positioned as not to obstruct the view.

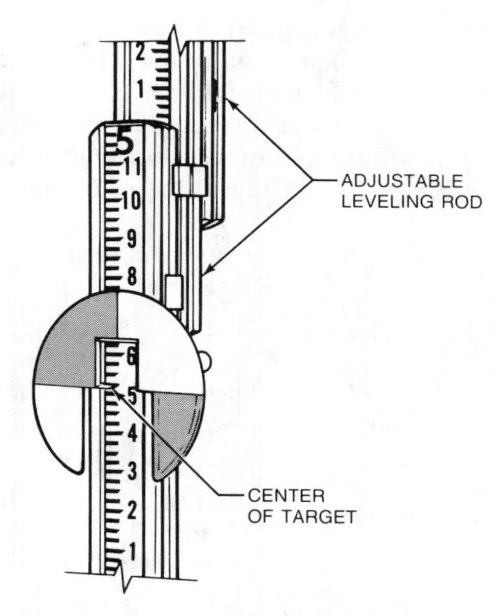

Figure E-11. A movable metal target is used to facilitate sighting over a long distance.

An *architect's rod* or *engineer's rod* is commonly used in construction. The architect's rod is graduated in feet, inches, and eighths of an inch and is used by carpenters and other construction workers. The engineer's rod is graduated in feet, and tenths and hundredths of a foot and is commonly used by surveyors and engineers. See Figure E-12.

A measuring tape or wood rule held against a unmarked wood rod may be used as a leveling rod when sighting over short distances. In some cases, an unmarked wood rod can be used as a leveling rod without using a measuring tape or wood rule. The line of sight is marked on the unmarked wood rod while it is positioned over an established reference point. The wood rod is moved to another location and the line of sight is marked on the rod. The distance between the two marks is the difference in elevation.

Verifying Grades and Elevations. A builder's level or transit-level is used to verify grade differences on a building site. The leveling instrument is positioned where all desired grade levels can be conveniently read through the telescope. The various grade levels are read and recorded. The larger the grade level reading is, the lower the actual grade level is. The difference in grade level between the highest and lowest grade levels is calculated by subtracting the smallest grade level reading from the largest grade level reading. See Figure E-13.

The builder's level or transit-level is used to establish and verify grade differences on steeply sloped lots. Stakes are positioned along the slope and at the top and bottom of the slope and driven flush with the surface. The leveling instrument is positioned between the first and second stakes and a rod reading is recorded at the first stake. The rod is positioned over the second stake and a rod reading is recorded. The leveling instrument is then positioned between the second and third stakes, and rod readings are recorded for the second and third stakes. This procedure is repeated until a rod reading is recorded at the top of the slope. Calculations are then performed to determine the total grade difference between the top and bottom of the slope. See Figure E-14.

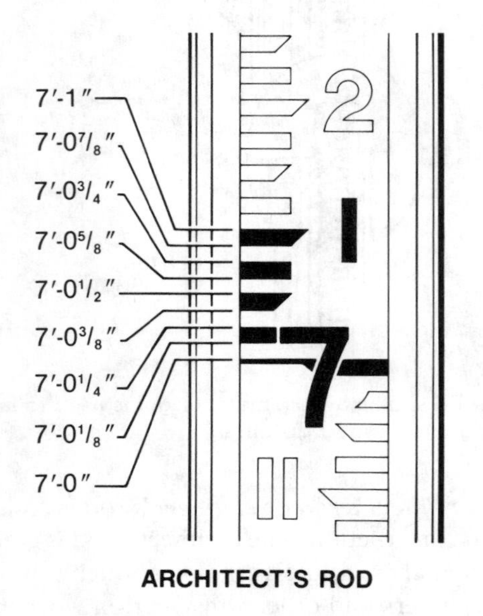

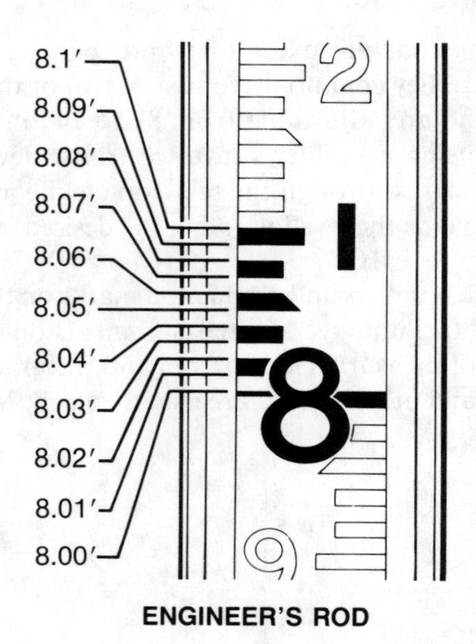

Figure E-12. An architect's and engineer's rod are commonly used with leveling instruments. The architect's rod is graduated in eighths of an inch and the engineer's rod is graduated in hundredths of an inch.

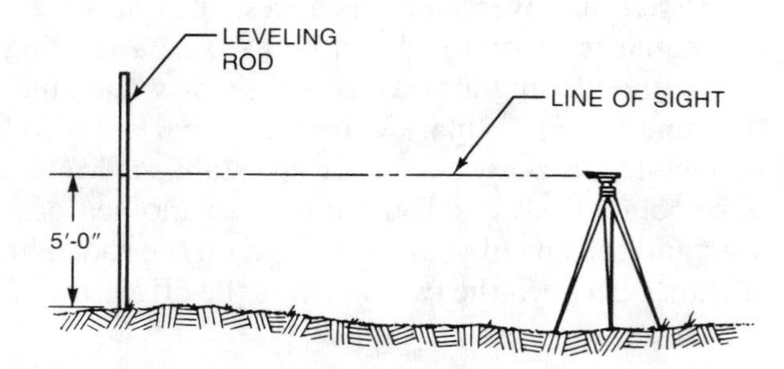

1. Set the leveling instrument up at a convenient location on the building site. Take a line of sight reading of the leveling rod held at a specific point. In this example, the line of sight reading is 5'-0".

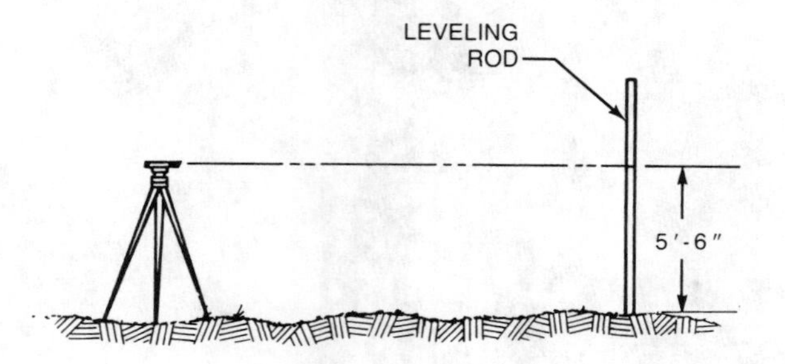

2. Move the leveling rod to another point and take a line of sight reading (5'-6"). Subtract the smallest reading from the larger reading to determine the grade difference (5'-6" – 5'-0" = 6").

Figure E-13. Grade differences are determined by subtracting the smallest grade level from the largest grade level.

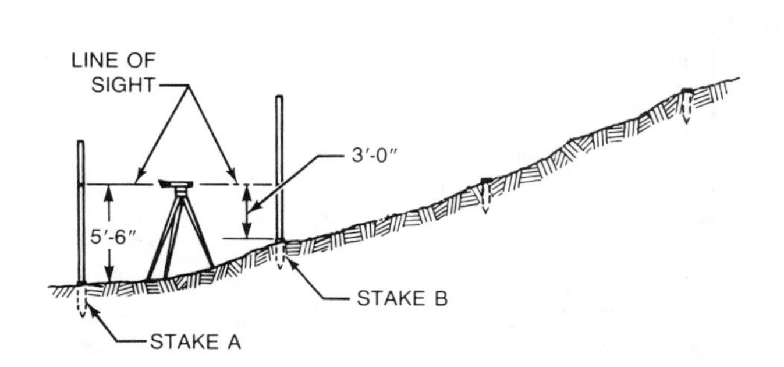

1. Drive stakes at equal intervals along the slope. Set up a leveling instrument and take a line of sight reading at stake A. (In this example the line of sight reading is 5'-6".) Take a line of sight reading at stake B (3'-0").

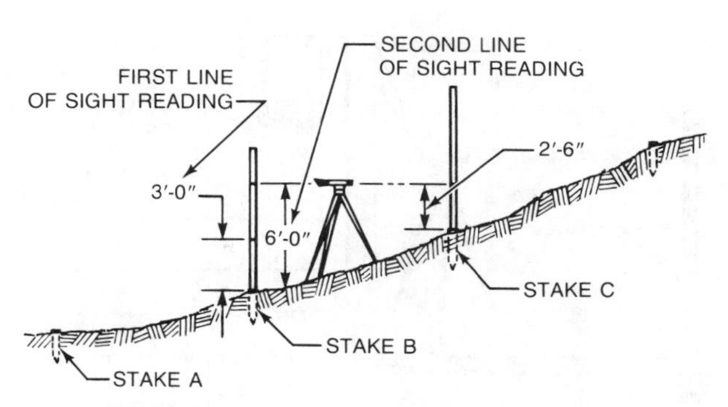

2. Move the leveling instrument to a point above stake B. Take a second line of sight reading at stake B (6'-0"). Subtract the first line of sight reading from the second reading (6'-0" − 3'-0" = 3'-0"). Add this figure to the line of sight reading at stake A (5'-6" + 3'-0" = 8'-6"). Take a line of sight reading at stake C (2'-6").

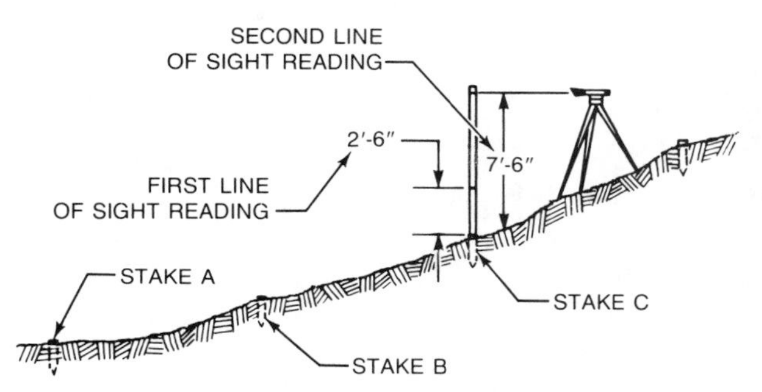

3. Move the leveling instrument to a point above stake C. Take a second line of sight reading at stake C (7'-6"). Subtract the first line of sight reading at stake C from the second reading (7'-6" − 2'-6" = 5'-0"). Add this figure to the sum of the line of sight readings in step 2 (5'-0" + 8'-6" = 13'-6").

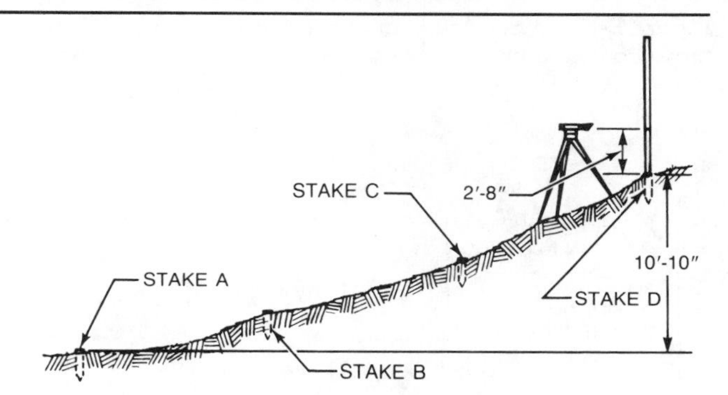

4. Take a line of sight reading at stake D (2'-8"). Subtract this figure from the sum in step 3 to determine the difference in grade level between stakes A and D (13'-6" − 2'-8" = 10'-10").

Figure E-14. The builder's level or transit-level is used to establish elevations on a steeply sloped lot.

Elevations are established for foundation footings and walls by using two methods: with an unmarked wood rod and rule, and with an unmarked wood rod. When using a unmarked wood rod and rule, the difference in elevation is determined by subtracting an initial reading from a second reading. For example, if the top of a footing form stake is to be located 6" below the bench mark or established reference point, an initial reading is taken at the bench mark or established reference point. The unmarked wood rod and rule are positioned on top of the footing form stake and a second reading is taken. The footing form stake is driven until the difference between the initial reading and the second reading is equal to the desired difference in elevation. See Figure E-15.

When using an unmarked wood rod, an initial reading is marked on the rod. The difference in elevation (6") is then measured and marked on the wood rod. The wood rod is positioned on top of the footing form stake and the stake is driven until the second mark aligns with the line of sight. See Figure E-16.

A leveling instrument and an unmarked wood rod are also used to establish level points over a long distance, such as corner stakes for foundation footing forms. After the desired elevation is established, an unmarked wood rod is positioned against the stake with the bottom of the wood rod even with the elevation mark. The wood rod is moved to another stake, and the elevation mark is aligned with the line of sight. The second stake is then marked

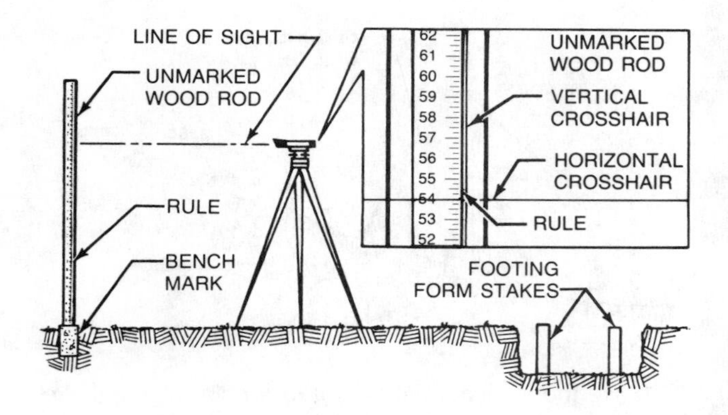

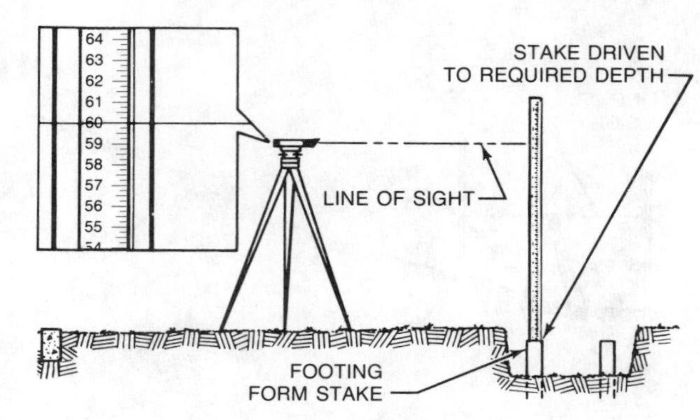

1. Hold an unmarked wood rod over a bench mark and take a line of sight reading on the rule.

2. Hold an unmarked wood rod on top of a stake. Drive the stake until the desired reading aligns with the horizontal cross hair. In this example, the top of the footing form stake is 6″ lower than the bench mark (60″ − 54″ = 6″).

Figure E-15. When using an unmarked wood rod and rule to determine elevations, the initial reading is subtracted from the second reading to obtain the difference in elevation.

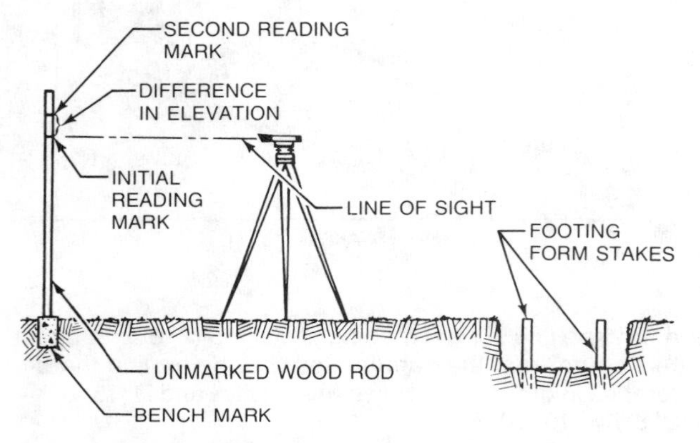

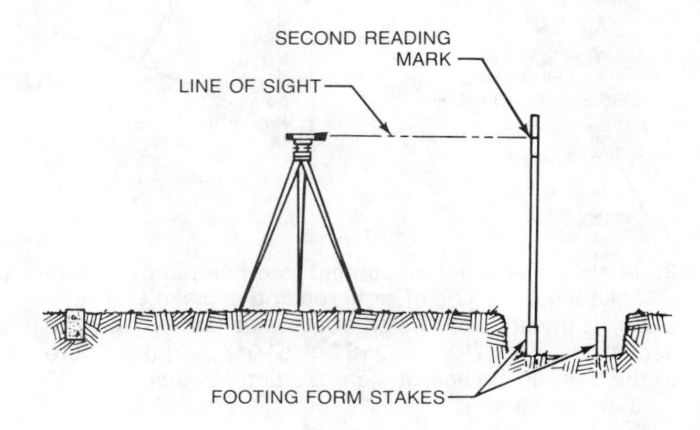

1. Hold the unmarked wood rod over the bench mark and mark the line of sight on the unmarked wood rod. Lay out the difference in elevation on the unmarked wood rod to indicate the second reading.

2. Place the unmarked wood rod on top of a stake. Drive the stake until the second reading mark aligns with the line of sight.

Figure E-16. When using an unmarked wood rod as a leveling rod, the elevations are measured and marked. The difference between the two marks is the difference in elevation.

at the bottom of the wood rod to indicate the same elevation as the initial stake. See Figure E-17.

Laying Out Right Angles. Buildings are commonly constructed in the shape of a rectangle or variation of a rectangle. The sides of a rectangle are at *right angles* (90°) to each other. A transit-level is the most efficient instrument to use to lay out the building lines for foundations of rectangular, L-, or T-shaped buildings. See Figure E-18. When squaring building lines with a transit-level, the instrument is set up and leveled over an established reference point, such as a nail driven into the top of a stake indicating the corner of a lot or building. A *plumb bob* is attached to a line that is hooked to the bottom of the leveling instrument. The plumb bob ensures that the transit-level is in its exact location when laying out the building lines. See Figure E-19. A procedure for setting up the transit-level follows.

1. Spread the tripod legs. Position the tripod so the tripod head is directly over an established reference point.

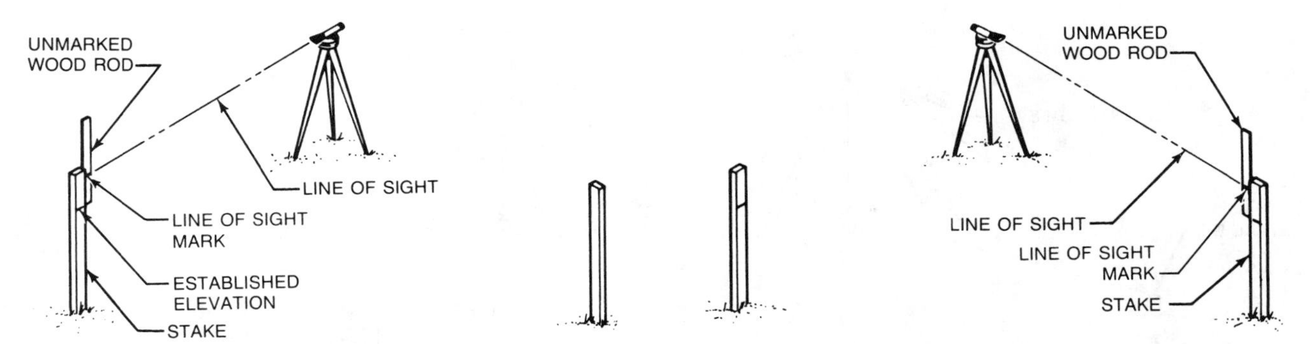

1. Establish the desired elevation on a stake. Position an unmarked wood rod along the stake, aligning the bottom with the mark on the stake. Mark the line of sight reading on the unmarked wood rod.

2. Move the unmarked wood rod to another stake and position it so the line of sight mark aligns with the line of sight of the leveling instrument. Mark the stake at the bottom of the unmarked wood rod.

Figure E-17. An unmarked wood rod and a leveling instrument are used to establish elevations over a long distance.

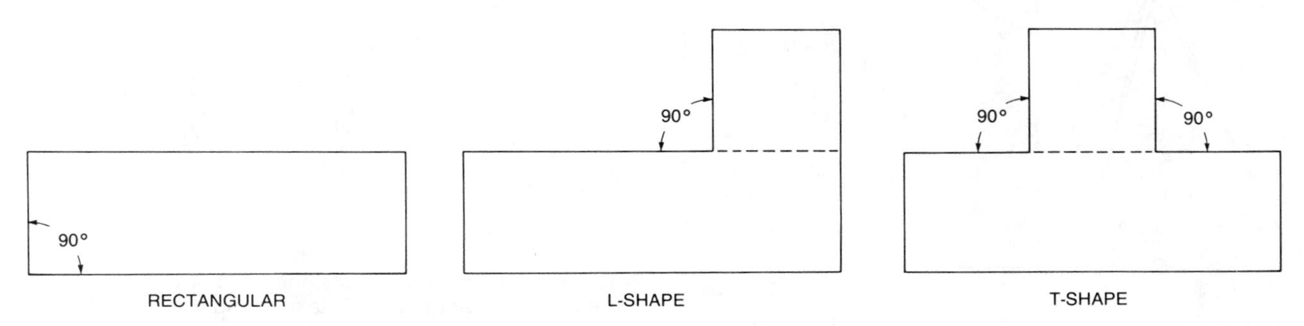

PLAN VIEW OF COMMON BUILDING SHAPES

Figure E-18. Building shapes commonly consist of lines that are at right angles to each other.

2. Attach the transit-level to the tripod head. Secure a line to the plumb bob hook at the bottom of the instrument. Adjust the line length so the plumb bob is approximately $\frac{1}{4}''$ above the reference point.
3. Roughly level the transit-level. Do not tighten the leveling screws.
4. Move the transit-level on the shifting center until the plumb bob aligns exactly with the reference point. If the transit-level does not have a shifting center, carefully shift the tripod until the plumb bob is in the correct position.
5. Adjust and tighten the leveling screws so the transit-level is level in all directions. A right angle is laid out by sighting back to another established reference point, such as the second corner of a lot or building. The instrument is then rotated 90° to establish a line at a right angle to the first line. See Figure E-20.

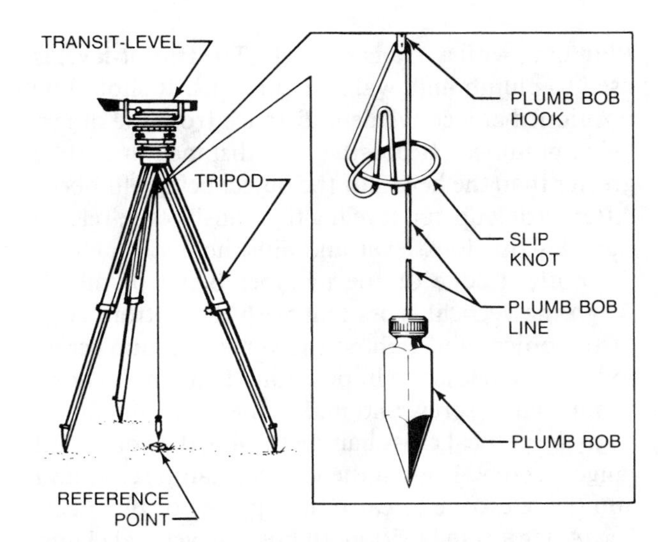

Figure E-19. A transit-level is placed directly over a reference point when establishing elevations. A slip knot is used to attach the plumb bob to the plumb bob hook to allow it to be raised or lowered.

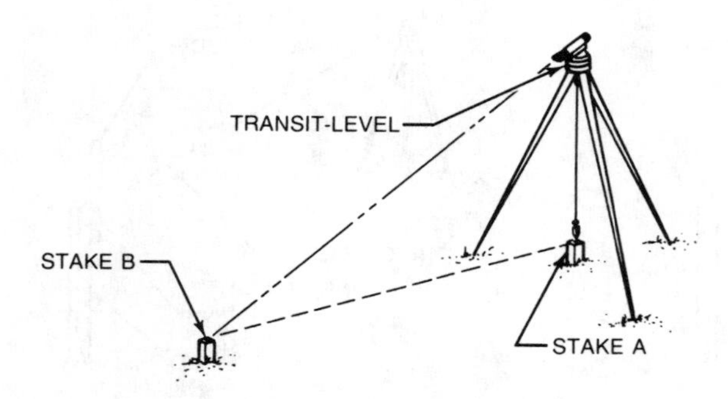

1. Set up the transit-level over stake A. Sight through the telescope and align the vertical crosshair with stake B.

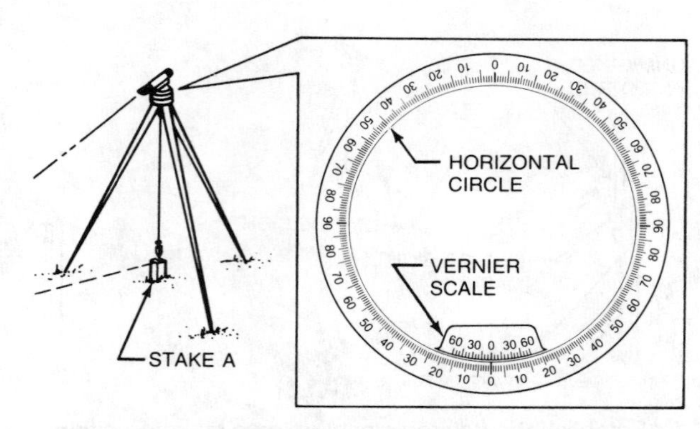

2. Turn the horizontal circle of the transit-level to align one of the "0" readings on the circle with the "0" on the vernier scale.

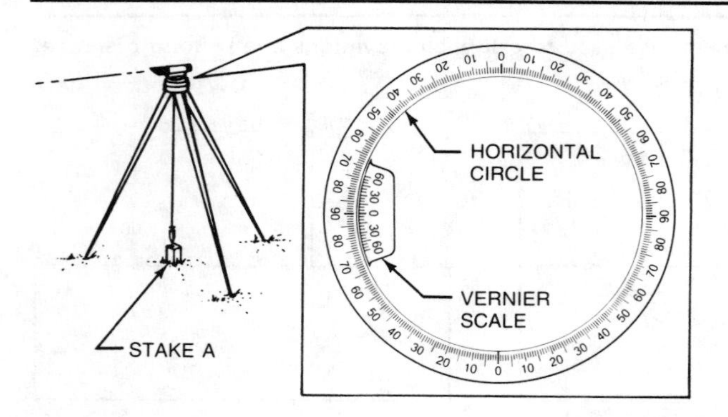

3. Rotate the transit-level until the "0" index on the vernier scale aligns with the 90° index on the horizontal circle.

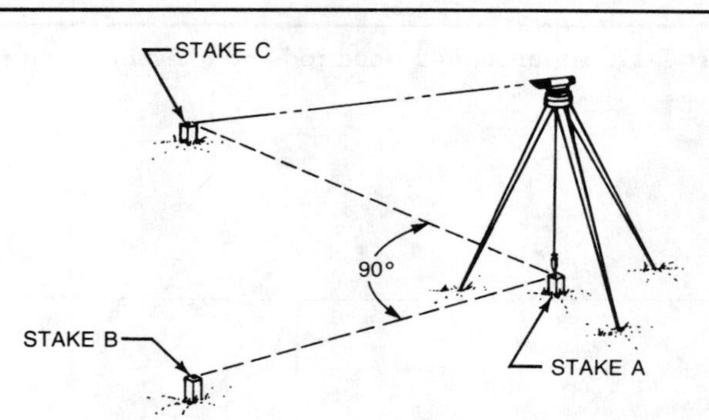

4. Measure the required distance and drive stake C. Aim the telescope at the top of stake C and drive a nail where the vertical crosshair and the exact measurement intersect. A right angle is formed between lines A-B and A-C.

Figure E-20. A transit-level is used to lay out right angles on a job site.

Plumbing with a Transit-Level. The transit-level is used to plumb high walls or columns. It should be positioned at a convenient distance from the object being plumbed. If possible, the distance should be greater than the height of the object being plumbed. After setting up and leveling the transit-level, release the telescope lock lever and aim the instrument at the bottom edge of the member being plumbed. Align the vertical cross hair with the bottom edge of the object and tighten the vertical clamp screw to hold the telescope in position. Tighten the horizontal clamp screw and make final adjustments to align the vertical cross hair by turning the horizontal tangent screw. Loosen the vertical clamp screw and aim the telescope toward the top edge of the member. After it is in position, tighten the vertical clamp screw and sight through the telescope. The member is plumb when the top edge of the member aligns with the vertical cross hair.

Establishing Points in a Straight Line. The transit-level is used to establish points in a straight line, such as a row of form stakes or piers. The transit-level is set up over one of the end stakes and aimed toward another end stake. The telescope is sighted and focused, and the vertical cross hair is aligned with the second end stake. Intermediate stakes are positioned and aligned with the vertical cross hair. The stakes are driven to the required depth and checked again for accurate layout. See Figure E-21.

LASER TRANSIT-LEVEL

The laser transit-level is a leveling instrument increasingly used in construction work for leveling and plumbing. It performs most of the operations of a transit-level, but only requires one person to set up and operate it.

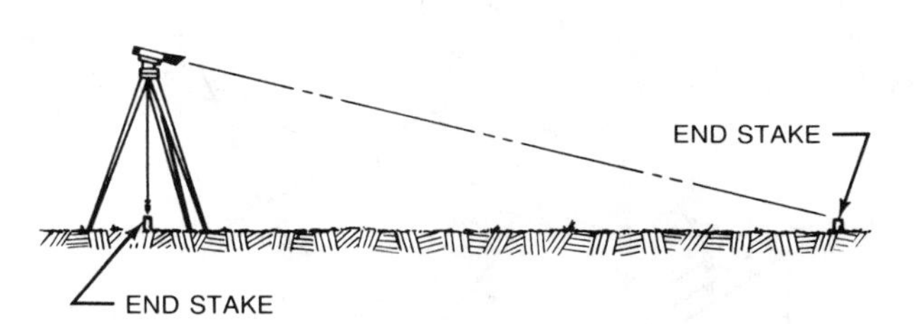

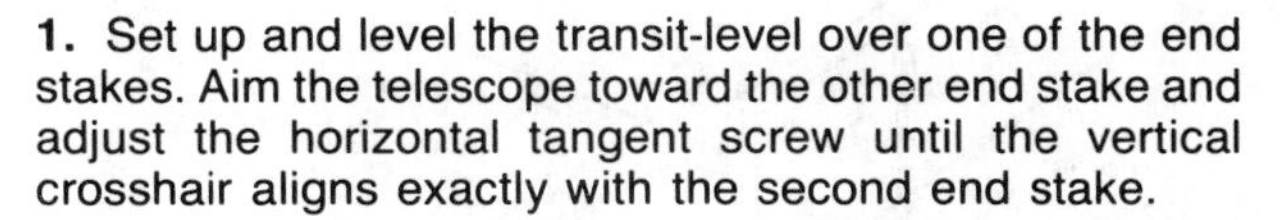

1. Set up and level the transit-level over one of the end stakes. Aim the telescope toward the other end stake and adjust the horizontal tangent screw until the vertical crosshair aligns exactly with the second end stake.

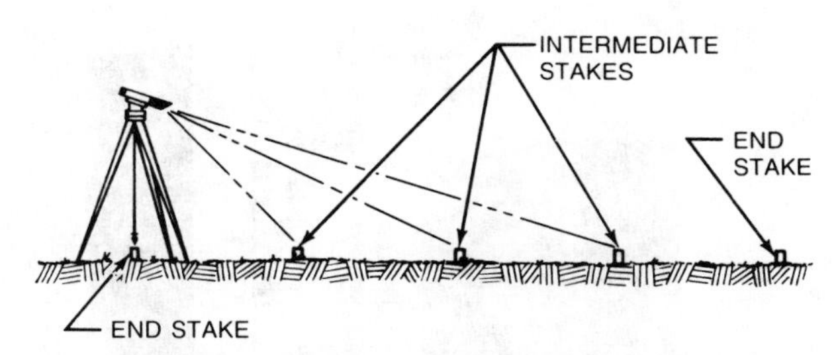

2. Position an intermediate stake and focus the telescope on the stake. Move the stake until the edge aligns with the vertical crosshair. Drive the stake to the required depth. Repeat the process until all stakes are driven.

Figure E-21. A transit-level is used to establish reference points in a straight line.

The main parts of the laser transit-level are the *rotating head, laser barrel, gimbel,* and *leveling screws*. The rotating head revolves at a maximum rate of 360 revolutions per minute, causing a beam projection. The laser barrel contains a helium-neon sealed-in tube that activates the laser beam. See Figure E-22. The laser beam is a concentrated red beam of light approximately 3/8″ in diameter. The instrument can be mounted on a tripod or stationary column with the gimbel. The leveling screws level the laser transit-level for accurate measurements. Some models of laser transit-levels have a self-leveling mechanism that keeps the instrument at its original setting despite changes in thermal conditions or minor jolts.

The laser beam is directed toward a *sensor* when grade levels or elevations are measured. A sensor is a light-sensitive target attached to a leveling rod that lights up when properly aligned with the laser beam. See Figure E-23. A laser transit-level uses various targets or sensors for different operations. A battery-operated sensor is commonly used for concrete construction layout. The sensor is attached to a leveling rod with brackets. See Figure E-24.

ROTATING HEAD
VARIABLE SPEED CONTROL
LASER BARREL
GIMBEL
LEVELING SCREWS

Laser Alignment, Inc.

Figure E-22. The rotating head of a laser transit-level revolves at a maximum rate of 360 revolutions per minute. The rate that the head rotates is adjusted using the variable speed control.

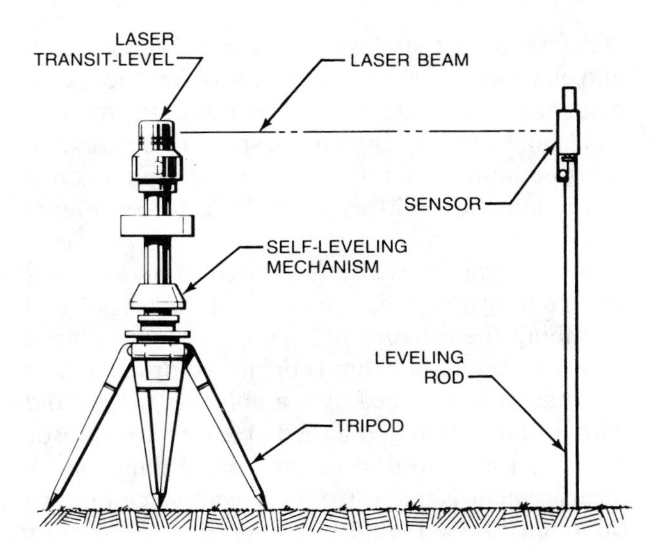

Figure E-23. A laser transit-level emits a concentrated beam of light.

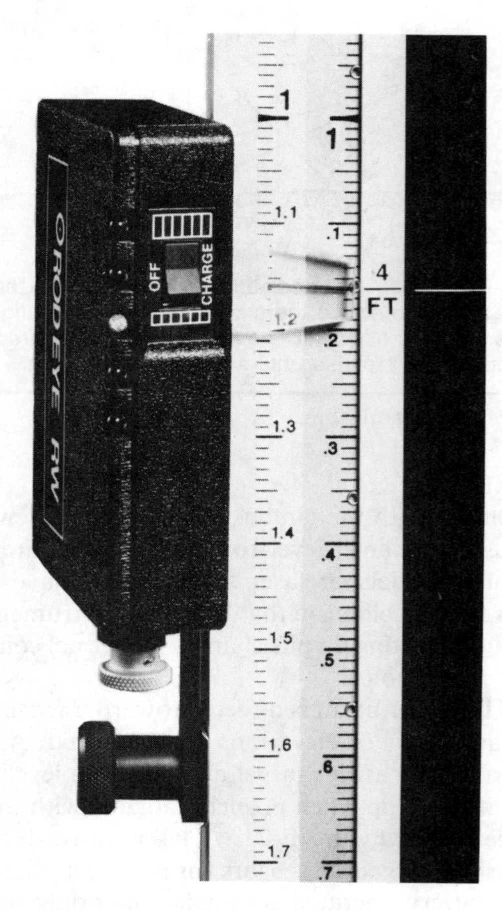

Figure E-24. A battery-operated sensor is attached to a leveling rod. When the laser beam is directly aligned with the sensor's eye, the indicator light comes on.

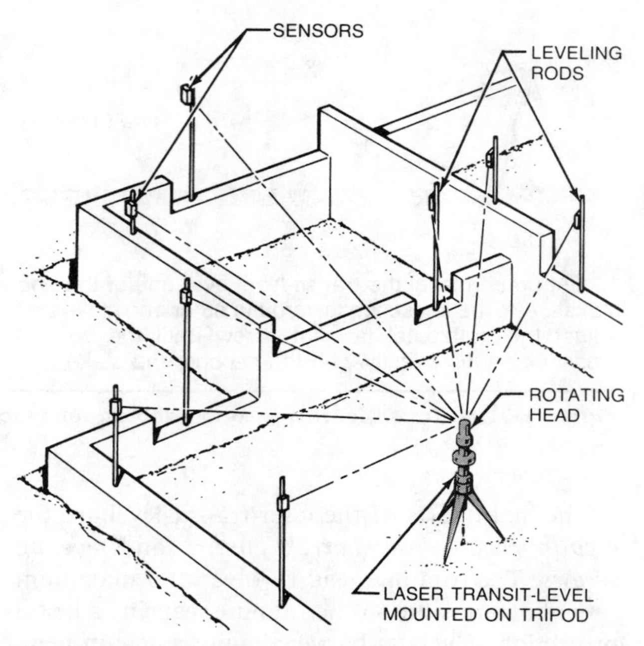

Figure E-25. A laser transit-level is used to establish the elevations at various points on the job site.

A laser transit-level is used for establishing grades and elevations over long distances. After the instrument has been positioned and adjusted, the rotating head can be set to a maximum speed of 360 revolutions per minute. Grade readings are then made at any point where sensors are positioned. See Figure E-25.

A laser transit-level may be used to plumb walls and columns by removing the rotating head and plumbing the instrument. With the rotating head removed, the laser beam is projected upward. The instrument is plumbed over a point on the ground with a plumb bob and aimed at a target or sensor attached to the wall or column being plumbed. A measurement is taken from the wall or column to the point of the plumb bob and from the wall or column to the sensor. When the two measurements are equal, the wall or column is plumb. See Figure E-26.

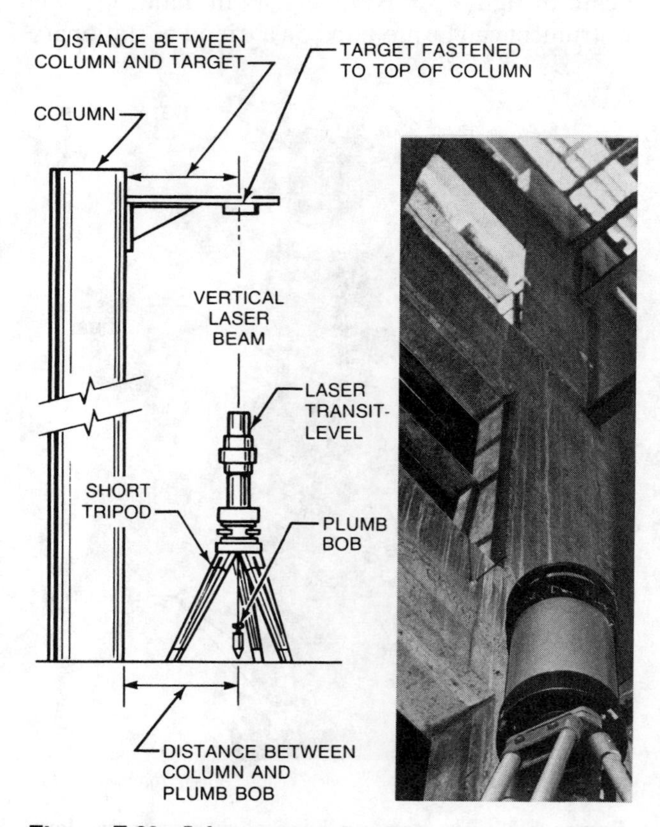

Figure E-26. A laser transit-level is used for plumbing walls and columns. When the distance between the column and target is equal to the distance between the column and plumb bob, the column is plumb.

Glossary

A

Abutment. End structure supporting beams, girders, and deck of a bridge.

abutment

Accelerator admixture. Substance that increases the rate of hydration, shortens setting time, or increases strength development.

American Concrete Institute (ACI). Association that sets standards for concrete construction.

Actual size. Thickness and width of lumber after shrinkage resulting from surfacing and seasoning.

Adjustable wood shore. Two-piece shore consisting of overlapping wood members that are secured in place with post clamps.

Admixture. Material other than cement, aggregate, and water that is added to a batch of concrete immediately before or during the mixing process.

Aggregate. Granular material, such as sand and gravel, used with cement to produce mortar or concrete.

Agitating truck. Truck with an agitator to transport freshly mixed concrete from batch plant to job site.

Air-entrained concrete. Concrete containing an admixture that produces microscopic air bubbles in the concrete. Used to improve workability and freeze resistance of concrete.

Anchor bolt. Bolt embedded in the top of a foundation wall or floor slab to secure a sill plate.

Anchor clip. Strap-type device embedded in the top of a concrete foundation wall that is folded over and nailed to the sill plate.

Architectural concrete. Permanently exposed concrete surface that features special designs or patterns such as textured finishes, and ribbed and fluted surfaces.

Auger. Earth-boring device attached to rig. Used to bore holes in the soil for deep piers or piles.

B

Backhoe. Excavating equipment used for small loading jobs and digging trenches for foundations.

Base course. Layer of selected material, usually gravel, placed beneath a concrete slab. Its main purpose is to control the capillary rise of water to the slab bed.

Batch. Quantity of mortar or concrete mixed at one time.

Batch plant. Location where concrete is mixed to specifications.

Batterboard. Horizontal board nailed to stakes driven into the ground. Used to hold a line or wire identifying property lines or building lines.

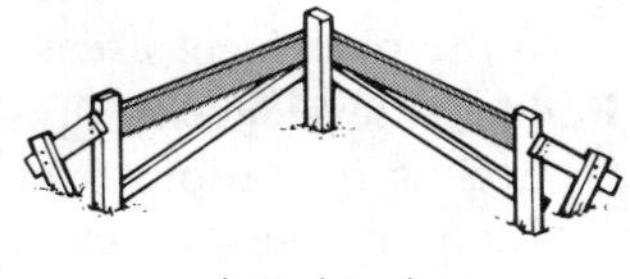
batterboard

Battered foundation. Foundation wall consisting of a vertical outside surface and an inclined inside surface that provides a wider base at the bottom of the wall.

Beam. Horizontal member that supports a bending load over a span, such as from column to column.

Beam bottom form. Bottom soffit of a beam form resting directly over the shore system.

Beam pocket. **1.** Opening in a concrete wall or column in which a beam rests. **2.** Opening in a column or girder form in which an intersecting beam form frames.

Beam side form. Vertical member of a beam form. Beam side is nailed against or rests on top of the beam bottom.

Bearing capacity. Maximum pressure that soil or other material can withstand without failure or a great amount of settlement.

Bearing pile. Pile that penetrates through layers of unstable soil until it reaches firm bearing soil.

Belled caisson. Concrete caisson flared at its bottom to provide a greater bearing area.

Bench mark. Point of reference for grades and elevations on a construction site.

Bleeding. Segregation in a concrete mix in which water rises to the surface of freshly placed concrete.

Bored caisson. Caisson constructed by placing a metal casing into a hole that is bored into the earth and filled with concrete after it is in position.

Brace. Diagonal or horizontal wood or metal member used to stiffen and support various parts of a form.

Bridge deck. The slab of the bridge superstructure that supports the traffic load.

buck

Buck. Frame placed inside a form to provide an opening for a door or window after the concrete has set.

Bucket. Large metal container into which concrete is discharged. The bucket is then raised by crane to the placement area.

Buggy. Manual or motor-driven cart used to move small amounts of concrete from hoppers or mixers to the placement area.

Builder's level. Telescope-like instrument used for leveling operations over long distances and establishing grades and elevations.

Building site. Location where construction occurs.

Built-in-place form. Form assembly built entirely in place.

Bulkhead. Vertical partition set in the formwork to stop fresh concrete from entering another area of the form.

Bulldozer. Earth-moving equipment used to start excavations and strip rocks and topsoil at the surface of the building site.

C

Caisson. Large cylindrical or rectangular casing placed in the ground and filled with concrete.

Capillary action. Physical process occurring in soil causing water and water vapor to rise from the water table toward the earth's surface.

capital

Capital. Flared section at the top of a concrete column that supports the floor slab above.

Casting bed. Base and support for casting precast structural members.

Cast-in-place concrete. Concrete placed in forms where it is required to set as part of the structure.

Catch basin. Area excavated and filled with gravel that receives surface water runoff. Also called *dry well.*

Cement. Ingredient that binds the sand and aggregate together in concrete after water is added.

Chamfer strip. Piece of stock placed in an inside corner of a form to produce a beveled edge.

Chute. Trough-like device used to move concrete from a higher point to a lower point.

Clay. Fine-grained, natural mineral that is plastic when moist and hard and brittle when dry.

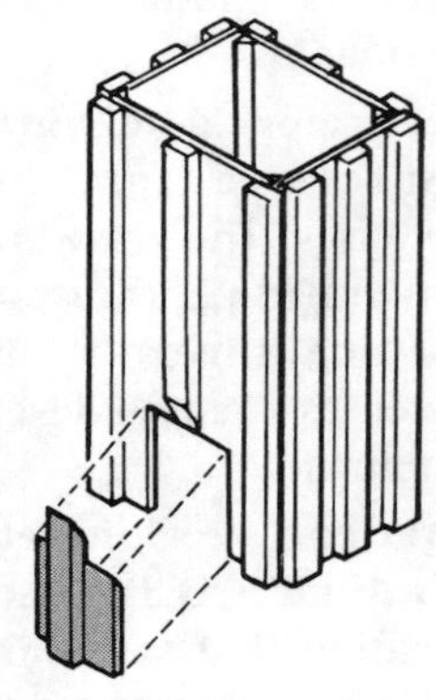
cleanout

Cleanout. Opening in the bottom of a form wall or column form that allows debris to be removed before the concrete is placed.

Climbing form. Large panel or ganged panel form that is lifted vertically for succeeding lifts.

Coarse aggregate. Crushed stone retained on a U.S. Standard #4 (4.75 mm) sieve.

Coarse-grained soil. Soil in which large particles such as sand and gravel are predominant.

Cohesive soil. Soil composed of very small particles such as silt and clay.

Coil tie. Internal disconnecting tie with external bolts that screw into an internal device consisting of metal struts with helical coils at each end.

Cold joint. Joint occurring where fresh concrete is placed adjacent to a previous placement of concrete that has already set.

Column. Vertical member supporting beams, girders, and/or floor slabs.

Column clamp. Device to hold column form sides together.

Common nail. Flat-headed nail used where appearance is not important, such as in form construction.

Compression. Stress caused by pushing together or a crushing force.

Compression test. Test conducted on a specimen of concrete to determine its compressive strength.

Compressive strength. Maximum resistance of a concrete or mortar specimen to vertical loads.

Concrete. Material consisting of aggregate combined with a binding medium of portland cement and water.

Concrete joist. Narrow, closely spaced beam that supports a floor or roof slab.

Concrete mix. Proportion of cement, sand, and aggregate that is combined with water to produce concrete.

Consistency. Degree of plasticity of fresh concrete and its ability to flow at the time it is being placed in forms.

Consolidation. Process of working fresh concrete so that a closer arrangement of particles is created and the number of voids is decreased or eliminated.

Construction joint. A cold joint between two adjacent placements of concrete.

Contour line. Line drawn on a survey or plot plan that passes through points having the same elevation.

Control joint. Shallow groove in the surface of the concrete that controls and confines cracking of the concrete resulting from expansion and contraction. Also called *relief* or *contraction joint.*

corner tie

Corner tie. Wood kicker or metal device used to brace the corners of forms.

Course. Horizontal layer of concrete. Several courses of concrete make up a *lift.*

Crawl space foundation. Low foundation featuring a narrow accessible space between the first floor joist and the ground.

Curing. Process of maintaining concrete moisture content and temperature in its early stages to allow desired properties to develop.

D

Dead load. Total weight of the superstructure of a building, including all construction materials.

Deck form. Form upon which concrete for a floor or roof slab is placed.

Decking. Sheathing material used for a deck form.

Deflection. Amount of bending or distortion of a material due to direct pressure.

dome pan

Dome pan. Metal or plastic square form used in a two-way concrete joist system.

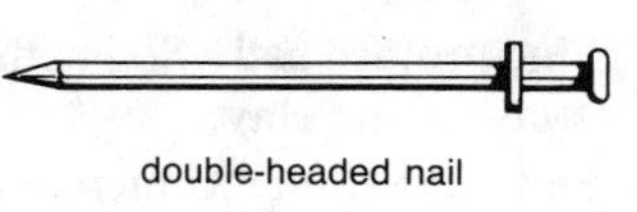
double-headed nail

Double-headed nail. Nail with two heads that permits easy removal. Commonly used in temporary construction. Also called *duplex nail.*

Double-post shore. Wood shore consisting of a head supported by two vertical posts.

Doubling-up walls. Trade term referring to placing the second (opposite) wall form.

Dowel. Deformed or plain round steel bar extending into adjoining portions of separately placed sections of concrete.

Drain pipe. Plastic or clay pipe used to convey water away from an area, such as a foundation wall.

Drain tile. Pipe constructed of clay, plastic, or concrete sections. Commonly placed alongside the foundation footing to carry water away from the foundation.

Drop panel. Thickened structural area over a column or capital that supports a flat slab floor.

E

Early strength. Concrete or mortar strength during the first 72 hours after placement.

Earth-formed footing. Concrete footing formed by a trench dug into firm and stable soil.

edge form

Edge form. Low wall form positioned around the perimeter of the placement area to contain fresh concrete. Used in flatwork, precast, and tilt-up construction.

Elephant trunk. Flexible tube extending from the bottom of a hopper.

Elevation. Grade level established to indicate vertical distance above or below a reference point.

Entrance platform. Low stoop or porch located at an entrance to a building.

Excavation. Removal of earth to allow for construction of a foundation.

Expansion joint. Separation between parts of a concrete structure. Used in locations where expansion and contraction forces are anticipated.

F

Fill. Soil or other material brought in from another location and deposited at the building site to raise the grade level.

Fine-grained soil. Soil composed of fine particles such as silt or clay.

Flat plate floor. Concrete floor slab system supported by columns tied directly to the floor.

Flat slab floor. Concrete floor slab system supported by columns and reinforced in two or more directions without beams or girders.

Flatwork. Construction of floor slabs, patios, sidewalks, or other horizontal surfaces.

Floating foundation. Thickened, reinforced slab placed monolithically with walls and/or footings that transmits the load over a large area. See *mat* and *raft foundations.*

Footing. Part of a concrete foundation that spreads and transmits the load over soil or piles.

Form. Temporary structure constructed to contain concrete while it sets.

Form anchor. Device embedded in concrete during placement to fasten formwork that will be constructed later.

Form hanger. Metal device used to support formwork suspended from structural framework such as steel or precast beams and girders.

Form liner. Wood or plastic material placed against the inside of a form wall to produce a textured or patterned finish or to absorb moisture.

Form tie. Device used to space and tie opposite form walls and prevent them from spreading.

form tie

Formwork. System of supporting freshly placed concrete.

Foundation. Part of a structure that rests on and extends into the ground and provides support for the load of the superstructure.

Free fall. Distance of descent of freshly placed concrete into forms without using a dropchute or other means of control.

Friction pile. Pile that receives its support from friction between the surrounding soil and exterior surface of the pile.

Front setback. Distance between the front of a structure and the front property line.

Front walk. Walk that extends from the front entrance of a building to a driveway or sidewalk.

Frost line. Depth to which soil freezes in a particular area.

Full basement foundation. Foundation that provides living and/or storage areas below the superstructure.

G

Ganged panel form. Large prefabricated forms constructed by joining a series of smaller panels.

ganged panel form

Girder. Large horizontal member that supports a bending load over a span, such as from column to column.

Grade. Existing or proposed ground level of a building site.

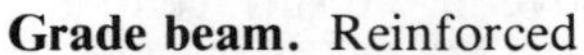

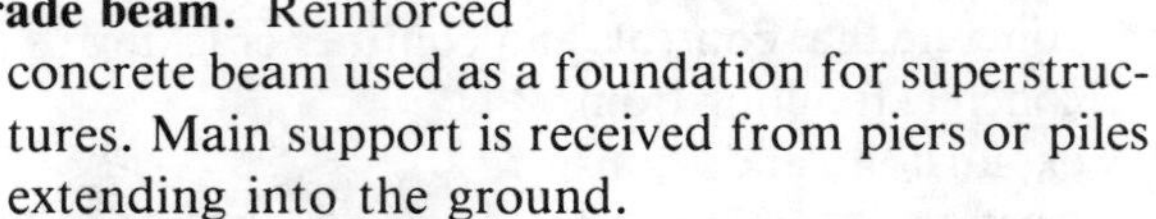

Grade beam. Reinforced concrete beam used as a foundation for superstructures. Main support is received from piers or piles extending into the ground.

Green concrete. Concrete that has set but has not appreciably hardened.

Ground beam. Concrete beam placed horizontally at ground level that ties wall and/or column footings together.

Groundwater. Water beneath the earth's surface that is primarily affected by the water table.

Grout. Mixture obtained by combining cement, sand, and water.

H

Heave. Upward thrust due to frost or moisture absorption and expansion of soil.

Heavy construction. Concrete or heavy timber construction methods used to erect structures such as factories, bridges, freeways, and dams.

Honeycomb. Voids in concrete resulting from segregation and poor consolidation at the time concrete is placed.

Hook points. Locations in a precast member providing access for crane attachment.

Hopper. Funnel-shaped box used to place concrete into a form.

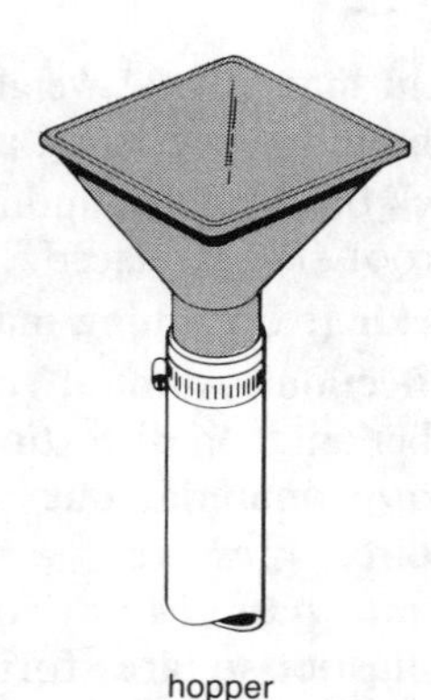
hopper

Hub. Stake used to indicate a corner of property.

Hydration. Chemical reaction between water and cement in concrete mix.

I

Insert. Anchoring device placed in precast concrete members that provides a crane attachment point.

Internal disconnecting tie. Form tie consisting of two external sections that screw into an internally threaded bolt and remain in place after the forms are stripped. Eliminates the use of spreader cones and reduces the size of holes remaining in the concrete.

Isolation joint. Separation created between adjoining parts of a concrete structure to allow for movement of the parts. Usually filled with caulking compound or asphalt-impregnated material.

J

J-Bolt. Type of anchor bolt embedded into the concrete at the time of placement. The threaded end projects from the concrete to allow for the attachment of a sill plate or other structural member.

K

Kerf. Saw cut made partially through a wood member.

Key strip. Chamfered piece pressed into the concrete immediately after placement; used to form a keyway.

Keyway. A groove formed in one lift or placement of concrete that is filled with concrete from the next lift.

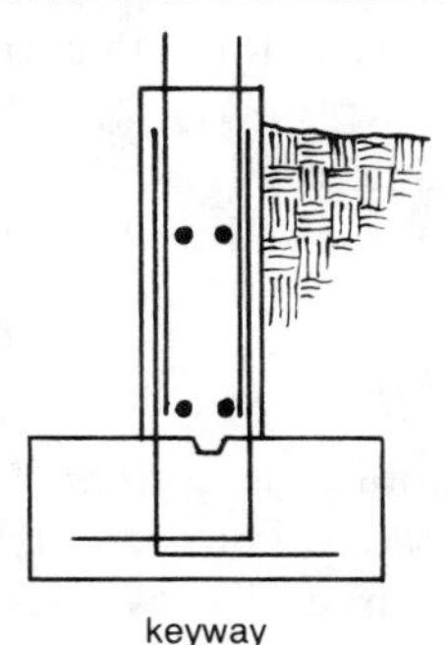
keyway

Kicker. A wood block or board that reinforces another form member against an outward thrust.

L

L-foundation. Foundation wall resting along the edge of a spread footing.

L-head shore. Shore formed with the horizontal member projecting from one side. Commonly used to support forms for spandrel beams.

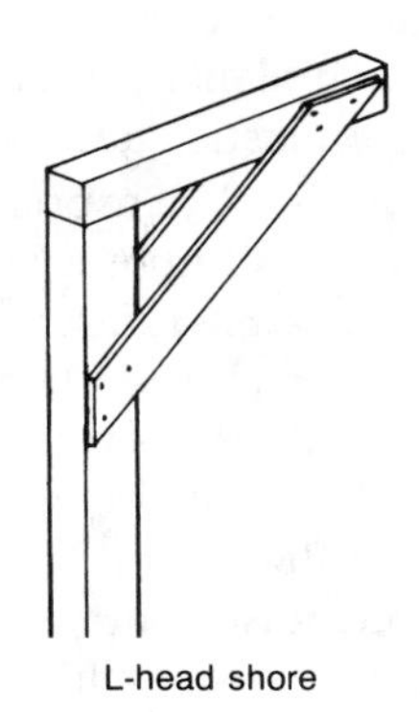
L-head shore

Lally column. Steel pipe column that rests on a concrete pier footing. Used to support wood or steel beams.

Lateral pressure. Horizontal pressure such as the force of soil against the side of a high foundation wall.

Ledger. Horizontal member that supports permanent or temporary structural members.

Leveling rod. Graduated rod used in conjunction with a builder's level or transit-level.

Lift. Layers of concrete placed in a wall and separated by horizontal construction joints.

Lift bar. Horizontal bar used with cranes that are equipped with cables threaded over pulleys. Cables are attached to precast panels to be raised into position.

Lift plate. Metal plate that is bolted to inserts embedded in precast concrete members.

Line of sight. Imaginary straight line extending from a builder's level or transit-level to object being sighted.

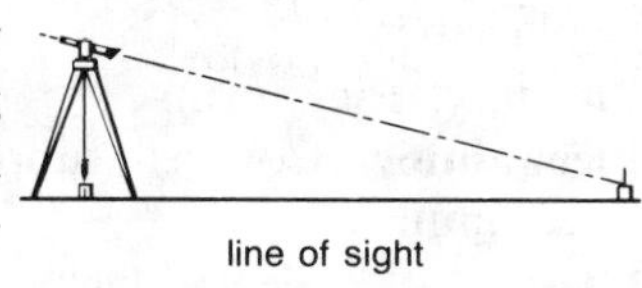
line of sight

Live load. Non-constant loads, such as people, furniture, or equipment, supported by a structure.

M

Mat foundation. Spread foundation consisting of a solid slab of heavily reinforced concrete that supports loads.

Minimum bending radius. Bending capacity of wood such as plywood used in construction of curved forms.

Mix. Mixture of aggregate, cement, water, and required admixtures.

Monolithic concrete. Concrete placed in forms without construction joints.

Motor grader. Earth-moving machine used for final grading operations.

Mudsill. 1. Plank or timber placed on top of soil to support shores. **2.** Wood member fastened to the top of foundation walls to which joists or studs are nailed.

N

Nominal lumber size. Dimension of sawed lumber before it is surfaced and seasoned.

O

Occupational Safety and Health Administration (OSHA). U.S. government safety regulations establishing guidelines for construction trade and other industries.

P

Pan. Metal or plastic prefabricated form unit used in construction of concrete floor joist systems.

Panel. Form section consisting of sheathing and stiffeners that can be erected and stripped as a unit.

Panel form. Prebuilt form section made up of panel sheathing, studs, and top and bottom plates.

parapet

Parapet. 1. Short wall along the edge of a bridge deck. **2.** Part of wall that extends above roof level.

Patented tie. Patented device used to secure and space opposite walls of forms during concrete placement.

Pier box. Pier form.

Pier footing. Concrete supporting base for a post or column.

Pilaster. Rectangular column incorporated with a concrete wall to strengthen the wall and provide support for the end of a beam.

Pile. Long steel, wood, or concrete member penetrating deep into the soil to support grade beam foundation walls or columns.

Pipe pile. Round steel pile. Hollow interior is filled with concrete after pile has been driven into ground.

Placement. Process of placing and consolidating concrete. *Pour* is a term used interchangably with placement.

Plank. Lumber over 1″ thick and 6″ or more in width.

Plasticity. Property and state of freshly mixed concrete that determines its shaping qualities and workability.

Plate. Flat horizontal member placed at the top and bottom of panel studs.

Plot plan. Drawing in a set of prints that shows size of lot, location of building on the lot, grades, and other information required to perform preliminary groundwork and foundation construction.

Plumb. 1. Vertical. **2.** To make vertical.

Plyform®. Plywood product specifically designed for form construction.

Plywood. Manufactured panel product consisting of veneers that are glued together under intense heat and pressure. Used extensively as form sheathing.

Polyethylene film. Thin sheet of plastic frequently used as vapor barrier.

Portland cement. Product obtained by pulverizing and mixing limestone with other products to produce cement required for concrete mix.

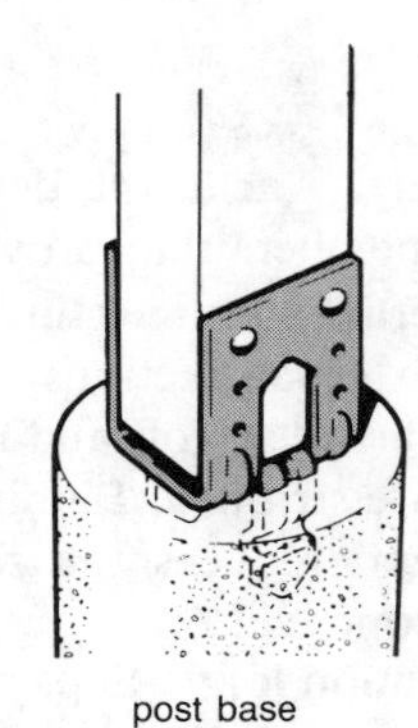
post base

Post base. Metal device embedded in concrete piers or walls. Used to secure the bottom of wood posts.

Post-tensioning. Method of prestressing concrete in which the tendons are tensioned after concrete has set.

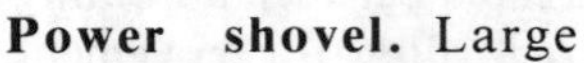

Power shovel. Large earth-moving equipment used for excavation.

Pozzolan. Admixture having cementing properties in the presence of moisture and ordinary temperatures.

Precast concrete. Concrete structural member formed somewhere other than its final position.

Prefabricated forms. Forms constructed from prebuilt panel sections.

Prestressed concrete. Concrete reinforced by stressing steel cables before or after concrete placement. Stressing cables places concrete in a state of compression.

Pretensioning. Method of prestressing reinforced concrete in which steel is stressed before the concrete sets.

Property line. Recorded legal boundaries of a piece of property.

Public walk. Walk that runs along a street bordering the building lot. Also called *sidewalk*.

R

Radius, minimum bending. Bending capacity of wood, such as plywood, used in construction of curved forms.

Raft foundation. Continuous, heavily reinforced slab of concrete placed monolithically with walls and/or columns over soil with poor bearing capacity or where heavy loads must be supported.

Ready-mixed concrete. Concrete manufactured at batch plants and delivered by truck to the job site in a plastic state.

Rebar. Steel reinforcing bar.

Reshore. Temporary shore firmly placed under concrete beams, girders, or slabs after form shores have been removed. Used to avoid deflection of the shored member or damage to concrete that is partially cured.

Ribbed floor slab. Thin floor slab integrated with concrete joists that tie into supporting girders and columns. Also called *one-way joist system*.

Ribbon. Narrow strip of material, usually wood, used in formwork.

Riser. Vertical member between two stair steps.

Rock pocket. Visible rock formations and voids in set concrete due to improper consolidation and segregation.

Rodding. Compaction of concrete with tamping rod.

S

Scaffold shoring. Shoring that consists of tubular steel frames assembled to support beam and floor forms.

scaffold shoring

Screed. System used to level and strikeoff the concrete when placing concrete for slabs.

Segregation. Varying concentration of concrete ingredients resulting in nonuniform proportions in the concrete mix.

Seismic risk zone. Area where conditions of earthquakes exist.

Separation. Coarse aggregate separating from rest of mix during placement occurring because of improper placement and consolidation.

Service walk. Walk that is located between a driveway or public sidewalk and rear entrance of building.

Set. The stiffening of concrete after it has been placed. *Initial set* refers to first stiffening. *Final set* occurs when concrete has attained significant hardness and rigidity.

Sheathing. Material used to form the face of wall forms or the deck of floor forms.

She bolt. See *internal disconnecting tie.*

Sheet piling. Interlocking metal piles designed to resist lateral pressure. Frequently used in deep excavations.

Shore. Temporary support for formwork and fresh concrete that has not developed full strength.

Shore jack. Metal device fastened to the bottom of a wood shore to allow height adjustment to be made.

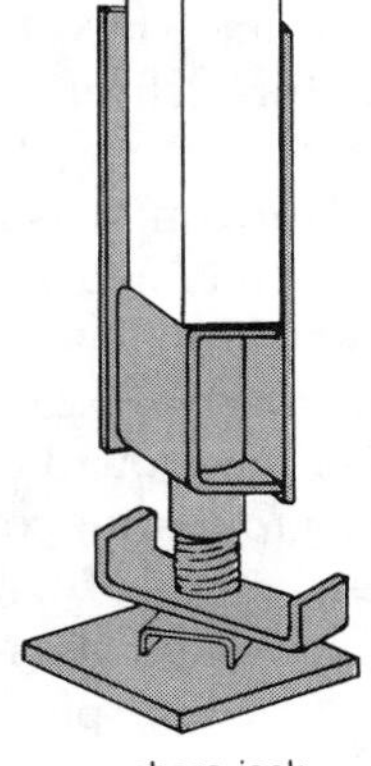
shore jack

Shore post. Vertical member used in shoring systems.

Shoring. System used to prevent sliding or collapse of earth banks around an excavation.

Shut-off. Vertical bulkhead placed at the end of wall forms to contain fresh concrete.

Sill plate. Plate fastened to the top of foundation walls to provide a base for floor joists or studs. Also called *mudsill.*

Silt. Granular material consisting of fine mineral particles that range from 2 to 50 μm in diameter.

Single-post shore. Shore consisting of a single post placed under stringers supporting floor forms.

Single wall form. Wall form system consisting of one constructed form wall and the other wall composed of solid rock, extremely hard soil, or an existing foundation of an adjoining building.

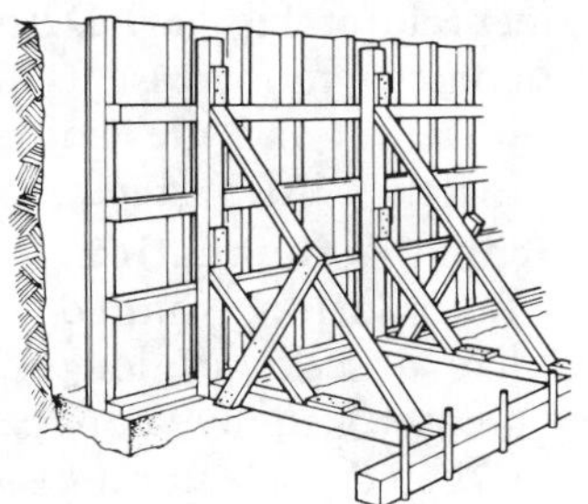
single wall form

Site cast. Precast structural concrete members cast in casting beds on the job site.

Site work. Preliminary layout, excavation, and other preparations required before construction can begin.

Slab. Flat horizontal layer of concrete supported by ground, beams, columns, or walls.

Slab-on-grade foundation. System that combines concrete foundation walls with a concrete floor slab resting directly on a bed of gravel.

Sleeve. Metal or fiber cylinder set inside a form to shape a passage for pipes or other objects through the finished concrete wall.

Slump test. Test performed at the time of concrete placement to measure consistency of concrete.

Snap tie. Patented wall tie device with cones acting as form spreaders. Ends of ties extend from the set concrete wall and are broken off at the breakbacks.

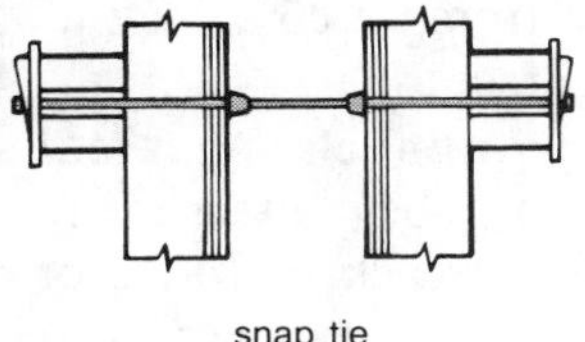
snap tie

Soldier pile. H-shaped pile driven into the ground that is commonly used for shoring high vertical earthen banks. Three-inch thick planks are placed between the flanges of the beams. Also called *soldier beam.*

Spading. Consolidation of concrete with a narrow wood rod or spading tool.

Spandrel beam. Beam located in outer wall of a building usually to support floors or roof.

Spread footing. Generally rectangular base placed beneath foundation walls to distribute the building load over a greater area.

Stack cast. Precast method in which a series of panels are cast on top of each other.

Stairway, closed. Stairway enclosed by walls.

Stairway, open. Unenclosed stairway running between two levels.

Stake. Wood or metal piece sharpened at its lower end and driven into the ground to either anchor the lower ends of braces or to hold the sides of footing forms in place.

Steel reinforcing bar. Deformed steel bar placed in concrete to increase its ability to withstand lateral pressure and tie adjoining concrete members together. Also called *rebar.*

Stepped foundation. Foundation shaped like a series of long steps. Usually constructed on sloping lots.

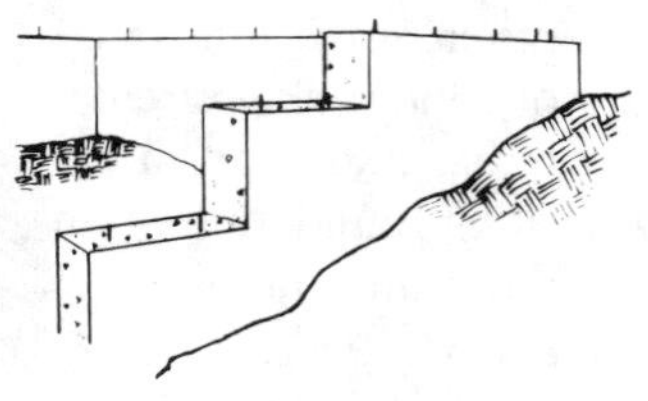
stepped foundation

Stepped pier. Pier consisting of two or more rectangular piers decreasing in size and placed on top of one another.

Strike board. Wood or metal straightedge used for screeding concrete.

Strikeoff. To level concrete to its correct finish grade.

Stringer. Horizontal timber placed on top of shores. Commonly used as part of a floor form system.

stringer

Strip. To remove forms from set concrete surfaces.

Strongback. Vertical member attached to the back of forms or precast members to reinforce or stiffen them.

Stud. Vertical member used to stiffen and support form sheathing.

Superstructure. **1.** Part of a building above the foundation. **2.** In bridge construction, concrete deck or traffic surface of the bridge.

Suspended formwork. Formwork suspended from a structural member and supported with U-shaped snap ties or coil hangers that are positioned over the beam or girder.

T

T-foundation. Foundation system consisting of a wall placed in the center of a spread footing resting on soil.

T-head shore. Shore formed with the horizontal member projecting equally on both sides of post.

Tapered pier. Pier footing with inclined sides.

Template. **1.** Frame used in positioning formwork members. **2.** Wood piece used to lay out and secure anchor bolts during concrete placement.

template

Tension. Pulling or stretching force.

Test pit. Shallow excavation dug to examine soil conditions on the job site.

Textured plywood. Form sheathing producing special surface effects such as wood grain, boards, and other designs.

Tilt-up construction. Concrete construction method in which wall panels of a building are cast horizontally at a location adjacent to its eventual position and lifted into position by crane.

Toenail. **1.** Nail driven at an angle. **2.** To drive a nail at an angle.

Total rise. Vertical distance from the bottom surface of the lower level of the stairway to the top surface of the upper level of the stairway.

Tower crane. Crane consisting of a high tower and gib. Sections are added to the tower to achieve greater heights.

Transit-level. Surveying instrument used for leveling over long distances as well as establishing grades and elevations. Telescope can also be moved vertically for plumbing and other layout work.

Transit-mixer truck. Truck equipped with a large drum to mix concrete for delivery to job site.

Tread. Horizontal surface of the step in a stairway.

Trench. Narrow excavation in the ground extending to bearing soil.

Tubular fiber form. Round column forms constructed of spirally wound fiber plies.

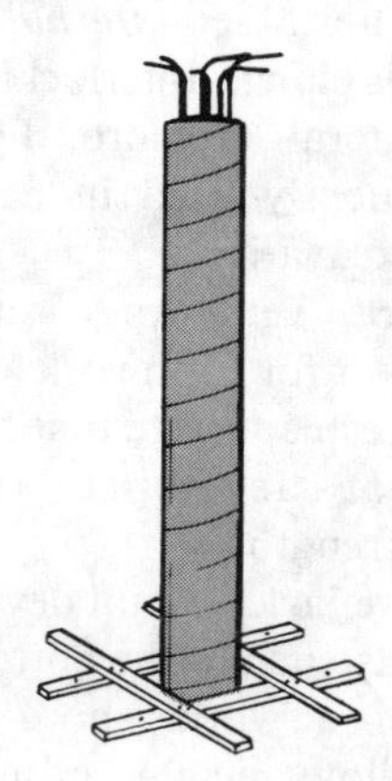
tubular fiber form